Formaldehyde and Other Aldehydes

Formaldehyde and Other Aldehydes

Committee on Aldehydes

Board on Toxicology and Environmental Health Hazards

Assembly of Life Sciences

National Research Council

NATIONAL ACADEMY PRESS
Washington, D.C.   1981

NOTICE: The project that is the subject of this report was approved by the Governing Board of the National Research Council, whose members are drawn from the councils of the National Academy of Sciences, the National Academy of Engineering, and the Institute of Medicine. The members of the Committee responsible for the report were chosen for their special competences and with regard for appropriate balance.

This report has been reviewed by a group other than the authors according to procedures approved by a Report Review Committee consisting of members of the National Academy of Sciences, the National Academy of Engineering, and the Institute of Medicine.

The National Research Council was established by the National Academy of Sciences in 1916 to associate the broad community of science and technology with the Academy's purposes of furthering knowledge and of advising the federal government. The Council operates in accordance with general policies determined by the Academy under the authority of its congressional charter of 1863, which establishes the Academy as a private, nonprofit, self-governing membership corporation. The Council has become the principal operating agency of both the National Academy of Sciences and the National Academy of Engineering in the conduct of their services to the government, the public, and the scientific and engineering communities. It is administered jointly by both Academies and the Institute of Medicine. The National Academy of Engineering and the Institute of Medicine were established in 1964 and 1970, respectively, under the charter of the National Academy of Sciences.

The work on which this publication is based was performed pursuant to Contract 68-01-4655 with the Office of Research and Development of the Environmental Protection Agency (Dr. Alan P. Carlin, Project Officer).

Library of Congress Catalog Card Number 81-81738

International Standard Book Number 0-309-03146-X

Available from:
NATIONAL ACADEMY PRESS
2101 Constitution Ave., N.W.
Washington, D.C. 20418

Printed in the United States of America

First Printing, May 1981	Third Printing, October 1983
Second Printing, November 1981	Fourth Printing, December 1984

COMMITTEE ON ALDEHYDES

JACK G. CALVERT, Ohio State University, Columbus, Ohio, <u>Chairman</u>

LYLE F. ALBRIGHT, Purdue University, West Lafayette, Indiana

EILEEN BRENNAN, Cook College, New Brunswick, New Jersey

STUART M. BROOKS, University of Cincinnati School of Medicine, Cincinnati, Ohio

CRAIG D. HOLLOWELL, University of California, Berkeley, California

DAVID H. W. LIU,* Woodward-Clyde Consultants, San Francisco, California

CHARLES F. REINHARDT, Haskell Laboratory, E. I. du Pont de Nemours & Company, Wilmington, Delaware

JAMES A. FRAZIER, National Research Council, Washington, D.C., <u>Staff Officer</u>

NORMAN GROSSBLATT, National Research Council, Washington, D.C., <u>Editor</u>

LESLYE B. GIESE, National Research Council, Washington, D.C., <u>Research Assistant</u>

JEAN E. PERRIN, National Research Council, Washington, D.C., <u>Secretary</u>

---

*Early in the preparation of this report, Dr. Liu was with SRI International, Menlo Park, California.

BOARD ON TOXICOLOGY AND ENVIRONMENTAL HEALTH HAZARDS

RONALD W. ESTABROOK, University of Texas Medical School, Dallas, Texas, Chairman

THEODORE CAIRNS, Greenville, Delaware

VICTOR COHN, George Washington University Medical Center, Washington, D.C.

JOHN W. DRAKE, National Institute of Environmental Health Sciences, Research Triangle Park, North Carolina

A. MYRICK FREEMAN, Bowdoin College, Brunswick, Maine

RICHARD HALL, McCormick & Company, Hunt Valley, Maryland

RONALD W. HART, National Center for Toxicological Research, Jefferson, Arkansas

PHILIP LANDRIGAN, National Institute of Occupational Safety and Health, Cincinnati, Ohio

MICHAEL LIEBERMAN, Washington University School of Medicine, St. Louis, Missouri

BRIAN MacMAHON, Harvard School of Public Health, Boston, Massachusetts

RICHARD MERRILL, University of Virginia, Charlottesville, Virginia

ROBERT A. NEAL, Vanderbilt University, Nashville, Tennessee

IAN NISBET, Massachusetts Audubon Society, Lincoln, Massachusetts

CHARLES R. SCHUSTER, JR., University of Chicago, Chicago, Illinois

GERALD WOGAN, Massachusetts Institute of Technology, Cambridge, Massachusetts

ROBERT G. TARDIFF, National Research Council, Washington, D.C., Executive Director

ACKNOWLEDGMENTS

This document is the result of individual and coordinated efforts by the members of the Committee on Aldehydes. Although, as detailed below, individual members were responsible for specific sections, the entire report was reviewed by the full Committee. The summary (Chapter 2) and the recommendations (Chapter 3) represent a consensus of the Committee members.

The introduction (Chapter 1) was prepared by the Chairman, Dr. Jack G. Calvert. Chapter 4, on the properties, production, and uses of the aldehydes, was prepared by Dr. Lyle F. Albright, Dr. Calvert, and staff; the Appendix, which contains details on many specific aldehydes, was prepared by staff. Chapter 5, on sources and concentrations, was a joint effort of Dr. Calvert, Dr. Albright, Dr. Eileen Brennan, Dr. Craig D. Hollowell, and Dr. David H. W. Liu, assisted by Dr. William R. Mabey. Chapter 6, on measurement methods, was written by Dr. Hollowell, assisted by his associate, Dr. Robert R. Miksch, and by Dr. Brennan and Dr. Liu. Dr. Stuart M. Brooks and Dr. Charles F. Reinhardt prepared Chapter 7, on health effects of formaldehyde, and Chapter 8, on health effects of some other aldehydes, with the assistance of Dr. Francis N. Marzulli, Mr. Richard C. Graham, and Dr. Joel Bender. Chapter 9, on the effects of aldehydes on vegetation, was written by Dr. Brennan, and Chapter 10, on the effects on aquatic organisms, was written by Dr. Liu.

We acknowledge the contributions of Dr. Robert Frank, who served as chairman of the Committee in its formative period.

Special recognition should be given to Dr. Albright, who chaired a subcommittee meeting at which resource information was presented by industrial representatives of the Formaldehyde Institute, the American Textile Manufacturers, Inc., the Manufactured Housing Institute, the Hardwood Plywood Manufacturers Association, and Aerolite SPE Corp. Special thanks should also be extended to Dr. Brooks, who chaired a subcommittee on the health effects of formaldehyde, which included Dr. Donald Proctor and Dr. Edward Emmett of Johns Hopkins University, who assisted in the evaluation of the health effects, and representatives of the Formaldehyde Institute, who made available resource information on the health effects. The efforts of Dr. John Clary and Mr. Kip Howlett in organizing representation and the presentations made at these subcommittee meetings are greatly appreciated, as well as those of Dr. Leon Starr of the Celanese Corporation for his consultation and information resources furnished to the Committee.

For providing resource material and other information, we express our gratitude to the following:

- Dr. Eric R. Allen, Atmospheric Sciences Research Center, State University of New York, Albany, New York
- Mr. Douglas Anthon, Lawrence-Berkeley Laboratory, University of California, Berkeley, California
- Dr. Charlotte Auerbach, University of Edinburgh, Edinburgh, Scotland
- Mr. W. L. Benning, Manufactured Housing Institute, Arlington, Virginia
- Mr. Charles A. Campbell, Aerolite SPE Corporation, Florence, Kentucky
- Dr. T. Cooke, Textile Research Institute, Princeton, New Jersey
- Mr. Maynard Curry, U.S. Department of Housing and Urban Development, Washington, D.C.
- Dr. John W. Drake, National Institute of Environmental Health Sciences, Research Triangle Park, North Carolina
- Dr. Julius Fabricant, New York State College of Veterinary Medicine, Ithaca, New York
- Ms. Mary Leah Fanning, Lawrence-Berkeley Laboratory, University of California, Berkeley, California
- Mr. Robert B. Faoro, U.S. Environmental Protection Agency, Research Triangle Park, North Carolina
- Mr. R. David Flesh, Del Green Associates, Foster City, California
- Dr. Alfred L. Frechette, Massachusetts State Department of Public Health, Boston, Massachusetts
- Dr. A. Myrick Freeman II, Department of Economics, Bowdoin College, Brunswick, Maine
- Mr. Ernest Freeman, Department of Energy, Washington, D.C.
- Dr. Thomas E. Graedel, Bell Laboratories, Murray Hill, New Jersey
- Mr. Serge Gratch, Ford Motor Company, Dearborn, Michigan
- Mr. William Groah, Hardwood Plywood Manufacturers Association, Arlington, Virginia
- Dr. Daniel Grosjean, Environmental Research and Technology, Inc., West Lake Village, California
- Mr. Jan Heuss, General Motors Research Laboratory, Warren, Michigan
- Mr. Richard C. Holmquist, American Mining Congress, Washington, D.C.
- Dr. Nelson S. Irey, Armed Forces Institute of Pathology, Washington, D.C.
- Dr. Howard Johnson, SRI International, Menlo Park, California
- Ms. Nancy S. Johnston, Consumer Product Safety Commission, Bethesda, Maryland
- Dr. Philip Landrigan, National Institute for Occupational Safety and Health, Cincinnati, Ohio
- Dr. Charlotte R. Lindley, Department of Chemistry, Ohio State University, Columbus, Ohio
- Dr. Gunnar Lundqvist, Arhus University, Arhus, Denmark

- Dr. Geoffrey Meadows, E. I. du Pont de Nemours and Company, Wilmington, Delaware
- Mr. Andrew Micula, New Jersey Department of Environment Protection, Trenton, New Jersey
- Dr. Dominick J. Mormile, Consolidated Edison, New York, New York
- Mr. Charles Morschauset, National Particle Board Association, Silver Spring, Maryland
- Dr. Demetrios J. Moschandreas, GEOMET Technologies, Inc., Gaithersburg, Maryland
- Mr. John F. Murray, Formaldehyde Institute, Scarsdale, New York
- Ms. Laura A. Oatman, Minnesota Department of Health, Minneapolis, Minnesota
- Mr. Edward Owens, Aberdeen Proving Grounds, Maryland
- Mr. James Paine, Texas Air Control Board, Austin, Texas
- Dr. G. J. Piet, National Institute for Water Supply, Voorburg, The Netherlands
- Dr. James M. Ramey, Celanese Corporation, New York, New York
- Dr. Yuko Sasaki, Tokyo Metropolitan Research Institute for Environmental Protection, Tokyo, Japan
- Dr. Eugene N. Skiest, Borden, Inc., Columbus, Ohio
- Dr. Ronald J. Spanggord, SRI International, Menlo Park, California
- Dr. Karl J. Springer, Southwest Research Institute, San Antonio, Texas
- Dr. Edgar R. Stevens, California Statewide Air Pollution Research Center, Riverside, California
- Mr. William Stockwell, Department of Chemistry, Ohio State University, Columbus, Ohio
- Mr. John G. Tritsch, American Textile Manufacturers Institute, Washington, D.C.
- Dr. Michael P. Walsh, U.S. Environmental Protection Agency, Washington, D.C.
- Mr. Frank Walter, Manufactured Housing Institute, Arlington, Virginia
- Mr. Ralph C. Wands, Mitre Corporation, McLean, Virginia
- Dr. Jonas Weiss, CIBA-GEIGY Corporation, Ardsley, New York
- Dr. Katherine W. Wilson, Pacific Environmental Services, Santa Monica, California
- Ms. Mary Ann Woodbury, Wisconsin State Department of Health and Social Services, Madison, Wisconsin

Free use was made of the resources of the Toxicology Information Center of the Board on Toxicology and Environmental Health Hazards, National Research Council; the National Library of Medicine; the National Agricultural Library; the Library of Congress; and the Air Pollution Technical Information Center of the Environmental Protection Agency. Also acknowledged is the assistance given to the Committee by the National Academy of Sciences Library and various other units of the National Research Council.

CONTENTS

1	Introduction	1
2	Summary	3
3	Recommendations	11
4	Commercial Production, Properties, and Uses of the Aldehydes	20
5	Aldehyde Concentrations, Emission, and Environmental Generation and Transformation Reactions	36
6	Analytical Methods for the Determination of Aldehydes	132
7	Health Effects of Formaldehyde	175
8	Health Effects of Some Other Aldehydes	221
9	Effects of Aldehydes on Vegetation	256
10	Effects of Aldehydes on Aquatic Organisms	276
Appendix:	Properties, Uses, and Synonyms of Selected Aldehydes	289

CHAPTER 1

INTRODUCTION

This report was prepared at the request of the Environmental Protection Agency (EPA) by the Committee on Aldehydes, which was appointed by the National Research Council in the Board on Toxicology and Environmental Health Hazards, Assembly of Life Sciences. The Clean Air Act requires that from time to time the Administrator of EPA evaluate the need for air-quality criteria on pollutants that may have adverse effects on man or the environment. This report is to be used by EPA in assessing the need for such criteria on some of the aldehydes. It is intended to identify and characterize the more important aldehydes that pollute the environment, the sources of their emission, their concentrations, their transformation and transport, their effects on the health of animals and humans, and their effects on the aquatic and terrestrial environments. It is not intended to recommend concentrations of polluting aldehydes for use in developing regulations, but rather to evaluate the available data for EPA to use in judging the need for regulatory strategies to control aldehyde pollution. It is hoped that wide dissemination of this report will inform physicians and other health professionals about the health effects of aldehydes and how they may be encountered at hazardous concentrations in the environment.

The Committee had hoped to address the economics of the options for abatement of aldehyde pollution, and it chose formaldehyde as the model of the aldehydes because of its perceived importance and because it is used in a wide variety of products. Techniques for abating formaldehyde emission are still evolving and being tested; their value has not been proved, and their costs and cost-benefit relationships are not known. Therefore, on the grounds of a lack of usable information, the economic analysis of control options was abandoned.

Chapters 2 and 3 summarize the Committee's findings and set forth the Committee's recommendations, respectively.

Chapter 4 describes commercial methods of production of the aldehydes and their uses. Chapter 5 reviews the reported atmospheric concentrations of the aldehydes in clean and urban environments, in indoor environments, and in surface and drinking waters; considers the sources of direct emission from industrial operation, combustion, consumer products, natural vegetation, and indoor environments; and evaluates current theories of the mechanism of aldehyde generation in the atmosphere, the aldehyde removal processes that operate in the environment, and the secondary effects of aldehydes in the chemistry of the polluted atmosphere. Chapter 6 reviews and evaluates the methods of analysis of formaldehyde and selected higher aldehydes and methods of sampling and of preparing standards.

Formaldehyde has prominence throughout the report, because it is ubiquitous, is used in very large quantities, and is mutagenic in microorganisms and insects and carcinogenic in Fischer 344 rats. In Chapter 7, the health effects of formaldehyde in terrestrial animals and humans are discussed in detail; Chapter 8 covers the health effects of selected other aldehydes; Chapter 9 discusses the effects of selected aldehydes on vegetation; and Chapter 10 discusses the effects of some aldehydes on aquatic organisms.

The Appendix summarizes in tabular form several of the important physical and chemical properties of a number of the aldehydes that have been found in the environment. This compilation is not intended to be exhaustive, nor should the importance of these aldehydes be inferred from their listing.

This report cites references available to the Committee up to July 1, 1980, with one exception: a study that was made public in November 1980 and provided more recent information on an earlier study that was cited. The work of the Committee from July 1, 1980, onward was devoted to an analysis of the information in hand. Thus, scientific papers and analyses published after that date were not considered.

CHAPTER 2

SUMMARY

Although many of the aldehydes are minor components of the natural environment, we now recognize the potential impact of some of them on the urban and indoor environments. Thus, there is a need to study their sources, concentrations, transport, and transformations and their effects on various environmental and biologic systems.

PRODUCTION, USES, AND PROPERTIES OF THE ALDEHYDES

The aldehydes are produced at the rate of several billion pounds per year in the United States. Formaldehyde is the most important aldehyde produced commercially; about 9 billion pounds per year of the 37-50% aqueous solution, called formalin, are prepared. The production methods depend on the catalytic oxidation of methanol. About half the formaldehyde produced is used in the preparation of urea-formaldehyde and phenol-formaldehyde resins, which are applied in the manufacture of plywood, particleboard, foam insulation, etc. Another 25% is used to make other high polymers and resins. Its use is so diversified that there is a potential for exposure in a number of occupational, environmental, and consumer settings. About 1 billion pounds of acetaldehyde are prepared commercially each year in the United States, about 80% of it by the catalytic oxidation of ethylene in aqueous solution. This aldehyde is the major raw material for the preparation of acetic acid and other important chemicals. Smaller amounts of acrolein and the higher-molecular-weight aldehydes are prepared commercially. The simplest aldehydes are volatile compounds with characteristic pungent odors. These compounds are readily oxidized and polymerized.

ALDEHYDE EMISSION, CONCENTRATIONS, AND ATMOSPHERIC TRANSFORMATIONS

The aldehydes are introduced into the atmosphere through a variety of natural processes and as a result of human activity. In the atmosphere, they are generated through the photooxidation of both naturally occurring and anthropogenic hydrocarbons. They are injected directly into the atmosphere in the exhaust gases from automobiles and other equipment in which hydrocarbon fuels are incompletely burned.

Aldehydes are emitted from various industrial and manufacturing operations, power plants that burn fossil fuels, forest fires and open burning of wastes, and vegetation.

The accumulation of aldehydes in the atmosphere as a result of their direct release and photochemical generation is counterbalanced by several important removal paths. The aldehydes absorb the ultraviolet component of sunlight and decay into free-radical and molecular products. They also react rapidly with the highly reactive free radicals, largely the hydroxy free radical, present in the sunlight-irradiated atmosphere. Because of the high water solubility of formaldehyde and the other low-molecular-weight aldehydes, one expects the efficient transfer of aldehydes into rainwater, the oceans, and other surface waters. The high reactivity of the aldehydes leads to rather short half-lives, of around a few hours, in the sunlight-irradiated lower atmosphere. Thus, the atmospheric transport of the aldehydes over long distances is probably a less likely source in remote areas than their local generation from transported, longer-lived precursors, such as the less reactive hydrocarbons. The lifetime of formaldehyde in aqueous media may be somewhat greater than that of the gas-phase species, because the hydrated form of formaldehyde, which dominates in these conditions, does not absorb sunlight appreciably. The higher aldehydes do not have this protective mechanism, because of the lower degree of hydration. Microorganisms appear to play an important role in the degradation process, which may take 30-72 h in these cases (in natural conditions).

Many people may be exposed to aldehydes at high concentrations (i.e., above ambient) in the indoor environment of the home. Sources of aldehydes in conventional residential buildings and mobile homes include building materials, insulation, combustion appliances, tobacco smoke, and various consumer products. These sources emit aldehydes in substantial amounts; as a result, indoor aldehyde concentrations almost always exceed outdoor concentrations.

In any environment, the ambient concentrations of aldehydes depend on the rates of the formation and removal reactions. In a clean environment, aldehyde concentrations at ground level are commonly 0.0005-0.002 ppm (0.6-2.5 $\mu g/m^3$). In ambient urban air, the concentrations are much higher, usually an hourly average of 0.004-0.05 ppm (5-61 $\mu g/m^3$) during the daylight hours. Formaldehyde is the dominant aldehyde present, usually constituting 30-75% of the total aldehydes. Acetaldehyde may be present at about 60% of the formaldehyde concentration, with smaller amounts of the higher aliphatic aldehydes. Acrolein may be present at 10-25% of the formaldehyde concentration, and the aromatic aldehydes usually make up only a few percent of the total aldehydes. In most indoor environments, 24-h average formaldehyde concentrations of 0.05-0.2 ppm (61-246 $\mu g/m^3$) are not uncommon today. However, in some indoor environments concentrations of a few parts per million or higher have been reported. In the aquatic environment, aldehyde concentrations are generally less than 1 ppb. Concentrations of some aldehydes in the parts-per-million range have been reported in industrial effluents.

The aldehydes affect the chemistry of the chemically polluted atmosphere in a variety of complex ways. An increased aldehyde concentration decreases the induction period for the generation of the products of photochemical smog and increases the maximal concentration of ozone. The aldehydes other than formaldehyde are precursors of an important class of secondary pollutants, the peroxyacylnitrates and peroxybenzoylnitrates, which have been identified as highly active eye irritants and plant-damaging agents. Through several atmospheric reaction pathways, formaldehyde may be converted to formic acid. The interaction of formaldehyde and hydrogen chloride can lead to chloromethylethers, which are potential carcinogens, although current knowledge (which may be incomplete) indicates that the concentrations formed in the atmosphere at ambient concentrations of reactants would be so low that there is little probability of an impact on health. Thus, formaldehyde affects the quality of the ambient air not only directly, but also indirectly by way of its chemical transformations, involvement in photochemical smog reactions, and interaction in combination with other pollutants.

ANALYTICAL METHODS FOR THE DETERMINATION OF THE ALDEHYDES

The techniques for quantitative analysis of the aldehydes have not been investigated adequately with respect to the necessary reliability, sensitivity, and specificity. Accurate analysis for the aldehydes in the environment is essential to an assessment of their qualitative and quantitative influence on the environment and on the health of those exposed. The solution-phase spectrophotometric methods are the most commonly used analytical techniques. Although the individual aldehydes can be selectively measured with methods now available, the methods require improvement and standardization. Measurements of "total" aldehyde provide little help in the assessment of the impact of these compounds in the environment, because the variation in toxicity among the individual aldehydes is large--for example, the current Occupational Safety and Health Administration standards for exposure to acetaldehyde, formaldehyde, and acrolein are 200, 3, and 0.1 ppm, respectively. It is recognized that future analytical methods for aldehydes should provide an accurate determination of the specific aldehydes present in a given sample. Several methods appear to offer this potential. Many involve the derivatization of the aldehydes and the use of gas- or liquid-chromatographic separations and analysis. Further development is necessary to establish these methods for general use.

HEALTH EFFECTS OF FORMALDEHYDE

Formaldehyde has been the subject of numerous complaints regarding irritation of the eyes and respiratory tract, nausea, headache, tiredness, and thirst. These symptoms have been reported mainly by residents of homes in which formaldehyde has been identified as a result of off-gassing from urea-formaldehyde foam insulation, particleboard, or plywood. Studies of employees exposed to formaldehyde in the workplace and in controlled exposures have further indicated that the skin, eyes, and respiratory tract are the target organs affected.

Aqueous solutions of formaldehyde are damaging to the eye and irritating to the skin on direct contact. Repeated exposure to dilute solutions may lead to allergic contact dermatitis. There are some documented cases showing that formaldehyde is a cause of skin responses in sensitized persons using cosmetic formulations that contain formaldehyde at very low concentrations (0.01%). There are few documented cases showing that formaldehyde is a cause of hypersensitivity in persons with bronchial asthma; more commonly, asthma is aggravated by the irritating properties of formaldehyde. Systemic poisoning from ingestion is uncommon, because the irritancy of formaldehyde makes ingestion unlikely.

Numerous studies have shown that formaldehyde is irritating to the eyes and upper respiratory tract of laboratory animals. Preliminary results of a chronic-inhalation study sponsored by the Chemical Industry Institute of Toxicology (CIIT) have shown that formaldehyde induces nasal cancer in Fischer 344 rats exposed at 15 ppm 6 h/d, 5 d/wk for 18 mo, but not yet in B6C3F1 mice similarly exposed. (However, the CIIT reported at the Formaldehyde Symposium on November 20-21, 1980, in Raleigh, N.C., that nasal cancer had been observed in rats exposed at 6 ppm for 24 mo and in mice exposed at 15 ppm for 24 mo.) Fischer 344 rats have also shown dose-related histologic changes (epithelial dysplasia and squamous metaplasia) of the nasal mucosa after exposure at 2, 6, and 15 ppm. Although there is no evidence of the carcinogenicity of formaldehyde in humans, the results of these studies showing carcinogenicity in rats require that serious attention be given to an evaluation of the carcinogenic potential of formaldehyde in exposed humans. Formaldehyde has not altered reproduction or shown evidence of teratogenicity in animals, but it has exhibited mutagenic activity in several nonmammalian animal or cell systems. The human mutagenic and teratogenic potential of formaldehyde is not known.

The presence of environmental agents other than formaldehyde, smoking history, variability of health status, age, and genetic predisposition may modify responses to formaldehyde. These factors have not been adequately evaluated; that makes it difficult to assess accurately the health risks attributable solely to formaldehyde. However, the complaints of residents of homes with formaldehyde-containing products have been shown to be similar to complaints made by persons studied in the laboratory at similar formaldehyde concentrations; hence, these subjective complaints about health effects may be related to formaldehyde exposure in the home, although the presence of other pollutants causing the same symptoms must not be overlooked. Accordingly, a substantial proportion of the U.S. population may be likely to develop symptoms of irritation, if exposed to formaldehyde at low concentrations. As discussed in detail in Chapter 7, on the basis of laboratory tests and various kinds of population surveys, it has been estimated that some 10-20% of the general population may be susceptible to the irritant effects of formaldehyde at low concentrations. For example, most people report mild eye, nose, and throat irritation at a concentration of 1 ppm, whereas some note symptoms at concentrations below 0.5 ppm. In laboratory investigations, under controlled conditions, responses have been reported at formaldehyde concentrations as low as 0.01 ppm when formaldehyde was present in combination with other air pollutants. Low concentrations may also cause bronchoconstriction and asthmatic symptoms in some susceptible persons. The specific effects of continuous exposure on other susceptible populations--such as infants, young children, pregnant women, and the infirm--are not known. The exact numbers of susceptible people residing in environments where formaldehyde concentrations could produce adverse responses cannot be determined. Millions of persons live in mobile homes or conventional homes that have particleboard, plywood, or urea-formaldehyde foam (resin) insulation. On the basis of monitoring of a fairly large number of houses in these categories, significant formaldehyde concentrations were detected in several hundred American homes, and these concentrations were caused in large part by outgassing from these building materials. Much of this monitoring was done as a direct result of customer complaints. Yet other homes in these categories, including some with customer complaints, demonstrated formaldehyde concentrations basically comparable with those in homes that did not contain such building materials. On the basis of estimates of susceptibility of the general population to formaldehyde, it may be anticipated that a substantial number of persons are at risk of adverse health effects (upper and lower respiratory tract effects, eye irritation, etc.). Because of the incompleteness of the data, no conclusions can be drawn about the carcinogenic risks to humans exposed to formaldehyde.

HEALTH EFFECTS OF SOME OTHER ALDEHYDES

The principal effect of human exposure to other aldehydes, particularly acrolein and acetaldehyde, at low concentrations, is irritation of the eyes, skin, and mucous membranes of the upper respiratory tract. It has been demonstrated that several environmental irritants are ciliotoxic and mucus-coagulating agents. The aldehydes--which include acetaldehyde, propionaldehyde, and acrolein--may thus facilitate the uptake of other atmospheric contaminants by the bronchial epithelium.

Acetaldehyde, the least toxic of the atmospheric aldehydes, is slightly toxic when administered orally. The effect of direct contact with liquid acetaldehyde has not been studied, but industrial experience suggests that there is little hazard. Repeated-exposure studies indicated that significant toxic effects appear only at high concentrations. An 18-wk inhalation study in hamsters showed no adverse effects at 390 ppm (7.0×10^5 $\mu g/m^3$). Acetaldehyde is thought to be an important contributor to the health effects of cigarette smoke. It does not appear to have substantial mutagenic or carcinogenic effects, but more extensive studies are required to test this possibility. A major source of acetaldehyde in the body is the metabolism of ethanol. Acetaldehyde has shown embryotoxic and teratogenic effects in mice similar to those produced by ethanol.

Acrolein is the most acutely toxic of the atmospheric aldehydes. It is highly toxic by the oral and skin-absorption routes. It produces severe injury on contact with the skin and eyes. Inhalation of acrolein vapors by cats and rats produces severe eye and respiratory tract irritation at concentrations as low as 12 ppm (2.8×10^4 $\mu g/m^3$) and death in rats after 4-h exposure at 8 ppm (1.8×10^4 $\mu g/m^3$); its vapors produced little or no effect at up to 0.2 ppm (458 $\mu m/m^3$). Exposed animals appear to develop tolerance within a few weeks. Higher concentrations cause species- and dose-dependent histopathologic changes in both the upper and the lower respiratory tract. Although acrolein has been shown to be mutagenic in nonmammalian systems, it has not been shown to be carcinogenic in hamsters. In a single study, it was found not to be embryotoxic in rats.

Crotonaldehyde produces symptoms similar to those described for acrolein. Eye and respiratory tract irritation is seen with propionaldehyde, n-butyraldehyde, isobutyraldehyde, and chloral. Chloral is unique, in that its inhalation toxicity puts it in the highly toxic category for acute exposures. Other high-molecular-weight aldehydes—such as chloroacetaldehyde, valeraldehyde, furfural, the butyraldehydes, glyoxal, malonaldehyde, benzaldehyde, synapaldehyde, and the naturally occurring aldehydes—appear to be less toxic than formaldehyde and acrolein, although studies of these compounds are incomplete.

EFFECTS OF ALDEHYDES ON VEGETATION

Several studies concerning aldehyde phytotoxicity have been reported. Manifestations of injury include visible symptoms on foliage and effects on growth, photosynthesis, respiration, transpiration, seed germination, and pollen-tube elongation. Early California studies demonstrated that exposure of five smog-sensitive field crops to formaldehyde vapors (uncontrolled fumigations at 2 ppm, or 2.5×10^3 $\mu g/m^3$, for 2 h) caused no noticeable effect. Some doses of acrolein (0.1 ppm, or 229 $\mu g/m^3$, for 9 h) and trichloroacetaldehyde (0.8 ppm, or 4.8×10^3 $\mu g/m^3$, for 4 h) induced smog-like damage to alfalfa leaves, but higher doses of acrolein (0.6 ppm, or 1.4×10^3 $\mu g/m^3$, for 3 h and 1.2 ppm, or 2.8×10^3 $\mu g/m^3$, for 4.5 h) caused injury in spinach, endive, and beet leaves unlike that caused by smog. Visible injury in pinto bean leaves occurred after 70 min of exposure to acrolein at 2.0 ppm (4.6×10^3 $\mu g/m^3$). Products of the irradiated aldehydes in air have also been tested, in a 4-h exposure at 0.5 ppm. Irradiated formaldehyde and acetaldehyde caused no damage to petunias and pinto beans, but propionaldehyde and butyraldehyde caused a glazing of the lower leaf surface of both plants.

Slightly reduced rates of photosynthesis and respiration were measured when an alga (Euglena gracilis) was exposed to formaldehyde at 0.075 ppm (92 $\mu g/m^3$) for 1 h, and photosynthesis was significantly reduced after exposure to propionaldehyde at 0.1 ppm (123 $\mu g/m^3$) for 1 h. In fasted cells, the effects were minimized.

A rather large concentration (10^{-6} M, or 24 ppm) of a single higher aldehyde (trans-2-hexenal, pentanal, hexanal, heptanal, octanal, or nonanal) decreased the transpiration rate in wheat seedlings to less than that observed in complete darkness.

Aldehydes have been observed to inhibit pollen-tube elongation in lily. Although exposure to formaldehyde at 0.37 ppm (454 $\mu g/m^3$) for 1-2 h had no effect, a 5-h exposure at this concentration resulted in inhibition. Acrolein was more injurious, causing a 40% decrease in tube length when the lily was exposed at 0.4 ppm (917 $\mu g/m^3$) for 2 h.

Various other detrimental effects of aldehydes on plants have been observed. For example, oat, wheat, corn, barley, tomato, bean, lettuce, and radish showed a marked reduction in seedling growth and seed germination after exposure to polymer-treated woods. Presumably, the formaldehyde vapors that escaped from the wood were responsible.

On the basis of available information, one might expect to find some response of sensitive plants to aldehydes in ambient air. This will probably be seen first in the fast-growing herbaceous plants, rather than the woody, slow-growing species.

Present data suggest that aldehyde phytotoxicity itself is a minor pollution problem. However, in combination with the more common air pollutants, nitrogen dioxide and sulfur dioxide, phytotoxicity may be increased. The aldehydes may also contribute to the generation of the phytotoxic oxidants ozone and peroxyacyl nitrates, or PAN. Thus, the vegetation problem could become more serious if aldehyde and other pollutant concentrations rise substantially.

EFFECTS OF ALDEHYDES ON AQUATIC ORGANISMS

Thirty-six aldehydes have been identified in water, including industrial and sewage-treatment plant discharges, surface waters, and drinking water. Although the concentrations of many of these aldehydes in water are unknown, the concentrations of 22 in natural bodies of water or in drinking water have been determined to be less than 0.012 mg/L. The concentrations of five aldehydes that have been identified in aqueous waste discharges range up to 0.24 mg/L. Although the water-sampling sites have been limited, they are probably representative, and the results show that in general the aldehyde concentrations in the aquatic environment are relatively low.

Only seven of the 36 aldehydes (acrolein, formaldehyde, acetaldehyde, furfuraldehyde, crotonaldehyde, propionaldehyde, and vanillin) have been fully evaluated for acute toxicity in at least two aquatic species. The lowest reported median lethal concentrations

(LC_{50}s) for various exposure times and organisms range from about 0.05 mg/L for acrolein to 112 mg/L for vanillin and 130 mg/L for propionaldehyde. Acute-toxicity screening tests on 13 aldehydes showed most to be nontoxic to fish at 5 ppm and all to be nontoxic at 1 ppm. Only acrolein has been evaluated for chronic effects. From its evaluation of the data, the Environmental Protection Agency has determined the chronic LC_{50} values of acrolein to be 0.024 mg/L for the cladoceran Daphnia magna and 0.021 mg/L for the fathead minnow, Pimephales promelas.

On the basis of the method that uses calculated octanol-water partition coefficients (P), most of the aldehydes will probably not bioaccumulate substantially. However, six of them (capraldehyde, caprylaldehyde, 3,5-di-tert-butyl-4-hydroxybenzaldehyde, mesitaldehyde, nonylaldehyde, and undecylaldehyde) have log P values of at least 3.0; this suggests that they could accumulate appreciably in the tissues of aquatic organisms in the absence of rapid removal reactions.

Although little is known about the persistence of aldehydes in aqueous systems, it appears that a variety of aliphatic and aromatic aldehydes--including formaldehyde, acrolein, benzaldehyde, salicylaldehyde, syringaldehyde, and vanillin--can be biodegraded relatively rapidly.

The little information available now suggests that aldehydes (except acrolein) have low to moderate toxicity in aquatic organisms. We can conclude that the concentrations of aldehydes found in water are in most cases lower than those shown to have toxic effects in toxicity tests. There is some evidence that aldehydes do not persist for long periods in water that contains microorganisms; hence, the probability of occurrence of long-term effects appears to be low. However, many of the aldehydes have not yet been evaluated for toxicity in aquatic organisms, so our conclusion must be regarded as tentative.

CHAPTER 3

RECOMMENDATIONS

This document reviews the present knowledge on formaldehyde and some important higher aldehydes with respect to their production, properties, ambient and indoor concentrations, potential sources and sinks, and effects on humans, aquatic and terrestrial animals, and plants. The Committee recognizes that the first priority in its consideration is the determination of the effects of specific aldehydes on human health. However, serious deficiencies in its current knowledge prevent the immediate attainment of this primary goal. It is necessary to have unambiguous methods of analysis of specific aldehydes, comprehensive emission inventories, atmospheric generation and destruction rates, and measured concentrations of the individual aldehydes in indoor and outdoor environments to which human populations are exposed, as well as definitive health studies related to the specific aldehydes. This chapter identifies the missing scientific information that is needed if sound strategies for the control or abatement of aldehyde pollution are to be formulated and offers specific recommendations for obtaining the needed information. The Committee has not studied the direct or indirect economic impact of the implementation of these recommendations. The order of presentation reflects a suggested priority of the needed studies in each section, although we believe that all the recommendations deserve serious consideration by those concerned with the effects of the aldehydes on humans and the environment.

CHEMISTRY OF THE ALDEHYDES IN THE ENVIRONMENT

Many aspects of the sources, sinks, and transformation mechanisms of the aldehydes in the atmosphere, in outdoor and indoor environments, on the land, and in the surface waters remain ill-defined. A variety of further studies are required to permit the development of useful models of the potential ambient environmental and indoor concentrations of the aldehydes and their concentrations in natural surface and ground waters to which humans, terrestrial animals, aquatic organisms, and plant life will be exposed. Present information on the aldehyde exposure of the human population is at best incomplete.

There are some serious deficiencies in our present knowledge of indoor aldehyde sources and concentrations. In particular, studies on the following issues are required, to allow a careful assessment of the indoor-aldehyde problem: studies of building materials (particleboard, plywood, urea-formaldhyde foam insulation, etc.) from the point of view of their aldehyde emission rates and intervening factors (such as ventilation rate, temperature, and humidity); studies to measure the emission of other indoor sources of formaldehyde, such as gas-fired appliances, tobacco smoke, consumer products, and outdoor air; studies on the type and effectiveness of various schemes to reduce the indoor concentration of aldehydes; and monitoring studies that use reliable analytical techniques to assess aldehyde concentrations in a broad spectrum of occupied indoor environments.

Manufacturers of indoor plywood and particleboard should be able in the future to produce materials that have substantially lower emission of formaldehyde; however, it is not clear whether completely satisfactory solutions to the problem are possible. The Committee recommends that manufacturers continue work on the following promising approaches: Somewhat reduced amounts of formaldehyde should be used in preparation of the resins; improved polymerization recipes and better control of reaction conditions may result in less unreacted formaldehyde in the resin product. Techniques should be developed to remove excess or unreacted formaldehyde from the final product; for example, controlled heating and extended storage of the product before sale to the consumer will certainly promote escape of formaldehyde from the product. The surfaces of the final products should be sealed to minimize the escape of unreacted formaldehyde to the atmosphere; a specific sealing agent has recently been reported to reduce formaldehyde escape by about 70%, and paints and varnishes would presumably have some similar effect, but information on their use has not been reported; such sealers may also minimize moisture absorption and subsequent hydrolysis reactions. If suitable solutions to the formaldehyde problem of urea-formaldehyde resins cannot be obtained, the alternative is to select other resins, such as phenol-formaldehyde, melamine-formaldehyde, or epoxy; it is recognized that cost, appearance of the final consumer product, and somewhat poorer physical properties may militate against some or all of these alternative resins, and other building materials may be required as replacements for the present types of plywood or particleboard.

The combustion of fossil fuels--including natural gas, gasoline, diesel fuel, oil, and coal--and of wood, trash, etc., produces exhaust gases that contain both aldehydes and unburned hydrocarbons. Unburned hydrocarbons are transformed in part to aldehydes as intermediate compounds in atmospheric oxidation reactions. Ambient concentrations of these compounds in polluted urban areas are increased appreciably by exhausts from transportation vehicles. The controls used on new vehicles appear to offer reasonable regulation of both hydrocarbons and aldehydes. However, careful continued study of the emission from all internal-combustion engines is required as fuel composition and engine design are altered in the years ahead. In view of the high probability of the increased use of gasohol and methanol fuels and the expectation that formaldehyde and acetaldehyde are products of the incomplete combustion of these fuels, aldehyde emission from the exhaust of new and old vehicles of all kinds should be monitored regularly as these fuels increase in use. The aromatic-hydrocarbon content in liquid fuels, such as gasoline, may be of concern, for at least two reasons: first, there will be direct emission or vaporization losses, and, second, these compounds produce aromatic aldehydes in their atmospheric photooxidation. Benzaldehyde, the methylbenzaldehydes, and other aromatic aldehydes are precursors of the highly irritating peroxybenzoylnitrates, which would be formed in the atmospheric photooxidation reactions expected to occur. The increasing use of diesel fuels will cause somewhat new emission control problems. The emission of the higher aldehydes, as well as the common low-molecular-weight aldehydes, may be expected, and suitable controls may need to be investigated to address this potential problem.

The search for synergistic effects involving the aldehydes must continue. Thus, there should be special research efforts to investigate the ambient concentrations of bis(chloromethyl)ether in regions of high formaldehyde and hydrogen chloride concentrations. Further quantitative work is required to delineate the thermodynamic and kinetic properties of the formaldehyde-hydrogen chloride-bis(chloromethyl)ether system, to allow a quantitative assessment of the potential formation of the chloromethylether to which human populations may be commonly exposed.

Outdoor air concentrations of formaldehyde, the higher aliphatic and aromatic aldehydes, and acrolein should be monitored in the air on a continuing basis in a large number of heavily populated, rural, and remote areas, so that a reasonable data base on ambient aldehyde concentrations can be established.

The expected theoretical relation of the peroxyacylnitrates and peroxyarylnitrates and ozone to the precursor aldehydes should be tested in a continuing effort. Ambient concentrations of ozone and peroxyacetylnitrate and higher homologues should be measured at the same sampling sites used for the aldehyde determinations. Correlation checks will require continuous monitoring of nitrogen dioxide, nitric oxide, nonmethane hydrocarbons, carbon monoxide, methane, sulfur dioxide, and possibly other contaminants.

Industrial-plant manufacture or use of aldehydes will always result in the release of some aldehydes to the environment. Although the present industrial control methods appear to be well conceived and efficient removal of aldehydes is theoretically possible, continued measurement of this emission is advised. This applies not only to plants preparing formaldehyde and acetaldehyde, the most important commercial aldehydes, but especially to the manufacture of the highly toxic aldehyde, acrolein.

Installation of urea-formaldehyde foams as insulation material has often resulted in excessive formaldehyde emission. At least a substantial portion of the emission can be attributed to poor installation techniques or improper use of materials. It is recommended that companies supplying the materials develop improved reactants, training procedures, and installation procedures for local contractors. Potential customers should also be made aware that formaldehyde will be emitted for some period after installation, even with the best combination of installation procedures and materials. Such emission to the atmosphere and the aquatic environment is not well characterized in most cases.

The aldehydes in the aquatic and terrestrial environments are potential sources of human exposure. Thus, it is recommended that pollution of the aquatic and terrestrial environments be studied to determine the hazardous concentrations of aldehydes to which humans and other organisms are likely to be exposed.

QUANTITATIVE ANALYSIS OF ALDEHYDES IN THE ENVIRONMENT

The ultimate value of any research related to the environmental effects of the aldehydes depends on the reliability, reproducibility, and accuracy of the analytical data that demonstrate the nature and amount of the aldehydes. Although many analytical procedures have been used in previous aldehyde studies, there are substantial problems associated with most of those in use today.

No technique common to the analysis of all aldehydes can yet be recommended. A series of more limited techniques that are widely used, generally for an individual aldehyde like formaldehyde, are discussed in Chapter 6. Very few of these are without fault in one or more respects: calibration, sampling procedure, or method of analysis. These limitations prevent their recommendation. Many of the techniques have common procedures, and hence common faults. Improved procedures for calibration, sampling, and analysis that are now recognized must be coupled to produce a series of refined, although still limited, techniques that can be recommended as standard measurement procedures and thus be applied immediately. However, emphasis should be on developing new techniques to secure the greatest benefit in the shortest time.

Wet-chemical spectrophotometric methods of analysis are the most practical and best-established methods for determining aldehydes, and their continued use is recommended for the immediate future, with some stipulations. First, there must be recognition that the information provided by these methods is limited; they generally measure either an individual aldehyde or the aldehydes as a class without discrimination. The accurate assessment of specific aldehydes as environmental pollutants may require the application of several methods. Second, recommendation of these methods of analysis should not impede investigations of promising alternatives. Possible improvements--such as increased sensitivity, decreased analysis time, or the simultaneous quantitative determination of several aldehydes--should be actively sought. With these two stipulations, we recommend the seemingly optimal wet-chemical spectrophotometric methods for aldehydes, of which the most widely used for the determination of formaldehyde are based on chromotropic acid and pararosaniline reagents. However, there are serious problems with methods that use chromotropic acid: determination of optimal analysis conditions, interfering substances, and lower sensitivity relative to alternative reagents. The pararosaniline method appears to be a suitable replacement for the chromotropic acid method. Its sensitivity is high, and interferences are minimal. Extensive testing should be continued, to confirm its use as a standard method for the analysis of formaldehyde in air. For the near future, the acetylacetone method shows the greatest promise, by virtue of its greater sensitivity, and should be evaluated for use in the analysis of air.

In view of the limitations of information obtained by measuring the total aldehyde content of mixtures that may have various ratios of aldehydes with substantially different toxicity, it is recommended that the methods of analysis for "total" aldehyde not be used in future studies involving atmospheric aldehydes.

Techniques for the quantitative analysis of a large number of specific aldehydes in the environment are highly desirable for maximizing information. Such techniques probably will rely on gas-phase or liquid-phase chromatography for separation. Because such techniques also rely on derivatization of the aldehydes, one possible approach would involve the use of passive monitors containing a derivatizing trapping agent in conjunction with a chromatographic separation method and analysis.

Techniques that can provide real-time measurements at field sampling sites are extremely desirable. It may be possible to develop continuous monitors that use established wet-chemical spectrophotometric methods of analysis; this would require little chemical research, but considerable engineering development. A number of alternative direct spectroscopic measurement techniques have been used in laboratory and atmospheric studies, but it is difficult to recommend these for a large number of sites without promise of future reduced cost and increased portability. However, it should be recognized that such direct techniques can provide analytical information based on characteristic spectral line structure and position; this allows an excellent check on possible unforeseen interferences that may be present in the less direct aqueous-phase spectrophotometric methods.

There are two methods of analysis to be considered for measuring the highly toxic and environmentally important aldehyde, acrolein. The method using 4-hexylresorcinol is well established, and its continued use is recommended. Field mishaps may be minimized and sample stability improved by collection in a bisulfite solution. A second fluorimetric method using m-aminophenol shows promise and could offer substantially improved sensitivity. Further tests of this second system are recommended.

To assess the health impact of aldehydes as environmental pollutants, it is desirable to expedite measurements so that a maximal number of samples may be analyzed. With this in mind, we note that passive monitors offer a great potential for expediting large-scale sampling. Therefore, it is recommended that research be directed toward perfecting a passive-monitor trapping agent consistent with one or more of the methods of analysis currently available. The collection of aldehydes on solid sorbents and later removal with an appropriate solvent represents one avenue of research. If they are consistent with available methods of analysis, passive monitors could be deployed with minimal delay.

There is no wet-chemical spectrophotometric method of analysis now available for the specific determination of acetaldehyde. Because the toxicity of acetaldenyde is low, relative to that of other aldehydes, this analysis may seem unnecessary and unimportant. However, acetaldehyde is a major source of peroxyacetyl nitrate in the urban environment, so it is very important to develop methods that allow the monitoring of this major precursor of a highly toxic compound. It is evident that the development of techniques for the quantitative analysis of all the individual aldehydes present will permit the measurement of acetaldehyde.

HEALTH EFFECTS OF FORMALDEHYDE

There is an urgent need for research to resolve several important questions related to the health effects of formaldehyde. The most noteworthy needs that the Committee has identified are outlined here.

It is not known what fractions of persons with asthma, atopic subjects, nonatopic persons, and patients with chronic obstructive lung disease constitute susceptible populations. Quantitative information on the proportion of the general population that is susceptible to the effects of formaldehyde and on the extent of the variability in response among this population may be obtained with appropriate epidemiologic techniques. A practical means for identifying susceptible subjects in the population is needed. Whether children, infants, pregnant women, older persons, and persons with specific medical conditions (e.g., heart disease) are also susceptible to the effects of formaldehyde also needs to be explored.

Controlled studies of the range of irritation responses to formaldehyde at concentrations below 1 ppm are few. Both epidemiologic studies and human inhalation experiments are necessary to assess the risk more precisely. These studies should include several formaldehyde concentrations below 1 ppm.

More objective means for determining human eye, nose, and throat irritation responses are needed. The reported studies have relied on subjective complaints. Because small-airway involvement may be a manifestation of lung involvement, future studies should incorporate tests of small-airway function, as part of both epidemiologic and chemical inhalation studies with humans.

In general, to identify specific health effects associated with exposure, information is needed from extensive, long-term epidemiologic studies that include persons from selected occupational sites and residences (conventional homes and mobile homes) and that involve cohorts (especially pregnant women, neonates, older children, and the infirm) and proper controls. Investigations should explore ways of identifying exposure with biologic tests (e.g., on urine or blood) and comparing them with the concentrations of chemical contaminants in air. Data on dose-response relationships are needed for use in developing control strategies. In addition, there must be careful documentation to show the relationship of human exposure to complaints, particularly nonspecific symptoms (headaches, tiredness, thirst, drowsiness, etc.).

Human epidemiologic investigations assessing the carcinogenic potential of formaldehyde are lacking. Human studies should address carefully the magnitude and duration of exposure, cigarette-smoking habits, and the presence of other environmental contaminants, such as bis(chloromethyl)ether, or confounding factors. Animal studies should include a number of different species, including primates. The importance of hyperplasia and metaplasia of nasal mucosa in humans and animals requires clarification, including the natural history and sequence of changes, dose-response relationships, the regression of lesions after removal of formaldehyde exposure, and potential screening tests of value (such as nasal swabs for cytologic examination).

Long-term effects of continuous low-dose exposure to formaldehyde are not known; particularly needed is an assessment of the mutagenic, embryotoxic, and teratogenic potential through human epidemiologic and laboratory animal studies. The observation of the mutagenic potential of formaldehyde in a wide variety of organisms points to the need for new work to ascertain the mutagenic and carcinogenic potential of formaldehyde in mammalian germinal or somatic cells. This information is required to evaluate properly the hazard to persons exposed to formaldehyde.

The mechanism of the airway response to formaldehyde is not known. Controlled inhalation studies with histamine or methacholine challenge tests are needed for assessing formaldehyde's effects on airways. Tests can be performed before and after low-dose exposures. In addition, investigations should be made to identify how formaldehyde sensitizes the airways and to determine whether there is an immunologic or nonimmunologic basis.

Epidemiologic studies of dermatitis due to formaldehyde are needed in determining prevalence, clinical history, and other contributing factors. Epidemiologic studies evaluating the risk and nature of skin reactions should include formaldehyde patch tests. It is not known whether airborne formaldehyde can cause allergic skin reactions. Therefore, studies of this and other routes of exposure and of the skin metabolism of formaldehyde are needed.

The effects of formaldehyde on nasal and lung defense mechanisms have not been well studied. More investigations showing the relationship of formaldehyde exposure and resulting effects on nasal and bronchial ciliary motility, alveolar machrophage function, and other defense processes are needed.

Limited information is available on the interactions of formaldehyde with other air pollutants. These studies are best performed as inhalation experiments, in which important variables can be controlled better than in field studies. The persons studied should include those believed to be susceptible to the effects of formaldehyde.

HEALTH EFFECTS OF SOME OTHER ALDEHYDES

Some of the higher-molecular-weight aldehydes appear to have effects that demand confirmation and quantitative evaluation to provide the proper health-risk evaluation and development of control strategies. Acetaldehyde was reported to be both embryotoxic and teratogenic in a single study in mice. These effects were similar to those of ethanol in humans. Because of the metabolic relationship between ethanol and acetaldehyde, the effects on the embryo need to be examined more extensively in other animal models. Acetaldehyde was shown to have chromosome-breaking activity in mammalian cells; that indicates that it may have mutagenic potential. Epidemiologic evidence also indicates that alcoholics have a higher risk of cancer. Again, the close metabolic relationship of acetaldehyde and ethanol requires that the carcinogenic potential be assessed. None of the existing studies provides sufficient information for an analysis of risks to humans.

Acrolein is seemingly one of the most acutely toxic and highly irritating of the aldehydes commonly encountered in the environment. In a single study in rats, acrolein was not found to be embryotoxic. However, the fetuses were not examined for malformations. Therefore, no information on the teratogenic potential of acrolein is available, and this should be studied further. Acrolein was not shown to be carcinogenic in a study on hamsters; there was a minimal effect on the carcinogenicity of benzo[a]pyrene. Both the cocarcinogenic effect and the carcinogenic potential of acrolein need to be evaluated further in other animal models, to determine whether the hamster is refractory in acrolein exposure studies. The intense eye irritation that is induced in humans by acrolein at very low concentrations should be investigated to establish the mechanisms that cause the severity of the reactions.

Investigations are also needed to characterize further the effects of the common aldehydes (e.g., butyraldehyde and acrolein) on humans, especially at concentrations present in the workplace, home, and general environment. Studies are needed to assess the importance of low-dose chronic exposures and interactions with other atmospheric contaminants.

Further studies are required to establish the suggested role of formaldehyde, acrolein, and possibly other aldehydes in eye irritation associated with high concentrations of photochemical smog. Tests for possible relations of eye irritation and formic acid, peroxyacetyl nitrate, and other products derived from the aldehydes should be made.

Human exposure to atmospheric aldehydes may be repetitive, as in occupational situations, or continuous, as in a residential environment contaminated with aldehydes from cigarette smoke, automobile exhaust, or out-gassing from aldehyde-containing consumer products. Animal studies are needed to investigate the pathophysiologic effects, immunologic aspects, and elements of sensitivity associated with continuous chronic exposure to aldehydes, for use in assessing the potential hazards and the results of epidemiologic studies.

EFFECTS OF ALDEHYDES ON VEGETATION

Several overt and subtle effects of aldehyde phytotoxicity have been reported in the very few studies of aldehydes that have been conducted. To understand the phenomenon fully, systematic studies like those conducted with the major air pollutants--sulfur dioxide, ozone, and hydrogen-fluoride--are recommended. The more common aldehydes such as formaldehyde, acetaldehyde, propionaldehyde, and acrolein should be used to screen economically important crops for sensitivity. Plant factors such as genetic variability, age, and nutrition and climatic and edaphic factors that influence plant growth should be examined to determine whether they increase plant susceptibility. Dose-response data should be obtained for the major aldehydes singly and in combination with each other and with other pollutants. In addition to visible injury (obvious symptoms), biochemical and physiologic alterations in plants should be assessed. The discovery of an aldehyde-sensitive "indicator" plant would prove useful in detecting aldehyde pollution in the environment.

EFFECTS OF ALDEHYDES ON AQUATIC ORGANISMS

Available data on the toxicity of aldehydes to aquatic organisms show that the acute toxicity of aldehydes can vary considerably. No toxicity data are available on the majority of the aldehydes that have been identified in aquatic systems. Although chronic effects are unlikely because of the instability of aldehydes in water, it is recommended that a program be developed and implemented to assess systematically the probable hazard of the commonly encountered aldehydes to aquatic life and to identify those which should be controlled.

CHAPTER 4

COMMERCIAL PRODUCTION, PROPERTIES, AND USES OF THE ALDEHYDES

The aldehydes are a very important class of organic compounds; they are characterized by the presence of the <u>formyl</u> functional group,

$$-\overset{\overset{\displaystyle O}{\|}}{C}-H$$

which we represent in this report as -CHO. The general structural formula of the aldehydes can be written as

$$R-\overset{\overset{\displaystyle O}{\|}}{C}-H$$

The first member of the aldehyde family is formaldehyde (HCHO), in which the R group is a hydrogen atom. For the higher aldehydes, acetaldehyde (CH_3CHO), propionaldehyde (C_2H_5CHO), and <u>n</u>-butyraldehyde (<u>n</u>-C_3H_7CHO), the R groups are CH_3, C_2H_5, and <u>n</u>-C_3H_7, respectively. The physical properties of the aldehydes that have some potential importance in the environment are summarized in Table A-1 of the Appendix. Table A-2 summarizes the uses of selected aldehydes and presents the various synonyms for their names.

Formaldehyde is the most common and important aldehyde in the environment, and the properties of its several common forms are considered in some detail in the first section of this chapter. In subsequent sections we consider the methods of aldehyde production and the manufacture of aldehyde-containing consumer products.

PROPERTIES OF VARIOUS FORMS OF FORMALDEHYDE

MONOMERIC FORMALDEHYDE

Monomeric formaldehyde is a colorless gas that condenses to form a liquid of high vapor pressure that boils at -19°C (760 Torr); it forms a crystalline solid at -118°C. It has a pungent odor that is highly irritating to the exposed membranes of the eyes, nose, and upper respiratory tract. In the pure dry, liquid form at low temperatures (-80 to -117°C), it does not polymerize rapidly; its stability depends on its purity, and it must be held at a low temperature to avoid polymerization. It is not commercially available in this form, but can be prepared for laboratory use by the original method of Spence and Wild.[23]

The molecule of gaseous formaldehyde in ambient air is designated by the molecular formula HCHO or the structural formula,

$$\begin{array}{c} H \\ \diagdown \\ C=O \\ \diagup \\ H \end{array}$$

TRIOXANE

Trioxane is the cyclic trimer of formaldehyde (trioxymethylene). It has the molecular formula of $C_3H_6O_3$, with three HCHO units per molecule. Its structural formula is:

$$\begin{array}{c} CH_2 \\ OO \\ | | \\ CH_2 CH_2 \\ O \end{array}$$

In pure form, it is a colorless, crystalline solid that melts at 61-62°C, and it boils at 115°C. It has a chloroform-like odor, and it is not irritating. It is combustible and burns readily when ignited (flash point, 45°C). It is soluble in water, and saturated solutions contain approximately 21 g/100 cc at 25°C.

PARAFORMALDEHYDE

Paraformaldehyde is a colorless solid in a granular form with an odor characteristic of monomeric formaldehyde. It is prepared by condensation of methylene glycol ($HOCH_2OH$), and its composition is best expressed by the formula $HO-(HCHO)_8-H$. Commercial grades of paraformaldehyde usually specify not less than 95% formaldehyde by weight, and they may contain up to 99%. Paraformaldehyde melts over a wide temperature range (120-170°C), which depends on the degree of polymerization. At room temperatures, it gradually vaporizes largely as the monomeric formaldehyde with some water formation, and the rate is increased by heating. Thus, it is commonly used as a source of formaldehyde for disinfecting large areas. It dissolves in hot water, and a solution of approximately 28% can be obtained by agitating it with water at 18°C for 5 wk.

FORMALIN

Formalin is the principal form in which formaldehyde is marketed; it is an aqueous solution that ranges in concentration from 37 to 50% by weight. The National Formulary solution contains not less than 37% formaldehyde with methanol (usually 6-15%, depending on the usage requirements) to suppress polymerization. It is a clear solution with the strong pungent odor of formaldehyde. Cloudiness is usually due to polymers, which form at various rates that depend on methanol content and storage temperature. The solution is slightly acid--0.005-0.01 M, calculated as formic acid.

In aqueous solutions, the dominant form of the formaldehyde is methylene glycol; in concentrated solution, it is one of many polymer molecules, $HO-(CH_2O)_n-H$, polyoxymethylene glycol.

SOME CHEMICAL REACTIONS OF FORMALDEHYDE

Formaldehyde vapor is relatively stable with respect to thermal decomposition; at temperatures above 400°C, it decomposes to form carbon monoxide, hydrogen, and methanol in the overall reactions,

$$2HCHO \rightarrow CH_3OH + CO \qquad (1)$$

$$\text{and } HCHO \rightarrow CO + H_2 \qquad (2)$$

Reaction 1 is catalyzed on metal surfaces and must occur heterogeneously.[4,14] Reaction 2 may occur as written--i.e., a direct decomposition into two stable molecules--or it may occur by a free radical pathway initiated by a primary rupture of a carbon-hydrogen bond: $HCHO \rightarrow H + HCO$; $H + HCHO \rightarrow H_2 + HCO$; $HCO + M \rightarrow H + CO + M$.

The carbon-hydrogen bonds in the formaldehyde molecule are relatively weak, and the rate constants for the hydrogen-atom abstraction reactions by free radicals are large (see Chapter 5). For example, the HO-radical attack on formaldehyde, $HO + HCHO \rightarrow H_2O + HCO$, has a rate constant that is near the collision number and is independent of the temperature.

Formaldehyde monomer vapors at pressures above about 0.5 Torr show a tendency to polymerize at room temperature.[22] The equilibrium vapor pressure of monomeric HCHO over polymeric HCHO is much higher at high temperatures, and monomer pressures of several hundred Torr can be maintained readily for several hours without substantial polymerization if the containing vessel is heated to 100°C or higher.

In the aqueous phase, formaldehyde is oxidized readily by even mild oxidizing agents, such as $Ag(NH_3)_2^+$, and this property has been exploited in the development of several wet-chemical analytical methods for formaldehyde.

OXIDATION AND REDUCTION REACTIONS

On oxidation under controlled conditions in the gaseous or dissolved state, formaldehyde may be converted in part to formic acid, or under more highly oxidative conditions to carbon monoxide (with some carbon dioxide), and water. The photooxidation of formaldehyde in the gas phase leads to carbon monoxide, hydrogen, hydrogen peroxide, formic acid, and some other metastable products (see Chapter 5). Performic acid is produced under special conditions through the oxidation of formaldehyde solutions at low temperatures.

REACTIONS OF FORMALDEHYDE WITH FORMALDEHYDE

Cannizzaro Reaction

This reaction involves the reduction of one formaldehyde molecule with the oxidation of another. Although it is normally catalyzed by alkalies, the reaction can occur when formaldehyde is heated with acids at 40-60°C: $2HCHO(aq) + H_2O \rightarrow CH_3OH + HCO_2H$. At 70°C, the reaction may proceed through an aldol condensation, wherein carbohydrates are formed. Formaldehyde and other aldehydes that do not possess alpha-hydrogen atoms do not undergo ordinary aldol condensations, but can react almost quantitatively in alkaline solution and liberate hydrogen:

$$HCHO(aq) + NaOH \rightarrow HCO_2Na + H_2$$

$$H_2 + HCHO(aq) \rightarrow CH_3OH$$

Tischenko Reaction

Polymers of formaldehyde when heated with either aluminum or magnesium metal powder form methyl formate:

$$2HCHO(polymer) \rightarrow HCO_2CH_3$$

Polymerization Reactions

The formation of resinous products on reaction with other chemicals is one of the most useful characteristics of formaldehyde and is the reason for its immense importance in the synthetic resin industry. Under suitable conditions, the molecules of many compounds are linked together by methylene groups when subjected to the action of formaldehyde. Phenol- and urea-formaldehyde resins are polymethylene compounds of this type.

Two distinct mechanisms are probably involved in resin-forming reactions: the polycondensation of simple methyl derivatives and the polymerization of doubled-bonded methylene compounds. Although in some cases the mechanism is definitely one or the other of the two, it is often not clear which is followed, and both may play a part in some instances. Recent evidence indicates that the formation of urea-formaldehyde resins, which used to be regarded as a simple polycondensation of methylol ureas, may actually involve the primary formation of a methylene urea that then polymerizes to give a cyclic trimethylenetriamine whose methylol derivatives are finally cross-linked by condensation.

Thermoplastic resins are the result of simple linear condensations, whereas the production of thermosetting resins involves the formation of methylene cross-linkage between linear chains. Both types may be produced from the same raw materials by variations in the relative amounts of formaldehyde used, the conditions of catalysts, and the temperature. However, with compounds whose molecules present only two reactive hydrogen atoms, only thermoplastic resins can be obtained.

A diverse group of organic compounds--including alcohols, amines, amides, proteins, phenols, and hydrocarbons--form resins with formaldehyde.

INDUSTRIAL PRODUCTION AND USES OF THE ALDEHYDES

Aldehydes as a family are produced in the United States at a rate of several billion pounds per year.[25] An even greater quantity is produced in other parts of the world. The more important aldehydes (on the basis of production rates) are made with feedstocks obtained from petroleum or natural gas; hence, they are generally considered to be petrochemicals.

Several aldehydes find large and generally major uses as feedstocks for the production of other chemicals. Considerable amounts of several aldehydes are used captively in a given plant--i.e., they are produced and used in the same plant. Large quantities of aldehydes, however, are transported to other plants or locations for use. The following factors are considered here with respect to the most important aldehydes: industrial processes used for production, annual rates of production, end uses, and properties.

FORMALDEHYDE

Production

Formaldehyde is the most important aldehyde in the United States and in the remainder of the world, on the basis of rates of production.[25] Most formaldehyde is stored and transported as aqueous solutions containing 37-50% formaldehyde and 1-15% methanol. In 1978, total production capacity in the United States was about 9×10^9 lb of aqueous solution, or about 3.3×10^9 lb on an anhydrous basis. Actual production is estimated to be only 70% of capacity, or approximately 6.3×10^9 lb of formaldehyde solutions per year. It is estimated that 65% of the formaldehyde produced is used in the same plant in which it is produced.

Costs for transporting formaldehyde tend to be high, because water and methanol also need to be transported. Hence, as a general rule, formaldehyde solutions are transported only relatively short distances, and little formaldehyde is exported or imported. Several large formaldehyde plants are near lumber plants in the South and the far West, inasmuch as the two largest uses for formaldehyde solutions are in production of plywood and particleboard.

Methanol is the starting feedstock for commercial production of formaldehyde.[5,7,10,15,21,27] For some 40 or 50 yr, methanol has been produced almost exclusively by the reaction of carbon monoxide and hydrogen under high pressure in the presence of catalysts. Both carbon monoxide and hydrogen are generally obtained from natural gas or petroleum fractions. Other materials that, at least in theory, can be used are coal, shale oil, oil from tar sands, and cellulose. Coal has already found limited use, and it will probably increase in the future. For at least the next 10 or 15 yr, however, petroleum-based hydrocarbons will probably remain the preferred feedstock for production of methanol.

Methanol (wood alcohol) was produced in the early part of this century primarily from wood. But this process cannot compete economically with the process that uses petroleum-based feedstocks. Another process that is no longer economically feasible is a process in which propane and butane are partially oxidized to produce a wide variety of oxygenated products, including methanol, formaldehyde, and acetaldehyde.[2]

In the United States, 16 companies produce formaldehyde; capacities of formaldehyde plants vary widely from about 14 to 600 x 10^6 lb/yr.[15] Three companies (Celanese, Borden, and Du Pont) have over 50% of our national capacity. Two quite different processes are used.[5,7,10,19,20] In one, a mixture of methanol and oxygen is passed over a silver catalyst. The main reaction is the dehydrogenation of methanol:

$$CH_3OH \xrightarrow{Ag} HCHO + H_2$$

Part of the hydrogen is oxidized with oxygen to produce water vapor.

In the other process, a mixture of methanol and air is passed over a catalyst consisting of molybdenum and iron oxides. The main reaction is this oxidative dehydrogenation of methanol:

$$CH_3OH + O_2 \xrightarrow{Mo, Fe\ oxides} HCHO + H_2O$$

Relative advantages of the two processes have been discussed in considerable detail by Diem,[7] Sleeman,[21] and Chauvel et al.[5] The capital costs of the silver-catalyst process are higher, but its operating costs are lower. The ratio of formaldehyde to methanol in the product solution is normally higher in the oxide-catalyst process; this product is preferred for some end uses.[5]

Uses

Major uses for formaldehyde have been reported elsewhere.[9,24] About 50% of the formaldehyde produced is consumed in the production of urea-formaldehyde and phenol-formaldehyde resins. These resins are used in the production of plywood, particleboard, foam insulation, and a wide variety of molded or extruded plastic items. Another 20-25% is used in the production of other resins or high polymers, including polyacetals, melamine resins, and alkyd resins. Hence, 70-75% of the formaldehyde is used in the production of high-polymeric resins or plastics. Formaldehyde is also used to produce hexamethylenetetramine, pyridine, trioxane, paraformaldehyde, chelating agents, and nitroparaffin derivatives. Formaldehyde solutions (often referred to as formalin) are used as disinfectants, embalming fluids, and textile-treatment agents and in leather and dye manufacture.

ACETALDEHYDE (CH_3CHO)

Production

In 1978, production capacity for acetaldehyde in the United States totaled about 1.7×10^9 lb, but actual production was approximately 1.0×10^9 lb.[17,24] Acetaldehyde is generally stored and transported as a liquid. Because it has a normal boiling point of 20.8°C, storage vessels must be capable of withstanding high pressures.

About 80% of the world's aldehyde is produced by controlled oxidation of ethylene with an aqueous solution of palladium and cupric chlorides as catalysts.[1] The overall desired reaction is as follows:

$$C_2H_4 + 0.5O_2 \rightarrow CH_3CHO$$

This acetaldehyde process was first commercialized in about 1960;[12] two versions are now used industrially. A two-stage version was developed by Wacker Chemie, but Farbwerke Hoechst has developed a one-step version. These two versions are often referred to as the Wacker process and the Wacker-Hoechst processes, respectively.[11] In both, more than 93% of the ethylene feedstock is converted to and recovered as acetaldehyde. Carbon dioxide, water vapor, and chlorinated hydrocarbons are byproducts.

In the one-step process, oxygen is used as the oxidant. It is mixed with ethylene, and the mixture is bubbled upward through the catalytic solution. About 25% of the ethylene reacts per pass over the catalyst, and most of the unreacted ethylene is recovered and recycled to the reactor.

In the two-step process, ethylene is bubbled upward through the catalytic solution. The following reactions are the predominant ones in the first step of the process:

$$C_2H_4 + PdCl_2 + H_2O \rightarrow CH_3CHO + Pd + 2HCl$$

$$\text{and } Pd + 2CuCl_2 \rightarrow PdCl_2 + 2CuCl$$

In the second step, the catalyst solution is regenerated with air in a separate reactor, as follows:

$$2CuCl + 2HCl + 0.5O_2 \rightarrow 2CuCl_2 + H_2O$$

Almost all the ethylene reacts in a single pass through the reactor in the two-step version; hence, recovery and recycling of ethylene are not critical as a rule.

In both versions of the process, acetaldehyde is separated from the exit gas stream from the reactor by water absorption, and an aqueous solution of acetaldehyde is produced. Unreacted ethylene, if any, is recycled to the reactor. There is always a need for a vent stream, to remove chlorinated byproducts. This vent stream contains some ethylene and low amounts of acetaldehyde; the exact concentrations of these materials in the vent stream apparently have never been reported for any specific industrial plant. If necessary, however, an absorber could be designed and operated to remove essentially all acetaldehyde from the vent stream. To prevent most of the combustible hydrocarbons, including acetaldehyde, from escaping to the surroundings, it is generally more economical to send the vent stream to a flare or to the furnace. In the one-step process, the vent stream has a substantial fuel value.

Before development of the Wacker technology, the following two processes were of major importance:
- Hydration of acetylene. This process was commercialized in Germany during World War I. Several modifications have been reported, but the process has not been competitive with the Wacker-Hoechst process, because of the relatively high price of acetylene, compared with ethylene.
- Dehydrogenation of ethanol. In many respects, this process is similar to the one for production of formaldehyde from methanol. To make its use feasible in the future, the required ethanol reactant must be available at a much lower cost than ethylene.

Although these two process are still used to a limited extent, it is unlikely that they will be used in any new plants in the near future.

In transporting or storing acetaldehyde, extensive precautions must be taken to prevent leaks and ensure safe conditions, because this aldehyde boils at room temperature. When mixed with air, it is highly flammable and reacts to form acetic acid, highly explosive peroxides, and other products. It is transported in drums or insulated trucks or tank cars. Specific information was not found on acetaldehyde concentrations in the atmosphere in or near acetaldehyde plants.

Uses

About 60% of the acetaldehyde produced is used as feedstock for the production of acetic acid and acetic anhydride. The remaining 40% is used in the production of pentaerythritol, peracetic acid, pyridine, crotonaldehyde, 1,3-butylene glycol, and various other chemicals. Hester and Himmler[12] reviewed the numerous chemicals manufactured in 1958 from acetaldehyde.

ACROLEIN ($CH_2=CHCHO$)

Production

Acrolein is produced in the United States by Shell Chemical Co. and Union Carbide Corp.; the annual production in 1978 was estimated at about 45×10^6 lb.[25]

Acrolein has been produced by several processes in the past, including condensation of formaldehyde with acetaldehyde and the pyrolysis of diallyl ether.[26] The method currently used is the catalytic oxidation of propylene; a mixture of propylene, air, and steam in a mole ratio of approximately 1:10:2 is passed over a catalyst of mixed metal oxides. Acrolein yields, on the basis of inlet propylene feed, are about 70%, but substantial amounts of acetaldehyde and acrylic acid are also produced. A water absorption unit and distillation are used for recovery and separation of acrolein, acetaldehyde, and acrylic acid.

Acrolein is a colorless liquid; it is highly volatile and highly reactive. Because it is highly irritating, absorbers are used to minimize acrolein losses to the atmosphere. In addition, gaseous emission streams are generally sent to either a flare or a furnace, to destroy acrolein in any gas stream by combustion before it is exhausted to the atmosphere. Careful design and close attention to operating and maintenance procedures are necessary to minimize acrolein losses or leaks at pumps, valves, and storage vessels. Undesired reactions of acrolein, such as polymerization, are minimized by adding inhibitors and stabilizers to the liquid acrolein.

Uses

Approximately half the acrolein produced is used as a feedstock for production of glycerine,[26] and about 25% to produce the amino acid methionine, an essential protein added to various foods. The remaining 25% of acrolein is used in the production of many chemicals, including glutaraldehyde, 1,2,6-hexanetriol, quinoline, pentaerythritol, cycloaliphatic epoxy resins, oil-well derivatives, and water-treatment chemicals.

HIGHER ALIPHATIC ALDEHYDES

Production

The Oxo process is the application of a chemical reaction called oxonation, or more properly hydroformylation, for production of C_3-C_{16} aliphatic aldehydes.[13] Carbon monoxide and hydrogen are caused to react with the double bond of an olefin to produce an aldehyde with at least one more carbon atom than the olefin. In the case of ethylene, the overall reaction for production of propionaldehyde is as follows:

$$CH_2=CH_2 + CO + H_2 \rightarrow CH_3CH_2CHO$$

In the case of proplyene, both n-butyraldehyde and isobutyraldehyde are produced.

In the past, cobalt carbonyls were used almost exclusively as catalysts for the Oxo process, and relatively high pressures, often 200-400 atm, were required.[13] In the last few years, various catalysts have been proposed that offer a variety of advantages, including higher yields, improved product compositions, and lower operating pressures. Rhodium catalysts, for example, are now widely used in Oxo processes of at least several olefins.[16]

A portion of the aldehyde formed is hydrogenated to produce alcohols. For example, some propionaldehyde is hydrogenated to 1-propanol, some n-butyraldehyde to 1-butanol, and some isobutyraldehyde to 2-butanol.

Propionaldehyde is produced by two American manufacturers, Eastman Kodak Co. and Union Carbide Corp. Production in 1978 was estimated at over 190×10^6 lb.[25] About 750×10^6 lb of butyraldehydes were produced in 1976. Major American producers of butyraldehyde are Badische, Celanese Corp., Eastman Kodak Co., and Union Carbide Corp.

The Oxo process is used for the manufacture of several aldehydes that are consumed in the production of plasticizers. In such cases, the aldehydes are hydrogenated to produce alcohols; the alcohols then react with acids to produce the esters that serve as plasticizers.

Uses

Propionaldehyde is used primarily as a chemical intermediate; the percentages consumed in this country for different purposes are approximately as follows: 1-propanol, 40%; propionic acid, 37%; and trimethylolethane, 23%. Butyraldehydes are used as chemical intermediates in the production of 1-butanol, 2-butanol, 2-ethyl-1-hexanol, and a wide variety of specialty chemicals. Over 1×10^9 lb of plasticizers are used each year in the preparation of poly(vinyl chloride) plastics; C_4-C_{12} aldehydes are used for this purpose. Several of the higher aliphatic aldehydes, particularly C_{13}-C_{16} aldehydes, are used in the production of detergents.

BENZALDEHYDE (C_6H_5CHO)

Benzaldehyde is a colorless or yellowish, highly refractive oil with an odor resembling that of oil of bitter almonds; it is the simplest aromatic aldehyde.[6] Total world production is probably less than 10×10^6 lb/yr.

Toluene is the feedstock used for production of benzaldehyde. At least three processes have been used industrially:[6]

- Toluene is chlorinated to produce benzal chloride (or α,α-dichlorotoluene), which is then hydrolyzed to produce benzaldehyde.
- Liquid toluene is oxidized in the presence of a catalyst, such as manganese dioxide.
- Toluene vapors are oxidized on a catalyst, such as vanadium pentoxide.

Major U.S. producers of benzaldehyde are Benzol Products Co., Heyden, Newport Chemical Corp., and Tennessee Product and Chemical Corp.

Uses

Benzaldehyde has important uses in dyes, pharmaceuticals, perfumes, and flavoring agents.

FURFURAL

$$\begin{array}{c} \text{HC} \overset{O}{\frown} \text{C-CHO} \\ \| \quad \| \\ \text{HC} - \text{CH} \end{array}$$

Production

Furfural (2-furaldehyde, furfuraldehyde, furfurol, or furol), a colorless liquid aldehyde is produced from a variety of agricultural byproducts, including corncobs, oat hulls, rice hulls, bagasse, cottonseed hulls, and paper-mill wastes.[8] It is soluble in most organic solvents, but only slightly soluble in water. Furfural is essentially a substituted furan; the aldehyde group is attached to the five-member heterocyclic ring that contains one oxygen atom and two carbon-carbon double bonds.

The raw materials used for furfural production are typically brought together in dilute sulfuric acid, and the mixture is heated under pressure. On completion of the reaction the pressure is released, causing the furfural to vaporize, with considerable water. The crude furfural is then purified primarily by distillation.

Uses

Furfural is used in the manufacture of furan and several tetrahydrofuran compounds. It is used extensively as a selective solvent in the production of lubricating oils, gas oils, diesel fuels, and vegetable oils. It also finds uses in the production of modified phenol-formaldehyde resins and in the extractive distillation of butadiene.

MANUFACTURE OF ALDEHYDE-CONTAINING CONSUMER PRODUCTS

UREA-FORMALDEHYDE RESINS

Some urea-formaldehyde resins emit formaldehyde over extended periods. A brief discussion of the manufacturing (or polymerization) technique used to produce these resins will help to explain the emission problem and will suggest ways to eliminate or at least minimize it. The resins are prepared by causing urea to react with formaldehyde.[3] Each of the four hydrogen atoms in a urea molecule is potentially reactive. If urea and formaldehyde reacted on an exactly equimolar basis, the following reactions indicate the formation of a typical so-called thermoplastic resin (or high polymer).

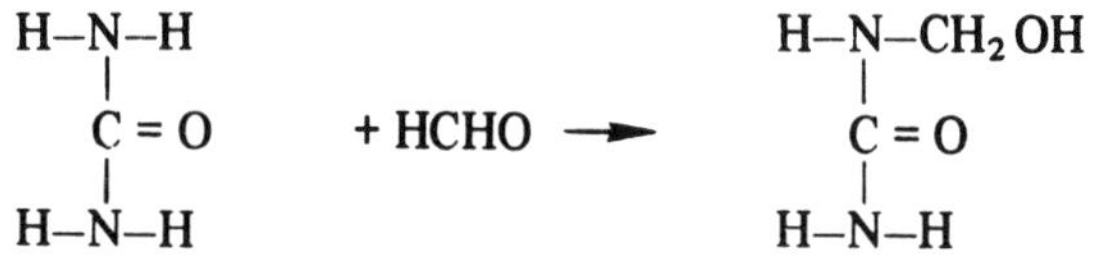

$$\begin{array}{c} H-N-H \\ | \\ C=O \\ | \\ H-N-H \end{array} + HCHO \longrightarrow \begin{array}{c} H-N-CH_2OH \\ | \\ C=O \\ | \\ H-N-H \end{array}$$

urea

The intermediate product formed is basically a monomer, and it polymerizes as follows to produce a thermoplastic resin and water:

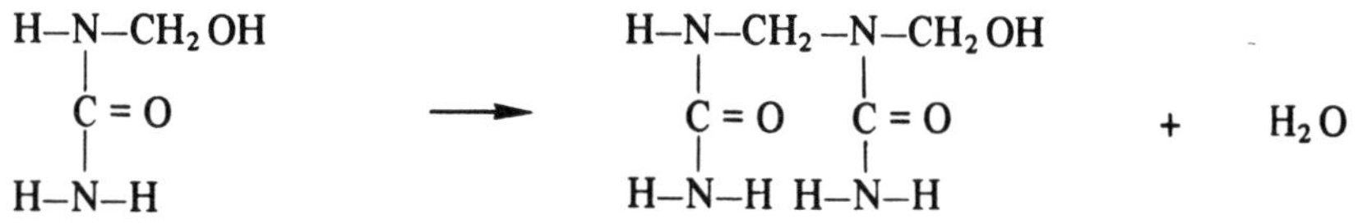

$$\begin{array}{c} H-N-CH_2OH \\ | \\ C=O \\ | \\ H-N-H \end{array} \longrightarrow \begin{array}{c} H-N-CH_2-N-CH_2OH \\ | \quad\quad\quad | \\ C=O \quad C=O \\ | \quad\quad\quad | \\ H-N-H \; H-N-H \end{array} + H_2O$$

Eventually:

$$n \begin{matrix} H-N-CH_2OH \\ | \\ C=O \\ | \\ H-N-H \end{matrix} \longrightarrow H \left[\begin{matrix} N-CH_2 \\ | \\ C=O \\ | \\ H-N-H \end{matrix} \right]_n OH \quad + \quad (n-1)H_2O$$

Some additional formaldehyde is, however, needed to react with at least a few of the unreacted $-NH_2$ groups and to provide chemical cross-links between polymer chains. When such cross-links occur, the desired thermosetting polymers or resins are produced. The amount of formaldehyde added to the reaction mixture is critical, for the following reasons:

- An excess of formaldehyde results in faster polymerization or cross-linking, which tends to lower manufacturing costs.
- Sufficient formaldehyde is needed to provide adequate cross-linking and to cause satisfactory properties in the final product.
- An excess of formaldehyde results in unreacted formaldehyde in the final consumer product, which slowly diffuses from the product and, especially in indoor applications, may result in increased formaldehyde concentrations.

In addition to unreacted formaldehyde in urea-formaldehyde resins, some formaldehyde may be formed by hydrolysis involving these resins. These hydrolysis reactions are essentially the reverse of the reactions shown above. When the resins are exposed to water or to a humid atmosphere, some moisture is adsorbed; this results in the slow formation and release of formaldehyde. Factors that affect the release of formaldehyde from UF resins are discussed in greater detail by Meyer.[18]

Urea-formaldehyde resins are a large and relatively old family of high polymers that have been used in the production of numerous molded plastic items. With respect to the release of formaldehyde to the air, definite problems have occurred in the following applications:

- Foams used in walls or attics of homes or other buildings for insulation.
- Particleboard.
- Indoor plywood.
- Paper products and some textiles.

In plywood and particleboard, the role of the resin is to act as an adhesive to bind the thin sheets of wood and wood particles together.

OTHER CONSUMER PRODUCTS

Several other high polymers that are prepared with formaldehyde probably contain unreacted formaldehyde that may eventually be emitted. Phenol-formaldehyde (or phenolic) resins are prepared by causing phenol and formaldehyde to react. Melamine resins are reaction products of melamine and formaldehyde. The amount of unreacted formaldehyde in the resin is obviously important. Phenolic and melamine resins can be used as adhesives in the production of plywood and particleboard. Phenolic resins are used because of their desirable physical and chemical properties: they are quite resistant to hydrolysis; they are relatively inexpensive, compared with alternative resins (but somewhat more costly than urea-formaldehyde resins); and there are often some problems with appearance. Although plywood produced with phenolic resins is often dark or somewhat stained, it is usually covered or coated in some way. Loss of formaldehyde in such plywood, if it does actually occur, would be less critical, because of outdoor application. There is little likelihood that aldehyde would ever build up to high concentrations in the ambient air.

Both phenolic and melamine resins are used in large quantities to fabricate numerous molded or extruded plastic products. Because fabrication is at high pressure, the final plastic product has essentially no porosity. Hence, in these products, diffusion of unreacted formaldehyde to the surface is extremely slow. There is no evidence that formaldehyde emission is a problem with phenolic or melamine plastic products.

Polyacetal resins are formed by polymerization of formaldehyde or trioxane. Ethylene oxide is sometimes used as a comonomer, and the polyacetal resin is a copolymer. At or near ambient conditions, polyacetals are highly stable. Polyacetals that are homopolymers of formaldehyde are thermally unstable at high temperatures, such as might be experienced during a fire. In such cases, they decompose quite rapidly and release formaldehyde.

REFERENCES

1. Aguilo, A., and J. D. Penrod. Acetaldehyde, pp. 115-162. In J. J. McKetta, Ed. Encyclopedia of Chemical Processing and Design. Vol. 1. New York: Marcel Dekker, Inc., 1976.
2. Albright, L. F. Commercial vapor phase processes for partial oxidation of light paraffins. Chem. Eng. 74:165, 14 August 1967.
3. Billmeyer, F. W. Textbook of Polymer Science, pp. 468-475. 2nd ed. New York: Wiley-Interscience Publishers, 1971.
4. Calvert, J. G., and E. W. R. Steacie. Vapor phase photolysis of formaldehyde at wavelength 3130A. J. Chem. Phys. 19:176-182, 1951.
5. Chauvel, A. R., P. R. Courty, R. Maux, and C. Petitpas. Select best formaldehyde catalyst. Hydrocarbon Processing 52(9):179-184, 1973.

6. Darshau, P., and A. P. Kudchadker. Benzaldehyde, pp. 171-182. In J. J. McKetta, Ed. Encyclopedia of Chemical Processing and Design. Vol. 4. New York: Marcel Dekker, Inc., 1977.
7. Diem, H. Formaldehyde routes bring cost, production benefits. Chem. Eng. 85:83-85, 27 February 1978.
8. Dunlop, A. P. Furfural and other furan compounds, pp. 237-251. In A. Standen, Ed. Kirk-Othmer Encyclopedia of Chemical Technology. 2nd ed. Vol. 10. New York: Wiley-Interscience Publishers, 1966.
9. Foster D. Snell Division. Preliminary Study of the Costs of Increased Regulation of Formaldehyde Exposure in the U.S. Workplace. Prepared for Formaldehyde Task Force, Synthetic Organic Chemical Manufacturers Association. Florham Park, N.J.: Booz, Allen and Hamilton, Inc., Foster D. Snell Division, 1979. 372 pp.
10. Gerloff, U. Compare BASF formaldehyde process. Hydrocarbon Proc. Petrol. Refin. 46(6):169-172, 1967.
11. Guccione, E. Acetaldehyde via ethylene oxidation gets tryout in single-stage design. Chem. Eng. 70:150-152, 9 December 1963.
12. Hester, A. S., and K. Himmler. Chemicals from acetaldehyde. Ind. Eng. Chem. 51:1424-1430, 1959.
13. Kyle, H. E. Oxo process, pp. 373-390. In Standen, Ed. Kirk-Othmer Encyclopedia of Chemical Technology. 2nd ed. Vol 14. New York: Wiley-Interscience Publishers, 1967.
14. Longfield, J. E., and W. D. Walters. The radical-sensitized decomposition of formaldehyde. J. Am. Chem. Soc. 77:6098-6103, 1955.
15. Lovell, R. J. Emissions Control Options for the Synthetic Organic Chemicals Manufacturing Industry. Formaldehyde Product Report. Knoxville Tennessee: Hydroscience, Inc., for U.S. Environmental Protection Agency, Office of Air Quality Planning and Standards, 1979. [150] pp.
16. Low-pressure Oxo process yields a better product mix. Chem. Eng. 84:110-115, 5 December 1977.
17. Ma, J. J. L. Acetaldehyde. Process Economics Program Interim Report No. 24A2. Menlo Park, California: Stanford Research Institute, December 1976. [114] pp.
18. Meyer, B. Urea-Formaldehyde Resins. Reading, Mass.: Addison-Wesley Publishing Company, Inc., 1979. 423 pp.
19. Sheldrick, J. E., and T. R. Steadman. Product/Industry Profile and Related Analysis on Formaldehyde and Formaldehyde-Containing Consumer Products. Part II. Products/Industry Profile on Urea-Formaldehyde. Columbus, Ohio: Battelle Columbus Division, for U.S. Consumer Product Safety Commission, 1979. [24] pp.
20. Sheldrick, J. E., and T. R. Steadman. Product/Industry Profile and Related Analysis on Formaldehyde and Formaldehyde-Containing Consumer Products. Part III. Consumer Products Containing Formaldehyde. Columbus, Ohio: Battelle Columbus Division, for U.S. Consumer Product Safety Commission, 1979. [39] pp.
21. Sleeman, D. G. Silver-catalyst process obtains high-strength formaldehyde solutions. Chem. Eng. 75:42-44, 1 January 1968.

22. Spence, R. The polymerisation of gaseous formaldehyde. J. Chem. Soc. 1933:1193-1197, 1933.
23. Spence, R., and W. Wild. The preparation of liquid monomeric formaldehyde. J. Chem. Soc. 1935:338-340, 1935.
24. Stanford Research Institute. Formaldehyde. In Chemical Economics Handbook. Menlo Park, Cal.: Stanford Research Institute, 1979.
25. Suta, B. E. Production and Use of 13 Aldehyde Compounds. Menlo Park, Cal.: SRI International, for U.S. Environmental Protection Agency, Office of Research and Development, 1979. [32] pp.
26. Weigert, W. M., and H. Haschke. Acrolein and derivatives, pp. 382-401. In J. J. McKetta and W. A. Cunningham, Eds. Encyclopedia of Chemical Processing and Design. Vol. 1. New York: Marcel Dekker, Inc., 1976.
27. Weimann, M. More-methanol formaldehyde route boasts many benefits. Chem. Eng. 77:102-104, 9 March 1970.

CHAPTER 5

ALDEHYDE CONCENTRATIONS, EMISSION, AND ENVIRONMENTAL GENERATION
AND TRANSFORMATION REACTIONS

The aldehydes are introduced into the environment through a variety of different pathways, which are considered in this chapter. They are injected directly into the atmosphere with exhaust gases from mobile sources and other equipment in which the incomplete combustion of hydrocarbon fuels occurs. They arise from various industrial and manufacturing operations and power generating plants that burn fossil fuels, from uncontrolled forest fires and the open burning of wastes, and from vegetation. Aldehydes are also generated in the atmosphere through the interaction of various reactive species (ozone, hydroxyl radicals, etc.) with hydrocarbons and some of their oxidation products. In recent years, it has been recognized that formaldehyde vapors may be released indoors, as well as outside, from various domestic activities and, more importantly, from particleboard and other building and insulation materials, chemically treated cloth, and other products that are formulated with formaldehyde-containing polymers. In fact, indoor concentrations of the aldehydes generally exceed those found in the outside air today.

The buildup of aldehydes in the atmosphere as the result of their direct release and their atmospheric generation is counterbalanced by many aldehyde removal paths. The aldehydes absorb the ultraviolet component of sunlight and undergo photodecomposition. They also react rapidly with the ubiquitous, highly reactive, transient hydroxyl (HO) free radical present in sunlight-irradiated atmospheres. Because of the high water solubility of formaldehyde and the other low-molecular-weight aldehydes, one expects the transfer of aldehydes into rainwater, the oceans, and other surface waters.

The rates of generation of ozone and the peroxyacylnitrates in the polluted atmosphere are strongly influenced by aldehyde photodecomposition and other reactions.

The combined effects of aldehyde injection, generation, and removal lead to a highly variable ambient concentration of the aldehydes. Their concentration can become high (about 0.05 ppm) in areas of high human activity and poor atmospheric ventilation. They are also present in the natural atmosphere, and concentrations of 0.002-0.006 ppm are commonly monitored in remote regions. Concentrations many times higher have been reported in some nonoccupational indoor environments. This chapter considers the aldehyde concentrations observed and then, in more detail, the many processes that control these concentrations. Throughout this document, the concentrations of gaseous aldehydes are usually given in parts per million (ppm) or micrograms per cubic meter ($\mu g/m^3$). "Parts per million" as used here refers to molecules of the species in question per million molecules of air at 25°C and 1 atm; for these conditions, concentrations expressed in the two units may be interconverted according to the following relations:

concentration in $\mu g/m^3$ = (concentration in ppm)(40.87)(M), and

concentration in ppm = (concentration in $\mu g/m^3$)(0.02447)/(M),

where M is the molecular weight of the specific aldehyde, e.g., 30.03 for formaldehyde and 44.05 for acetaldehyde. "Parts per hundred million" (pphm), "parts per billion" (ppb), and "parts per trillion" (ppt), which are used occasionally, refer to molecules of the species in question per hundred million, billion, and trillion molecules of air (at 25°C and 1 atm), respectively.

ENVIRONMENTAL CONCENTRATIONS OF THE ALDEHYDES

THE CLEAN ATMOSPHERE

The formation of formaldehyde and the other aldehydes in the natural unpolluted atmosphere is both anticipated in theory and observed experimentally. Reported ranges of concentration of total aldehydes in ambient clean air are as follows: Antarctica (1968), <0.0005-0.01 ppm; rural Illinois and Missouri (1973), 0.001-0.002 ppm; Panama (1966), <0.0002-0.0027 ppm; and Amazon basin (1970), 0.001-0.006 ppm[25] (Breeding et al.,[25] in reporting concentrations in the central United States, cited references to other measurements of formaldehyde in clean air). Spectroscopic measurements (high-resolution infrared absorption) have been used to identify formaldehyde in the atmospheric column over Reims, France.[15] From an analysis of the absorption line shapes at 2806.858 and 2869.871 cm^{-1} and reference to theoretical formaldehyde concentration-altitude profiles, Barbe et al. derived the approximate formaldehyde concentration-altitude profile shown in Figure 5-1 (dashed line).[15] The formaldehyde concentration decreased from about 10^{10} molecules/cc (about 0.0004 ppm) at ground level to about 10^7 molecules/cc (about 0.00002 ppm) at 26 km. These measurements are in reasonable accord with the theoretical estimates of Levy,[113] which are shown in Figure 5-1 as the solid curve.

URBAN ATMOSPHERES

Aldehydes are among the most abundant of the carbon-containing pollutant molecules in most urban atmospheres; only the hydrocarbons, carbon monoxide, and carbon dioxide are at higher concentrations. Shown in Figures 5-2, 5-3, and 5-4 are the aldehyde concentrations observed in the areas of Los Angeles, California, in 1968,[156]

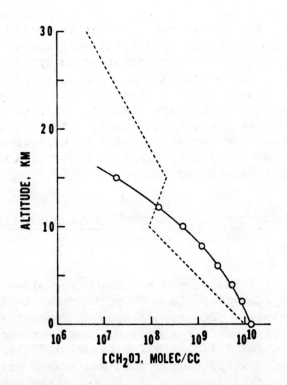

FIGURE 5-1 Atmospheric formaldehyde concentration versus altitude profiles; solid line, theoretical estimates by Levy[113] calculated for a clear summer day, noon, latitude 34°; dashed line, estimates of Barbe et al.[15] based on matching infrared line shapes and intensities observed in solar light transmitted through the atmosphere at Reims, France, November and December 1978, at sunset.

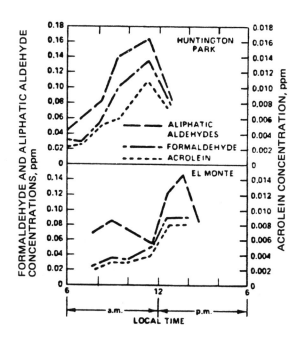

FIGURE 5-2 Hourly aldehyde concentrations at Huntington Park and El Monte, Calif., October 22, 1968. Reprinted with permission from Scott Research Laboratories, Inc.[156]

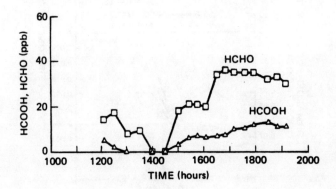

FIGURE 5-3 Concentrations of formaldehyde and formic acid measured in Riverside, Calif., at various times on October 14, 1977. Reprinted with permission from Tuazon et al.[178]

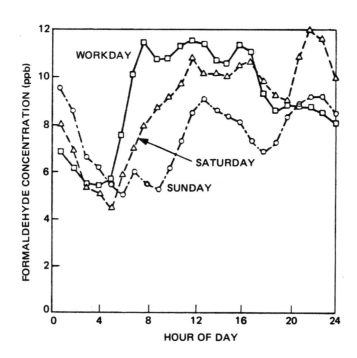

FIGURE 5-4 Diurnal variation of formaldehyde concentrations measured at Newark, N.J., for different days of the week; averaged from data taken from June 1 to August 31 for 1972, 1973, and 1974. Reprinted with permission from Cleveland et al.[46]

Riverside, California, in 1977,[178][179] and Newark, New Jersey, in 1972-1974.[46] The relatively high concentrations of the aldehydes observed in Los Angeles some years ago in 1968 are not observed today. More typical are those shown in Figures 5-3 and 5-4. The diurnal variations observed reflect the meteorologic influence of air transport and mixing, as well as the other chemical and physical processes that form and remove these compounds. Table 5-1 summarizes additional analytic data for total aldehydes, formaldehyde, and acrolein as determined by chemical methods and reported in studies made in several large metropolitan areas. The data from the first four studies shown were obtained in the Los Angeles area and are based on averages of hourly samples taken only during the daylight hours. Data from the other studies shown are from annual averages of samples taken once each 24 h during the year. From these data and the more extensive data from urban centers in 26 states and Washington, D.C.,[168] it appears that the 24-h average concentration of total aldehydes is frequently above 0.1 ppm (12 $\mu g/m^3$) in many urban areas, but with wide variations; hourly daytime averages may be near 0.05 ppm (60 $\mu g/m^3$) and infrequently above 0.1 ppm (120 $\mu g/m^3$).

The chemical nature of the mixture of aldehydes present in outdoor air is expected to be a function of the local emission sources and the meteorologic factors (sunlight intensity and wavelength distribution, temperature, etc.) near the sampling site. However, formaldehyde usually makes up some 30-75% of the total aldehyde observed in ambient urban air. The complete molecular speciation and composition are not now available for the fraction of the ambient-air samples determined to be "total aldehydes" by chemical methods, but all the normal aliphatic aldehydes containing 1-12 carbon atoms, the 14-carbon aldehyde, and a few of the common branched-chain aldehydes have been identified in ambient air.[73] Nine difunctional aldehydic compounds (C_5-, C_6-, and C_7-dialdehydes, hydroxyaldehydes, and aldehydic acids) have been detected in aerosols. A study carried out in California in 1972 included analysis for acetaldehyde, as well as formaldehyde.[33] Daily averages of formaldehyde were around 0.035 ppm; those for acetaldehyde, the only other aldehyde identified, were about 0.02 ppm.

Although Graedel[73] reported that only one of the aromatic aldehydes had been detected in ambient air (4-methylbenzaldehyde, at 50-280 ppt), benzaldehyde and 2-methylbenzaldehyde, 3-methylbenzaldehyde, and 4-methylbenzaldehyde probably are also present in at least minute amounts. Such a conclusion is based on two types of information: first, the latter aromatic aldehydes have been detected in the emission of automobiles burning gasolines that contain aromatic hydrocarbons (as is discussed in more detail later); second, on the basis of photooxidation investigations with toluene and xylenes (common ingredients of gasoline), such aldehydes are formed.[57] If the aromatic-hydrocarbon content of gasoline increases, the amounts of aromatic aldehydes present in the atmosphere will probably also increase.

TABLE 5-1

Measurement of Ambient Concentration of Aldehydes Observed in Several Urban Atmospheres Using Chemical Derivative (Spectrometric) Methods of Analysis

Date	Location	Total Aldehyde Concentration, ppb		Formaldehyde Concentration, ppb		Acrolein Concentration, ppb		Reference
		Range	Average[a]	Range	Average[a]	Range	Average[a]	
1960 (7/15-11/18)	Los Angeles, Calif.	10-360	130 (150)	0-130	36 (27)	1-11	8.8 (10.1)	151
1961 (9/25-11/15)	Los Angeles, Calif.	5-140	53 (38)	5-150	39 (25)	1-14	10 (13)	7
1968 (8/22-11/13)	Huntington Park, Calif.	3-173	44 (28)	4-136	29 (21)	0-8	3.4 (2.6)	156
1968 (8/23-11/14)	El Monte, Calif.	12-148	51 (27)	0-90	24 (18)	0-8	1.8 (2.3)	156
1973	Cincinnati, Ohio	--	11.7 (15.4)	--	--	--	--	93[b]
1974	Houston, Texas	--	7.8 (6.3)	--	--	--	--	93[b]
1974	St. Paul, Minn.	--	11.7 (6.0)	--	--	--	--	93[b]
1975	Cincinnati, Ohio	--	11.6 (11.9)	--	--	--	--	93[b]

[a] Derived from averages of hourly measurements made during daylight hours except that last four entries under "Total Aldehyde Concentration" are annual averages of 24-h samples; numbers in parentheses are standard deviations; validated data from recent years are not available to Committee.
[b] Ohio data derived by Jeffries and Kamens[93] from original data in Air Quality Data Ohio Counties--Hamilton, Clermont, Butler, and Warren, published by City of Cincinnati and Southwest Ohio Air Pollution Control District; Texas data from Texas Air Control Board in conjunction with City of Houston; Minnesota data from Minnesota Pollution Control Agency--Division of Air Quality.

Acrolein, the simplest unsaturated aldehyde, is of special interest, because of its effectiveness in inducing eye irritation and its general high toxicity. It appears to constitute a small but important fraction of the aldehydes in the urban atmosphere. Table 5-1 shows that the average acrolein concentration is 8-26% of the average formaldehyde concentration. Scientists in Tokyo have reported an average acrolein concentration of 7.2 ppb (0.0072 ppm), which is within the range shown in Table 5-1.[98a]

In summary, the present data suggest that--in a clean, unpolluted atmosphere--aldehyde concentrations at ground level are commonly about 0.0005-0.002 ppm. In polluted urban ambient air, the concentrations are much higher--commonly an hourly average of 0.01-0.05 ppm during the daylight hours. Formaldehyde is the dominant aldehyde present, and it usually makes up 30-75% of the total aldehyde present. Limited analytic data show that acetaldehyde may be present at about 60% of the formaldehyde concentration, and the higher aliphatic aldehydes are present at lower concentrations, decreasing rapidly with increasing molecular weight. Acrolein may be present at about 10-25% of the formaldehyde concentration; the aromatic aldehydes appear to make up only a few percent of the total aldehyde. Other dialdehydes or difunctional aldehydes presumably contribute to the aerosol mass in which they have been observed.

There is now no quantitative rationale that we can invoke to explain the large day-to-day or even year-to-year variations in aldehyde and hydrocarbon concentrations in the atmosphere. Experimental evidence from Houston, for example, obtained from 1973 through 1975 are most interesting in this regard. In some cases, the aldehyde concentrations were found to vary by a factor of 5-10 within several days. In one extreme variation, a reading of 52 $\mu g/m^3$ (0.042 ppm) occurred a week after and a week before readings of 6-7 $\mu g/m^3$ (0.005-0.006 ppm). In 1973, aldehyde values in Houston averaged higher than those in 1975 and especially those in 1974. No explanation was given for such differences. Atmospheric conditions must be responsible at least in part for day-to-day and year-to-year differences. The following are some of the factors that appear to be important in this variability:

- Wind conditions, including velocity and direction, strongly affect the dispersion of emission.
- Rain, standing water, or moist surfaces can be important sinks for formaldehyde.
- The extent of cloud cover and the position of the sun affect the sunlight intensity, which alters the rate of photochemical reactions.
- Air temperature affects the rate of chemical processes.
- Time of year may be important relative to atmospheric aldehyde concentrations; this is in part a result of temperature inversions that entrap the emission in the atmosphere near ground level. July through September, in general, had the highest concentrations in New Jersey,[46] Cincinnati,[93] and Houston.[93]

A direct comparison can be made between the ambient aldehyde concentrations in and near Houston and Cincinnati from 1973 through 1975 and between the concentrations in these two cities and St. Paul, Minnesota, in 1974.[93] The mean aldehyde concentrations in Cincinnati and St. Paul were in general slightly higher than those in Houston. These results and a detailed analysis of the Houston data suggest that the major refineries and chemical complexes of the Houston area do not contribute directly to the aldehyde concentrations. Baytown, in the Houston area, is the home of major refining and chemical plants; yet it had one of the lowest aldehyde concentrations in the entire Houston area. This may result in part from the time required for the conversion of hydrocarbons, perhaps the major impurity derived from the Baytown industry, to aldehydes through atmospheric reactions.

In 1973, the aldehyde concentrations in the Deer Park area of Houston were very high--rather consistently above those in any neighboring areas. There are several major refineries and petrochemical plants in this area. In 1974 and in 1975, aldehyde concentrations in Deer Park were similar to those in the remainder of the Houston area. The reason for the large reduction in 1974 and 1975 is unknown.

Aldehyde concentrations at or near Houston's and Cincinnati's major airports were similar to those in neighboring areas. Presumably, airports and planes contribute only a small fraction of the direct emission of aldehyde in metropolitan areas.

It is clear that transport of the air masses, the height of the mixing layer, and other meteorologic factors can be important in determining ambient aldehyde concentrations.

THE WORKPLACE

Workers in plants producing plywood or particleboard often use urea-formaldehyde, phenol-formaldehyde, or melamine-formaldehyde resins. Formaldehyde concentrations in several such plants have been reported[21,66,82,177,193] and may be as high as 10 ppm. The formaldehyde concentrations in the air in these plants obviously depend on the ventilating systems. Other key variables include the amount of free formaldehyde in the resin, the moisture content of the wood, the humidity of the air in the plant, and the processing temperatures. With current emission control technology, formaldehyde concentrations are or can be substantially lower.

Workers in a variety of other occupations are also exposed to formaldehyde, as shown in Table 5-2.

NONOCCUPATIONAL INDOOR AIR

The infiltration of outdoor air is one source of aldehydes in the indoor environment, but the primary sources are building materials, combustion appliances, tobacco smoke, and a large variety of consumer

TABLE 5-2

Formaldehyde Measurements in Occupational Environments

Sampling Site	Concentration, ppm		Method of Analysis
	Range	Mean	
Textile plants[161]	0-2.7	0.68	Sodium bisulfite, iodometric titration
Garment factory[22]	0.9-2.7	--	Collection in sodium bisulfite solution
Clothing store[128]	0.9-3.3	--	MBTH bubblers
Smog chamber[155]	0.01-unk	--	Chromotropic acid
Laminating plants[64]	0.04-10.9	--	Chromotropic acid
Funeral homes[102]	0.09-5.26	0.25-1.39	Chromotropic acid

products. Aldehydes can build up in buildings with greater insulation and tighter thermal containment intended to reduce infiltration (air exchange) and energy consumption.

Measurements of aldehydes in the indoor environment have typically focused on formaldehyde, whose indoor concentrations generally exceed those outdoors. Indoor monitoring data for U.S. homes are few, but limited monitoring data do exist for European homes, particularly in the Nordic countries. Table 5-3 summarizes the data that were recently described in detail by Suta.[174]

Several studies have concentrated on indoor formaldehyde emission from particleboard and plywood furnishings in houses. Measurements in Denmark,[9] Sweden (T. Lindvall and J. Sindell, personal communication), West Germany (Deimel;[54] B. Seifert, personal communication; Weber-Tschopp et al.[190]), and the United States (P.A. Breysse, personal communication) have shown that indoor concentrations often exceed 0.1 ppm and in some cases even exceed the 8-hr time-weighted average of 3 ppm for workroom air.[182,185] In 23 Danish houses, the average formaldehyde concentration was 620 µg/m^3 (about 0.5 ppm), and the range was 80-2,240 µg/m^3 (about 0.07-1.9 ppm).[9]

Over the last several years, complaints about indoor air quality have come from residents of mobile homes (constructed with formaldehyde-containing indoor plywood and particleboard). Since 1978, the U.S. Consumer Product Safety Commission (CPSC) has received hundreds of such complaints. Other federal agencies have also reported an increased number of complaints. In addition, dozens of lawsuits have been filed against UF-foam manufacturers and installers and mobile-home builders. It has been estimated that one of every 20 Americans--perhaps 11 million people--live in mobile homes that contain substantial quantities of particleboard, plywood, or both and are therefore potentially at risk of being exposed to formaldehyde. Thousands more live in homes insulated with UF foam. In August 1979, the CPSC issued two consumer advisories on UF insulation, citing possible health problems associated with this type of insulation.

As a result of occupants' complaints, formaldehyde was measured in more than 200 mobile homes in the United States; the concentrations reported ranged from 0.03 to 2.4 ppm (about 37-2,940 µg/m^3) (Breysse, personal communication). A study of formaldehyde emission in new office trailers with air-exchange rates as low as 0.16 air change per hour (ach) found formaldehyde concentrations in the range of 0.15 to 0.20 ppm,[60] in contrast with outdoor concentrations of less than 0.01 ppm.

Formaldehyde vapors are a concern in mobile homes, not only because the building materials used in their construction typically contain formaldehyde, but also because mobile homes are more tightly constructed than conventional homes and thus have less ventilation.

Aldehydes (measured by the MBTH method) were monitored in a study of 19 homes across the United States.[135] Outdoor concentrations were consistently lower than indoor concentrations--typically by a factor of 6 and quite often by an order of magnitude. Figure 5-5 is an illustration of the data collected in this study. The observed

TABLE 5-3

Summary of Aldehyde Measurements in Nonoccupational Indoor Environments

Sampling Site	Concentration,[a] ppm		Method of Analysis
	Range	Mean	
Danish residences[8]	1.8 (peak)	--	Unspecified
Netherlands residence built without formaldehyde-releasing materials[196]	0.08 (peak)	0.03	Unspecified
Residences in Denmark, Netherlands, and Federal Republic of Germany[196]	2.3 (peak)	0.4	Unspecified
Two mobile homes in Pittsburgh, Pa.[135]	0.1-0.8[b]	0.36	MBTH bubblers
Sample residence in Pittsburgh, Pa.[135]	0.5 (peak)[b]	0.15	MBTH bubblers
Mobile homes registering complaints in state of Washington[26]	0-1.77	0.1-0.44	Chromotropic acid (single impinger)
Mobile homes registering complaints in Minnesota[67]	0-3.0	0.4	Chromotropic acid (30-min sample)
Mobile homes registering complaints in Wisconsin[c]	0.02-4.2	0.88	Chromotropic acid
Public buildings and energy-efficient homes (occupied and unoccupied)[114]	0-0.21	--	Pararosaniline and chromotropic acid
	0-0.23[b]	--	MBTH bubblers

[a] Formaldehyde, unless otherwise indicated.

[b] Total aliphatic aldehydes.

[c] M. Woodbury, personal communication.

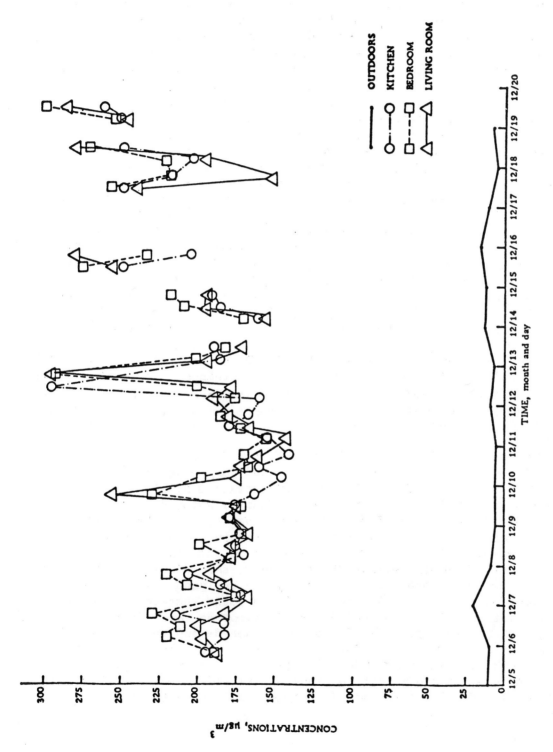

FIGURE 5-5 Diurnal variation in indoor and outdoor aldehyde concentrations. Single-family house, Chicago. Reprinted from Moschandreas et al.[135]

outdoor aldehyde concentration remained below 25 µg/m³ (0.02 ppm) at all times. The study determined that its field sample comprised two distinct classes of residences: those with high and those with low aldehyde concentrations (see Table 5-4). In all cases, however, indoor aldehyde concentrations exceeded outdoor. Although the source strengths were not determined in this study, the highest concentrations were observed in the mobile homes, and the plywood and particleboard generally appeared to be the primary source.

In a more recent study, formaldehyde and total aliphatic aldehydes (formaldehyde plus other aliphatic aldehydes) were measured at several energy-efficient research houses at various locations in the United States.[114] At low infiltration rates (<0.3 ach), indoor formaldehyde concentrations often exceeded 0.1 ppm (123 µg/m³), whereas outdoor concentrations typically remained at 0.016 ppm (20 µg/m³) or less. Normal air-exchange rates are about 0.75/ach. Figure 5-6 is a histogram showing the frequency of occurrence of formaldehyde and total aliphatic aldehyde concentrations measured at an energy-efficient house with an average of 0.2 ach. Data taken at an energy-efficient house in Mission Viejo, California, are shown in Table 5-5. As shown, when the house did not contain furniture, formaldehyde concentration was 80 µg/m³; when furniture was added, formaldehyde almost tripled. A further increase was noted when the house was occupied, very likely because of such activities as cooking with gas. When occupants opened windows to increase ventilation, the formaldehyde concentration decreased substantially. Although high, aldehyde concentrations observed in most of the energy-efficient dwellings that have been monitored were generally below 200 µg/m³.

Indoor and outdoor formaldehyde/aldehyde concentrations were found to be about the same at a public school in Columbus, Ohio, and a large medical center in Long Beach, California, and were well below 0.1 ppm (120 µg/m³). Both buildings have high ventilation rates, and that is probably why the indoor concentrations were low and essentially equivalent to outdoor concentrations.

Because many of the data reported from these field-monitoring studies involved houses whose occupants had complained of indoor air quality, these findings may not be representative of all homes. However, when data from the Washington sample, which was random, are compared with those from the mobile-home sample, which was based on occupant complaints, most of the differences in aldehyde concentration can be explained by differences in the age of the home. The mobile homes in the complaint sample are much newer than those in the random sample. Moreover, when Tabershaw et al.[175] analyzed the complaint data on mobile homes in Washington, it was found that there was no valid relationship between the degree of symptoms reported by occupants and the concentrations of formaldehyde and that, regardless of the actual exposure, all persons in the mobile-home sample reacted in substantially the same manner. Tabershaw Associates suggested that, because the study received substantial press coverage in Washington and other parts of the country, publicity may have

TABLE 5-4

Statistical Summary of Aldehyde Concentrations in Various Residential Structures (Outdoor Concentrations Very Low)[a]

Location and Type of Residence	Observed Range of 4-h Concentrations, µg/m^3	14-Day Monitoring Period Mean Concentration, µg/m^3	Standard Deviation, µg/m^3
Denver conventional	87-615	250	118
Chicago experimental I	140-300	200	38
Chicago experimental II	242-555	325	70
Pittsburgh mobile home 1	200-938	470	167
Pittsburgh mobile home 2	136-934	387	159
Washington conventional I	21-153	52	31
Baltimore conventional II	34-150	75	25
Washington experimental I	10-286	90	78
Baltimore experimental I	17-162	78	38
Baltimore experimental II	6-122	48	20
Pittsburgh low-rise 1	51-152	91	34
Pittsburgh high-rise 1	22-120	56	18
Chicago conventional I	20-190	54	29
Chicago conventional II	10-159	47	23
Pittsburgh low-rise 2	35-149	78	29
Baltimore conventional I	10-300	144	75
Pittsburgh high-rise 2	76-240	125	27
Pittsburgh high-rise 3	65-234	149	40
Pittsburgh low-rise 2	20-102	110	32

[a] Reprinted from Moschandreas et al.[135]

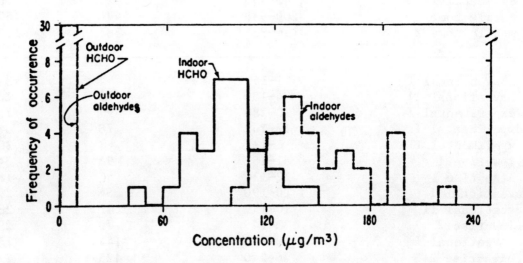

FIGURE 5-6 Histogram showing frequency of occurrence of formaldehyde and total aliphatic aldehydes at an energy-efficient house with 0.2 ach. Single-family house, Maryland. Reprinted with permission from Lin et al.[114]

TABLE 5-5

Indoor and Outdoor Formaldehyde and Aliphatic Aldehyde
Concentrations Measured at a Single-Family House (California)[a]

Condition	Number of Measurements	Sampling Time	Formaldehyde ($\mu g/m^3$)[b]	Aliphatic Aldehydes ($\mu g/m^3$)[c]
Unoccupied, without furniture	3	12	80 ± 9%	90 ± 16%
Unoccupied, with furniture	3	24	223 ± 7%	294 ± 4%
Occupied, day[d]	9	12	261 ± 10%	277 ± 15%
Occupied, night[e]	9	12	140 ± 31%	178 ± 29%

[a] Reprinted with permission from Hollowell et al.[86]

[b] Determined using pararosaniline method (120 $\mu g/m^3 \approx$ 100 ppb). All outside concentrations <10 $\mu g/m^3$.

[c] Determined using MBTH method, expressed as equivalents of formaldehyde. All outside concentrations <20 $\mu g/m^3$.

[d] Air exchange rate $\simeq$ 0.4 ach.

[e] Windows open part of time; air exchange rate significantly greater than 0.4 ach and variable.

motivated people with health problems, some of which were perhaps unrelated to formaldehyde, to call on the University of Washington to make an investigation.

Foreign houses (particularly Danish and Swedish) monitored for formaldehyde appear to have much higher concentrations than U.S. houses. These findings probably reflect differences in house construction and, hence, cannot be considered as representative of U.S. houses.

Although use of the Danish studies may not be appropriate for U.S. homes, the treatment of Andersen et al.[9] illustrates the many variables with which one must be concerned. He formulated a mathematical model that estimates the indoor air concentration of formaldehyde. In climate-chamber experiments, Andersen et al.[9] found the equilibrium concentration of formaldehyde from particleboard to be related to temperature, water-vapor concentration in the air, ventilation, and the amount of particleboard present. From this work, a mathematical model was established for room air concentration of formaldehyde.

When the mathematical formulation was applied to the room-sampling results, a correlation coefficient of 0.33 was found between the observed and predicted concentration--not a particularly good predictive ability. The authors then modified the value for the adjustable constants by calculating them for each room on the basis of monitoring results. The modified values led to a correlation coefficient of 0.88--a considerable improvement in predictability.

Formaldehyde release from interior particleboard occurs at a decreasing rate with an increase in product age. Eventually, the rate of formaldehyde evolution decreases to an imperceptible point. The length of time necessary for the phenomenon to occur (several years) depends on the atmospheric conditions to which the board has been subjected, as well as the degree of cure of the resin. The more unstable groups degrade first, and then the more stable free methylol groups.[101]

Field tests and a mathematical model were used in 1977 to determine the half-life of formaldehyde in particleboard typically used in Scandinavian home construction; it was about 2 yr when the ventilation rate in the home was 0.3 ach (C.D. Hollowell, personal communication). Suta[174] has analyzed the effect of home age on formaldehyde concentrations in Danish houses. The study indicated that the half-life of formaldehyde may be much longer than 2 yr. These data give the following relationship of formaldehyde concentration as a function of house age when no corrections are made for other pertinent factors, such as the amount of particleboard in the home, temperature, humidity, or ventilation: $C = 0.50e^{-0.012A}$, where C is formaldehyde concentration, in parts per million, and A is home age, in months. On the basis of this formula, the half-life of formaldehyde emission is 58 mo. The difference between half-life values derived from the test data and those from house-monitoring may result partly from the fact that particleboard is often added to older homes for repair and improvement.

Monitoring data for the 65 randomly selected mobile homes in Wisconsin show a similar decrease in formaldehyde concentration with increasing home age. The reported formaldehyde half-life was 69 mo, which is quite similar to that found in the Denmark study. Monitoring data on 45 complained-about mobile homes in Wisconsin also showed a decrease in formaldehyde concentration with increase in house age; the indicated half-life in this sample was 28 mo. When these data are combined, the formaldehyde half-life is 53 mo, or approximately 4.4 yr.

Not all residences are expected to have the same formaldehyde concentration. As suggested earlier, variation occurs even in homes of the same age, depending on the amount and type of particleboard and UF-foam insulation used in the construction, as well as on temperature, humidity, and ventilation. For this reason, monitored concentrations from a sample of similar homes will be characterized by a frequency distribution that can be approximated by a known statistical distribution, which, in turn, can be used to estimate the range of human exposures to formaldehyde in the residential environment.

The average ambient formaldehyde concentration appears to be approximately 0.4 ppm (490 $\mu g/m^3$) in both mobile homes and UF-foam-insulated homes. Few data are available on conventional houses that do not contain UF-foam insulation or that were not designed to be energy-efficient. The average formaldehyde concentration in conventional houses appears to range from 0.01 to 0.1 ppm (12 to 120 $\mu g/m^3$) and may be only slightly higher than outdoor concentrations. Houses containing larger amounts of particleboard would fall on the high side of this concentration range, and houses with no particleboard on the low side.

Average atmospheric formaldehyde concentrations are generally much lower than 0.1 ppm in U.S. cities, as indicated earlier. Examples of annual average concentrations are 0.05 ppm in Los Angeles,[6,168,180] 0.004-0.007 ppm in four New Jersey cities,[46] 0.04 ppm in Wisconsin cities (L. Hanrahan, personal communication), and less than 0.03 ppm in Raleigh, North Carolina, and Pasadena, California.[79] Formaldehyde concentrations at four Swiss locations ranged from 0.007 to 0.014 ppm; these concentrations are about one-fifth the corresponding indoor Swiss concentration.[189] In 1951, a mean value of 0.004 ppm was reported for mainland Europe.[43]

SURFACE WATERS AND DRINKING WATER

The high solubility of most aldehydes in water results in their accumulation in natural bodies of water. Sixteen aldehydes that have been identified in natural bodies of water, the names and locations of the bodies of water in which they have been found, and their concentrations are given in Table 5-6. Acrolein is not included, because it has been found only in surface water to which it was intentionally introduced.

Nineteen aldehydes that have been identified in drinking water are listed in Table 5-7. Quantitative information on most of these aldehydes is unavailable. Some aldehydes in drinking water may be produced during water treatment. Chloral, which has been identified

TABLE 5-6

Aldehydes Identified in Surface Water

Aldehyde	Body of Water and Location	Concentration,[a] μg/L
Acetaldehyde	Mobile River, Ala.[158]	NR
	Pacolet and Encoree River, So. Car.[158]	NR
	Mississippi River, New Orleans, La.[59a]	12
Benzaldehyde	Los Angeles River, Los Angeles, Calif.[59a]	1
	Unspecified river, Netherlands[b]	0.3
Butyraldehyde	Wisconsin River, Nekoosa, Wis.[59a]	6
Capraldehyde	Glatt River, Switzerland[72]	NR
	Unspecified reservoir, Netherlands[b]	0.1
Caproaldehyde	Unspecified reservoir, Netherlands[b]	0.03
Caprylaldehyde	Unspecified reservoir, Netherlands[b]	0.03
Cinnamaldehyde	Unspecified reservoir, Netherlands[b]	0.03
3,5-Di-tert-butyl-4-hydroxybenzaldehyde	Unspecified river, Netherlands[b]	0.1
Dichlorobenzaldehyde	Unspecified river, Netherlands[b]	0.03
Dimethylbenzaldehyde	Unspecified reservoir, Netherlands[b]	0.1
Enanthaldehyde	Unspecified reservoir, Netherlands[b]	0.1
Mesitaldehyde	Holston River, Kingsport, Tenn.[c]	3
2-Methylpropionaldehyde	Delaware River, Torresdale, Pa.[59a]	1
Paraldehyde	Lake Zurich, Switzerland[62]	NR
Undecylaldehyde	Unspecified reservoir, Netherlands[b]	0.03
Vanillin	Lake Superior, Ontario, Canada[62]	NR

[a] NR, not reported.

[b] G. J. Piet, personal communication.

[c] H. Boyle, personal communication.

TABLE 5-7

Aldehydes Identified in Drinking Water

Aldehyde	Location of Water Plant	Concentration,[a] µg/L
Acetaldehyde	Cincinnati, Ohio [47]	NR
	Miami, Fla. [47]	NR
	Ottumwa, Iowa [47]	NR
	Philadelphia, Pa. [100]	0.1
	Seattle, Wash. [100]	0.1
	Durham, No. Car. [126]	NR
Benzaldehyde	New Orleans, La. [100]	0.03
	Grand Forks, No. Dak. [183]	NR
	New York, N.Y. [183]	NR
	Voorburg, Netherlands[b]	0.3
Butyraldehyde	Voorburg, Netherlands [158]	0.1
Caproaldehyde	Voorburg, Netherlands[b]	0.03
Chloral	Cincinnati, Ohio [100]	2
	Grand Forks, No. Dak. [100]	0.01
	Philadelphia, Pa. [100]	5
	Seattle, Wash. [100]	3.5
	New York, N.Y. [100]	0.02
	Terrebonne Parish, La. [100]	1
	Kansas City, Kans. [107]	NR
Cinnamaldehyde	Voorburg, Netherlands[b]	0.005-0.03
Crotonaldehyde	Unspecified [158]	NR
Dimethylbenzaldehyde	Voorburg, Netherlands[b]	0.03-0.1
Enanthaldehyde	Voorburg, Netherlands[b]	0.03
2-Ethylbutyraldehyde	New York, N.Y. [100]	0.05
	Grand Forks, No. Dak. [100]	0.02
	Lawrence, Mass. [100]	0.04
	Terrebonne Parish, La. [100]	0.01
Furaldehyde	Chicago, Ill. [59a]	2
Isobutyraldehyde	Prague, Czechoslovakia [144]	0.13

TABLE 5-7 (continued)

Aldehyde	Location of Water Plant	Concentration,[a] µg/L
Isovaleraldehyde	Cincinnati, Ohio[47]	NR
	Miami, Fla.[47]	NR
	Ottumwa, Iowa[47]	NR
	Philadelphia, Pa.[47]	NR
	Seattle, Wash.[47]	NR
	Durham, No. Car.[47]	NR
	New Orleans, La.[184]	NR
Methacrolein	Unspecified[158]	NR
2-Methylpropionaldehyde	Cincinnati, Ohio[47]	NR
	Miami, Fla.[47]	NR
	Ottumwa, Iowa[47]	NR
	Philadelphia, Pa.[47]	NR
	Seattle, Wash.[47]	NR
	Durham, No. Car.[126]	NR
3-Methylvaleraldehyde	Ottumwa, Iowa[100]	1
Paraldehyde	Zurich, Switzerland[75]	NR
Propionaldehyde	Cincinnati, Ohio[47]	NR
	Miami, Fla.[47]	NR
	Ottumwa, Iowa[47]	NR
	Philadelphia, Pa.[47]	NR
	Seattle, Wash.[47]	NR
	Durham, No. Car.[126]	NR
Valeraldehyde	Ottumwa, Iowa[100]	0.5

[a] NR, not reported

[b] G. J. Piet, personal communication.

in the drinking water of seven U.S. cities, is thought to be formed during chlorination for water purification.[100]

SOURCES OF DIRECT EMISSION OF ALDEHYDES IN AMBIENT AIR

INDUSTRIAL OPERATIONS

Aldehyde Manufacturing Plants

The leakage of aldehydes into the atmosphere may occur in the operation of industrial plants that manufacture the aldehydes. Morris[133],[134] and Lovell[118] have estimated formaldehyde losses to the environment to be 0.4 g/1,000 g of product formed in solution. The atmospheric emission of formaldehyde from manufacturing processes in the United States can be roughly estimated at about 6×10^6 lb/yr. These losses usually occur at the following locations within the plants:

- The vent stream from the water absorber used for the recovery of formaldehyde and methanol usually contains carbon dioxide, carbon monoxide, hydrogen, water vapor, nitrogen (if air is used), and trace amounts of formaldehyde, methanol, and some byproducts. The absorber can be built and operated to recover formaldehyde so effectively that the exit gas stream is sometimes exhausted directly into the atmosphere. Other formaldehyde producers use slightly less efficient and cheaper absorbers with lower operating costs and use the exhaust streams as supplemental fuel in the furnaces of their power stations, because the streams often have appreciable fuel values. In another alternative, the vent stream from the absorber can be sent to a flare. Almost complete oxidation of formaldehyde, methanol, and carbon monoxide occurs in both a furnace and a flare, so the exhaust gas stream from the formaldehyde unit contains little formaldehyde.
- The vent stream from the top of the product fractionator used to prepare specification-grade product solution can be sent as needed to a flare, a furnace, or an absorber to reduce the formaldehyde content.
- Intermittent gaseous emission that occurs during plant startup or shutdown is sometimes sent to a flare, the furnace, or a small water absorber to reduce the formaldehyde content.
- Intermittent losses that occur at pump seals, from storage tanks, or at valves are sometimes controlled with portable gas blowers used in connection with small absorbers.

Each industrial plant uses various combinations of water absorbers, flares, furnaces, and catalytic incinerators to maintain formaldehyde concentrations in the ambient air of the plant at less than 3 ppm as the time-weighted average for an 8-h workshift. This is the maximal permissible concentration set by the Occupational Safety and Health Administration in 1979. The EPA does not have an ambient air standard for formaldehyde.

Information on emission from industrial plants producing the higher aldehydes is very limited. It is probably reasonable to assume that the percentages of the more volatile aldehydes--such as acetaldehyde, acrolein, and propionaldehyde--lost into the atmosphere are comparable with those reported for formaldehyde.

Other Industrial Sources

Shackleford and Keith[158] have reported that aldehydes occur in the effluent water streams from several types of industrial plants. Formaldehyde, acetaldehyde, acrolein, paraldehyde, sorbaldehyde, and syringaldehyde have been detected in unidentified chemical plants in rather scattered areas of this country. Aldehydes identified in the effluent from some sewage plants include acetaldehyde, benzaldehyde, crotonaldehyde, isovaleraldehyde, 2-methylpropionaldehyde, and salicylaldehyde. It is not known whether the aldehydes in these sewage plants were produced by microbial or chemical means or were in the feedstock to the sewage plants derived from various industrial plants.

Acrolein, anisaldehyde, benzaldehyde, salicylaldehyde, syringaldehyde, vanillin, and veratraldehyde have been detected in the effluent from paper mills. Benzaldehyde has been identified in the aqueous waste from textile mills. Some of these aldehydes probably form as a result of reactions involving wood or cellulose.

Fish-culture activities are also a source of formaldehyde in the aquatic environment. Formalin (aqueous solution of formaldehyde) is one of the most widely and frequently used chemical agents for treating fish with fungal or ectoparasitic infections. Treatment entails exposing the fish to formaldehyde at up to 250 mg/L of solution in ponds, raceways, or tanks. After use, these formaldehyde solutions are often discharged into the normal hatchery effluent stream from both private and government-owned fish hatcheries.

Anisaldehyde has been detected in the aqueous effluent of a pilot plant being used for coal gasification. This suggests that, when commercial coal-gasification plants are built, they may contribute to aldehyde effluent.

Table 5-8 shows reported aldehyde emission from various industrial sources, as collected by Stahl.[168]

Combustion

Combustion leads to both the direct and the indirect accumulation of aldehydes in the atmosphere of metropolitan areas. Aldehydes are present in at least trace amounts in the exhaust gases from combustion. In addition, there is often, if not always, some unburned hydrocarbon that escapes to the surroundings. As discussed later, this hydrocarbon oxidizes rapidly in the atmosphere to form aldehydes and other oxygenated products. Emission from transportation vehicles, power plants burning fossil fuels, home and industrial furnaces,

TABLE 5-8

Reported Aldehyde Emission from Various Sources[a]

Source	Aldehyde Emission (as formaldehyde)
Amberglass manufacture	
Regenerative furnace, gas fired	8,400 µg/m^3
Brakeshoe debonding	
(single-chamber oven)	0.10 lb/h
Core ovens	
Direct gas fired (phenolic resin binder from oven)	62,400 µg/m^3
Direct gas fired (linseed oil core binder from afterburner	<12,000 µg/m^3
Indirect electric (linseed oil core binder from oven)	189,600 µg/m^3
(from afterburner)	<22,800 µg/m^3
Insulated wire reclaiming, covering	
Rubber 5/8" o.d.	
Secondary burner off	126,000 µg/m^3
Secondary burner on	6,000 µg/m^3
Cotton rubber plastic 3/8-5/8" o.d.	
Secondary burner off	10,800 to 43,200 µg/m^3
Secondary burner on	4,800 µg/m^3
Meat smokehouses	
Pressure mixing burner	
Afterburner inlet	0.04 lb/h
Afterburner outlet	0.22 lb/h

TABLE 5-8 (continued)

Source	Aldehyde Emission (as formaldehyde)
Mineral wool production	
Blow chambers	109 µg/m^3
Curing ovens	
Catalytic afterburner inlet	1.90 lb/h
Catalytic afterburner outlet	0.90 lb/h
Direct flame afterburner inlet	2.20 lb/h
Direct flame afterburner outlet	0.94 lb/h
Wool coolers	32 µg/m^3
Litho oven inlet	120 µg/m^3
Litho oven outlet	32,880 µg/m^3
Litho oven outlet	4,680 µg/m^3
Paint bake oven	
Nozzle mixing burner	
Afterburner inlet	0.19 lb/h
Afterburner outlet	0.03 lb/h
Atmospheric burner	
Catalytic afterburner inlet	0.07 lb/h
Catalytic afterburner outlet	0.31 lb/h
Premix burner	
Catalytic afterburner inlet	0.3 to 0.4 lb/h
Catalytic afterburner outlet	0.2 to 0.5 lb/h
Phthalic acid plant	135,600 µg/m^3

TABLE 5-8 (continued)

Source	Aldehyde Emission (as formaldehyde)
Multijet burner	
Afterburner inlet	0.49 lb/h
Afterburner outlet	0.22 lb/h
Meat smokehouse effluent, gas fired boiler firebox as "afterburner"	
Water tube, 426 hp	
Afterburner inlet	0.22 lb/h
Afterburner outlet	0.09 lb/h
Water tube, 268 hp	
Afterburner inlet	0.39 lb/h
Afterburner outlet	0.40 lb/h
Water tube, 200 hp	
Afterburner inlet	0.39 lb/h
Afterburner outlet	0.30 lb/h
Locomotive, 113 hp	
Afterburner inlet	0.03 lb/h
Afterburner outlet	0.0 lb/h
HRT, 150 hp	
Afterburner inlet	0.03 lb/h
Afterburner outlet	0.18 lb/h
Meat smokehouse exhaust	
Gas fired afterburner inlet	104,400 µg/m^3
Gas fired afterburner outlet	40,200 µg/m^3
Electrical precipitation system inlet	88,800 µg/m^3
Electrical precipitation system outlet	56,400 µg/m^3

TABLE 5-8 (continued)

Source	Aldehyde Emission (as formaldehyde)
Phthalic anhydride production unit (multiple burner)	
Afterburner inlet	1.75 lb/h
Afterburner outlet	0.43 lb/h
Reclaiming of electrical windings (single chamber incinerator)	
100 hp generator starter	0.08 lb/h
14 pole pieces	0.08 lb/h
Auto armatures	0.13 to 0.29 lb/h
Auto field coils (multiple chamber)	0.49 lb/h
Auto field coils afterburner	0.08 lb/h
14 generator pole pieces	0.08 lb/h
Varnish cooking kettles	
Four nozzle mixing burner	
Afterburner inlet	0.30 lb/h
Afterburner outlet	0.11 lb/h
Inspirator burner	
Afterburner inlet	0.29 lb/h
Afterburner outlet	0.02 lb/h
Webb press	480 $\mu g/m^3$
	360 $\mu g/m^3$
	480 $\mu g/m^3$
	1,920 $\mu g/m^3$

[a]Reprinted from Stahl.[168]

garbage fires, and bonfires contributes to the rather high aldehyde concentrations in metropolitan areas. General reviews of the amounts and fate of aldehydes in the atmosphere have recently been issued.[115,173]

Transportation Vehicles

Transportation vehicles are important and possibly at times the predominant contributors to both aldehyde and hydrocarbon emission in some metropolitan areas. Much valuable information on emission from automobiles, trucks, buses, airplanes, etc., has accumulated in the last few years. The exhausts of various automobiles powered with gasoline engines have been collected by General Motors Corp. Tables 5-9 and 5-10 indicate the relative concentrations of the specific aldehydes identified in the exhaust gases from automobiles without and with catalytic converters, respectively. Formaldehyde is almost always the predominant aldehyde emitted, but at least 11 others have been identified, including at least three aromatic aldehydes. As discussed later, the amounts of aromatic aldehydes produced depend significantly on the aromatic hydrocarbon content of the gasoline used.

Several factors affect aldehyde and unburned-hydrocarbon concentrations in automotive emission. The most important ones are discussed below.

Operation of Gasoline Engines and Catalytic Converters.

Automotive manufacturers are under federal mandate to reduce the total hydrocarbon emission to 0.41 g/mile or less within the next several years. There is no question but that catalytic converters and other recent changes in motor-vehicle design and operation have resulted in substantial reductions in aldehyde and unburned-hydrocarbon emission. For example, higher air-to-fuel ratios are provided in modern engines. In 1971, the Los Angeles Air Pollution Control District (APCD) estimated that motor vehicles contributed about two-thirds of the total hydrocarbon emission inventories to the atmosphere of the Los Angeles area. By 1975, with the increased use of catalytic converters, the Southern California Air Pollution Control District (formerly the Los Angeles APCD) estimated that motor vehicles contributed less than half.

Newer cars equipped with catalytic converters often emit aldehydes at about 20-60 mg/mile; older cars without the more modern control devices emit aldehydes at about 70-300 mg/mile. Automobiles also emit a variety of paraffinic, unsaturated, and aromatic hydrocarbons. Jackson[91] found an average of 2.45 g/mile for 1970-1974 automobiles not equipped with catalytic converters. Hydrocarbons are emitted by 1974-1975 cars equipped with three types of catalytic converters at an average of 0.48-0.65 g/mile. Recently, Cadle, Nebel, and Williams[32] reported emission rates for catalytic and noncatalytic automobiles similar to these values.

As more vehicles become equipped with catalytic converters, motor vehicles will probably emit less aldehyde and unburned hydrocarbon.

The following factors also affect emission:

TABLE 5-9

Exhaust Aldehyde Composition: Gasoline Exhaust from Noncatalyst Cars[a]

Aldehyde	Concentration, mole %							
	Exxon[b]	Exxon[b]	Exxon[b]	Sun	Ford[c]	Ford[c]	CalPH	EPA
Formaldehyde	66.7	72.5	75.9	82.0	70.2	59.9	69.3	72.9	66.4
Acetaldehyde	9.3	8.7	7.8	8.9	7.2	14.3	7.5	8.5	12.3
Propionaldehyde[c]	15.7	--	--	--	0.4	7.0	0.7	6.4	16.3
Acrolein	3.2	4.3	--	--	9.8		2.6		
Butyraldehydes	--	--	--	--	0.4	3.0	1.0	1.7	1.1
Crotonaldehyde	--	--	--	--	0.4	1.4	0.4	0.4	1.1
Valeraldehydes	--	--	--	--	0.4	--	--		
Benzaldehyde	3.2	7.0	6.3	2.9	8.5	3.3	5.4	4.3	3.9
Toluenaldehydes	1.9	7.2	7.5	3.1	--	5.9	3.1	--	--
Others	--	0.3	2.5	3.1	2.7	5.2	10.0	5.8	--
Total	100	100	100	100	100	100	100	100	100
Driving cycle	72 FTP	72 FTP	72 FTP	7-mode	30 mph	72 FTP	idle	72 FTP

[a] J. M. Heuss, personal communication.
[b] Different gasolines.
[c] Different cars and driving cycles.
[d] May also include acetone.

TABLE 5-10

Exhaust Aldehyde Composition: Gasoline Exhaust from Catalyst Cars[a]

Aldehyde	Concentration, mole %							
	BOM[b]	BOM[c]			BOM[d]			EPA[e]
Formaldehyde	55.2	36.5	35.1	42.2	63.7	62.1	78.2	62.4
Acetaldehyde	25.5	34.9	40.8	23.5	14.9	17.5	13.3	16.4
Propionaldehyde	1.0	3.6	4.2	2.4	15.1	2.6	4.6	12.7
Acrolein	0	1.1	ND	ND	0.1	0.4	ND	
Crotonaldehyde	--	--	--	--	--	--	--	5.4
Methacrolein	0	4.1	7.1	2.4	ND	ND	ND	--
Benzaldehyde	10.1	11.3	7.5	17.0	3.7	12.2	2.3	3.0
Tolualdehydes	1.2	2.3	1.3	2.2	0.9	2.3	0.5	--
Ethyl benzaldehyde	2.5	3.2	2.9	6.0	0.7	0.6	0.3	--
$C_9H_{10}O$	4.5	3.0	1.1	4.3	0.9	2.3	0.9	--
Total	100	100	100	100	100	100	100	100
Driving Cycle	7-mode	72-FTP			72 FTP			75 FTP

[a] J. M. Heuss, personal communication.

[b] Pt mono. oxid. catalyst.

[c] Pt mono. oxid. catalyst; 3 gasolines.

[d] Dual RO catalysts; same 3 gasolines.

[e] 1975 Plymouth Fury.

• Age, condition, and degree of tuning or adjustment of the engine—engines require tuning and maintenance, if they are to have low emission rates.

• Converter replacement—catalytic converters age and must eventually be replaced, if they are to maintain sufficiently low emission rates.

• Design of engine—engine types in different cars sometimes result in quite different emission rates.

• Temperature of ambient air—75°F was found to result in lower aldehyde emission for some engines, compared with 25°F or 95°F; and starting a cold engine generally results in high aldehyde emission rates.

Composition of Gasoline or Liquid Fuel. The aromatic-compound content of gasoline has an important effect on the benzaldehyde content of the exhaust.[192] Little or no benzaldehyde was produced from a gasoline free of aromatic compounds, but the amount of benzaldehyde increased linearly with increased aromatic-compound content. With a 100% aromatic fuel, the ratio of formaldehyde to benzaldehyde was about 3:1; the total aldehyde content of the exhaust gases from the engine was, however, nearly constant, regardless of the aromatic-compound content of the gasoline. Ninomiya and Golovoy[143] found that rather large amounts of benzaldehyde were produced when toluene (an aromatic hydrocarbon) was blended into gasoline.

Gasolines produced by different refineries or produced at different times of the year often contain large variations in the amounts of specific hydrocarbons. These hydrocarbons may be grouped into three families: the alkanes (paraffinic hydrocarbons), the alkenes (olefinic hydrocarbons), and the aromatic hydrocarbons. Higher-quality or premium gasolines usually have more aromatic hydrocarbons and/or trimethylpentanes (highly branched C_8 alkanes). Especially in the recent past, they also generally contained more antiknock compounds, such as tetraethyl and tetramethyl lead. Furthermore, winter-grade gasolines as a rule contain more volatile hydrocarbons, such as butanes, to provide quicker starting. In tests by Bykowski[29] of gasohol (blends of 90% gasoline and 10% ethanol), summer-grade gasohol for two types of American automobiles emitted about 50-60% more aldehydes than winter-grade gasohol. The hydrocarbon composition of the fuel clearly has some effect on the type and amount of emission, but only preliminary data are available.

Within the last few years, there have been extensive efforts to develop blends of gasolines that contain various oxygenated hydrocarbons, including the following:

• Ethanol (ethyl alcohol) that is blended with gasoline at 5-20%—these blends are generally referred to as gasohol; the main objective in using ethanol is to develop liquid fuels obtainable from grains of cellulose-containing materials (such as wood, straw, and cornstalks) that are grown in this country.

• Methanol (methyl alcohol)—this has also been tested rather extensively and is sometimes confused with the gasohol approach;

methanol can be obtained from natural gas, petroleum, coal, and wood; it is now used as a fuel for some racing vehicles.

- Methyl-tert-butyl ether (MTBE)--MTBE is being blended with gasoline at 5-20% by several oil companies; it results in substantially higher octane ratings and is thought by many persons knowledgeable about gasoline to have a bright future.

Gasoline and gasohol have been compared in several types of automobiles, but the results are inconclusive. Bykowski[29] found lower aldehyde emission rates in one automobile when gasohol was used, but the opposite in another automobile. Chui, Anderson, and Baker[45] made a rather large number of tests with gasohol blends containing 20% ethanol in several Brazilian automobiles. There were small but significant increases in aldehyde emission from ethanol-gasoline blends when the engines operated at low load. Differences at normal load, however, were small and perhaps insignificant. A considerable number of investigations have compared methanol-gasoline blends and pure gasoline. In general, slightly more aldehyde was emitted from methanol-containing fuels;[13,27,148] some of the increase occurred in tests at higher compression ratios that simultaneously resulted in increased engine efficiencies. One of the major advantages of methanol is that it produces higher octane blends that can be burned at high compression ratios. It has been suggested[148] that aldehyde emission can be markedly reduced by proper adjustment of the air-to-fuel ratio and by spark-advance settings.

Preliminary information has also been published on the use of MTBE-gasoline blends. Emission from the burning of such blends is comparable with that from unblended gasolines, except for somewhat higher aldehyde and isobutylene emission from the blended fuel.[81] It is thought that emission problems of the blends can be made at least comparable with those of unblended gasolines by proper engine adjustment and minor changes in the operation of catalytic converters.

There is need for continued testing of aldehyde emission from automobiles as the use of gasohol and other new fuel blends increases, although present evidence suggests that proper use of catalytic converters and other devices may control this emission quite well.

Type of Engine. Several engines have been considered as alternatives to the conventional piston-cylinder gasoline engine used almost exclusively for many years in automobiles. Such engines include the diesel, stratified-charge (PROCO), and rotary engines. As with the conventional gasoline engine, rather large variations in performance occur from engine to engine, including aldehyde and other undesirable emission. Such differences are caused by numerous variables, including engine design and operation, fuel composition and quality, and use of or failure to use a catalytic converter. Comparisons of various engines have been conducted by most if not all major automobile manufacturers and oil companies, universities, research organizations, and government laboratories.

The following comparisons of diesel and gasoline engines are applicable:

- Diesel engines emit more hydrocarbon and aldehyde than gasoline engines equipped with catalytic converters. They also emit appreciably more particulate material. The relative importance of specific hydrocarbons and aldehydes in the emission from diesel engines tends to be quite different from the relative importance of those from conventional gasoline engines.
- In many cases, the aldehydes emitted from diesel engines have higher molecular weights than those from gasoline engines. Isobutyraldehyde is sometimes the most important aldehyde on a weight basis.[152,166] Over twice as much isobutyraldehyde as formaldehyde was emitted from one engine. In another and more typical case, formaldehyde emission was higher. The increased yield of the higher aliphatic aldehydes from diesel-fuel combustion probably results from the dominance of the higher-molecular-weight paraffinic hydrocarbons in this fuel. Benzaldehyde normally is either not detected in the emission from diesel engines or present in only small amounts; most diesel fuels contain little or no aromatic hydrocarbon.

Emission from a stratified-charge (PROCO) engine and emission from a conventional gasoline engine have been compared to at least a limited extent. In two comparisons, the stratified-charge engine emitted more aldehyde than a regular gasoline engine.[29,125] In two others, the opposite was reported (Bachman and Kayle;[12] J.M. Heuss, personal communication): two Honda CVCC engines emitted considerably less aldehyde. Insufficient information is available to draw any generalized conclusions relative to the aldehyde emission of the two types of gasoline engines.

There are limited data comparing a rotary engine such as was used at one time in Mazda cars with conventional gasoline engines (Bykowski;[29] Heuss, personal communication). The rotary engine emitted more aldehyde than the conventional engines in at least three comparisons; in two cases, the differences were large.

In summary, we may estimate that direct aldehyde emission from all vehicles in the United States amounts to about 2.6×10^8 lb/yr. Another 10^9 lb/yr is probably generated from the atmospheric oxidation of the hydrocarbon emitted by these vehicles. These estimates are very approximate and may be in error by as much as a factor of 3.*

Other Combustion Processes

Fossil-fueled power plants emit several undesirable materials to the atmosphere, including aldehydes, as shown in Table 5-11. The

---

*The following assumptions have been made in deriving these estimates: 120×10^9 gal of gasoline are consumed per year; travel amounts to 144×10^{10} miles/yr, with an average gasoline mileage of 12 miles/gal; hydrocarbon is emitted at 0.41 g/mile, as mandated for the future; and 20% of the emission is aldehydes, and 80%, hydrocarbon.

TABLE 5-11

Pollutant Emission Summary--Heat Generation Sources[a]

Source No.	Fuel Used	Firing Method	Flow, scfm	Temp, °F	H2O, %	Dry Basis CO2 %	O2 %	Total Particulates Lb Per 1000 Lb	Lb Per Million Btu	Ton of fuel[c]	Benzene-Soluble Organics	Carbon Monoxide Lb Per Million Btu	Ton of fuel[c]	Hydrocarbons (as Methane) Lb Per Million Btu	Ton of fuel[c]	Oxides of Nitrogen (as NO2) Lb Per Million Btu	Ton of fuel[c]	Oxides of Sulfur (as SO2) Ppm by Vol.	Lb Per Million Btu	Ton of fuel[c]	Formaldehyde Lb Per Million Btu	Ton of fuel[c]
1	Coal	Pulverized	415,000	260	5.4	12.3	6.9	0.50	0.59	14.0	0.7	0.004	0.1	0.007	0.16	0.47	11	1490	3.72	88	1.3x10⁻⁴	30x10⁻⁴
2		Chain grate stoker	32,300	235	6.2	12.3	6.1		2.23	61.6	0.3	0.10	2.8	0.004	0.11			405	1.00	28	0.9x10⁻⁴	26x10⁻⁴
3			45,000	430	6.7	12.1	7.7	0.99	1.31	31.0	0.3	0.51	12	0.005	0.11			2030	6.14	146	1.4x10⁻⁴	33x10⁻⁴
4		Spreader stoker	18,000	405	6.9	10.6	8.5	0.66	0.82	22.6	1.4	<0.1	<3	0.006	0.16						2.2x10⁻⁴	60x10⁻⁴
5		Underfeed stokers	3,340	380	2.1	3.0	17.2	0.68	0.62	17.0	1.1	0.16	4.5	0.116	3.2				21 x 10⁻⁴		590x10⁻⁴	
6		Hand-stoked	3,290	235	2.1	2.5	18.1	0.24	0.25	7.0	3.6	0.14	3.9	0.036	1.0	0.30	8.3	205	2.3	62	3.8x10⁻⁴	100x10⁻⁴
7			43	345	2.2	2.6	17.1	0.52	0.44	12	1.2	1.1	31	0.12	3.3	0.36	9.8	178	1.2	32		
8			78	220	2.1	2.8	17.7	1.80	1.29	37	17	3.5	99	0.73	21	0.11	3.2	80	0.54	15		
												<0.1	<4	0.013	0.51			1250	3.0	116	0.63 x 10⁻⁴	24x10⁻⁴
9	Oil	Steam-atomized	5,200	530	8.5	9.0	8.2	0.32	0.306	11.7	1.0					0.31	12	188	1.3	48	2.4x10⁻⁴	89x10⁻⁴
10		Low-pressure air-atomized	10,000	340	5.4	8.8	9.0	0.049	0.267	10.0	2.7	0.055	2.2	0.004	0.17			125	0.35	14	1.6x10⁻⁴	62x10⁻⁴
11			195	230	7.4				0.051	2.0	60											
12		Centrifugal-atomized	145	170	3.0	2.9	16.9	0.041	0.046	1.8	39	0.038	1.5			0.44	17	14	0.12	4.5		
13		Vaporized	115	175	2.6	1.8	18.3	0.070	0.080	3.1	9.4	0.075	2.9	0.021	0.82	0.03	1.3	35	0.46	18	6.4x10⁻⁴	250x10⁻⁴
14			49	185	1.9	1.2	19.3	0.067	0.071	2.8	11	0.25	9.8	0.030	1.2			4	0.08	3	5.8x10⁻⁴	230x10⁻⁴
15	Gas	Premix burners	3,640	380	8.4	3.6	14.3	0.026	0.021	1.0	11	0.013	0.6	0.003	0.14	0.14	6.4				0.89x10⁻⁴	41x10⁻⁴
16			325	210	11.4	5.9	10.0	0.030	0.032	1.5	8.0	3.00	140	0.082	3.8	0.16	7.3	0	0	0	2.2x10⁻⁴	100x10⁻⁴
17			92	170	4.8	2.4	16.9	0.010	0.006	0.3	33	0.02	0.9			0.35	16	0	0	0	2.4x10⁻⁴	110x10⁻⁴
18			82	140	4.1	2.2	17.3	0.011	0.007	0.3	23	0.026	1.2	0.022	1.0	0.09	4.1				1.1x10⁻⁴	53x10⁻⁴
19			11	295	4.4	2.0	17.5	0.027	0.026	1.2	19	0.030	1.4	0.016	0.74	0.06	2.8				26 x 10⁻⁴	1200x10⁻⁴

[a] Reprinted with permission from Hangebrauck et al.[77] Blanks indicate that no test was made.

[b] Pounds of particles per 1,000 lb of dry flue gas adjusted to 50% excess air.

[c] Density of natural gas = 0.0443 lb/ft³ (60°F, 1 atm).

amount and type of such emission vary greatly between plants, but the lowest rates of formaldehyde emission occurred in some coal-burning plants. In 1978, Natusch[137] reported rather limited data from power plants using coal, oil, and natural gas. Coal-fired furnaces emitted the most particles, the most carbon monoxide, and the least aldehyde (reported as formaldehyde). Aldehyde emission by coal-, oil-, and natural-gas-fired furnaces was reported to be 0.002, 0.1, and 0.2 lb from 1,000 lb of fuel. If the results of Natusch are average values and if the consumption data of the U.S. Department of Energy for coal, oil, and natural-gas consumption in power plants are used, the amount of aldehyde emitted from power plants in the United States is estimated to be approximately 50 million pounds per year (see Table 5-12). The unburned hydrocarbon emitted may lead to the eventual formation of 10-20 times more aldehyde. Large changes have occurred in the last 10-20 yr in the design and operation of fossil-fueled power plants. As a general rule, the major emphasis has been on increased energy efficiency. The results probably produce more complete combustion and decreased aldehyde concentrations, although specific data to support this hypothesis are not available.

Sulfur dioxide is considered by many to be a more obnoxious emission from oil- and coal-burning plans than aldehyde. Natusch[137] pointed out that the polycyclic organic materials formed to a small extent, particularly from coal-fired furnaces, were especially critical, in that such materials are generally considered to be carcinogens.

A variety of aldehydes have been identified as products of forest fires (Table 5-13). Bonfires and garbage fires also produce aldehydes and other undesirable byproducts. The amounts emitted obviously depend on the size of the fire, i.e., the amount of material burned (Table 5-14). In most cases, detrimental effects of such fires are limited to the immediate area of the fire.

The combustion of tobacco in the process of cigarette-smoking also generates a variety of aldehydes. This is an important source of aldehydes only in the indoor environment, and it is discussed in more detail later in this chapter.

VEGETATION

Plants in general have the ability to release volatile compounds into the air through their stomata and cuticle. Of these compounds, carbon dioxide, oxygen, and water have been studied in detail, owing to their metabolic relevance. Attention has recently turned to plants as a source of hydrocarbons important enough to affect air quality. Terpene diffused from forest trees has been the principal compound of concern;[150] aldehydes, in comparison, have received little attention.

There is ample evidence of the natural occurrence of aldehydes in plants. Schauenstein et al.[153] noted that aldehydes are widely distributed in fruits, imparting a characteristic aroma and flavor to pineapple, apple, grapefruit, lime, banana, pear, peach, lemon, blackcurrant, strawberry, orange, grape, and raspberry. Strawberry

TABLE 5-12

Estimated Aldehyde Emission from Power Plants in United States

Year	Fuel Consumption (Unrevised)[a]			Direct Aldehyde Emission, 10^6 lb[b]			
	Coal, 10^6 tons	Oil, 10^6 bbl	Natural Gas, 10^9 ft^3	Coal	Oil	Natural Gas	Total
1970	320.2	335.5	3,932	1.3	11.3	33.3	45.9
1973	384.2	560.2	3,660	1.5	18.8	31.0	51.3
1976	448.4	555.9	3,081	1.8	18.7	26.1	46.6
1978	481.2	635.8	3,188	1.9	21.4	27.0	50.3
1979	528.8	523.5	3,490	2.1	17.6	29.6	49.3

[a]U.S. Department of Energy, Federal Power Commission. Form 4.

[b]Based on following assumptions: coal, 0.004 lb/ton; oil, 0.2 lb/ton, 5.95 bbl of oil per ton; natural gas, 0.4 lb/ton, 23.6 x 10^3 ft^3 of gas per ton.

TABLE 5-13

Aldehydes Emitted by Forest Fires[a]

Aliphatic	Aromatic
Formaldehyde	Vanillin
Acetaldehyde	Coniferaldehyde
Propanal	Syringaldehyde
n-Butanal	Sinapaldehyde
Isobutanal	

Olefinic	Cyclic
Acrolein	Furfural
	5-Methylfurfural

[a]Data from Graedel.[73]

TABLE 5-14

Gaseous Emission from Open Burning[a]

Test No.	Material Burned	Gaseous Emission, lb/ton of material initially present				
		CO_2	CO	HC[b]	Formaldehyde	Organic Acids[c]
1	Municipal refuse	1,250	90	30	0.095	14
2		1,210	80	30	0.094	16
Avg.		1,230	85	30	0.095	15
3	Landscape refuse	860	80	35	0.005	18
4		550	50	25	0.006	8
Avg.		700	65	30	0.006	13
5	Automobile components	1,500	125	30	0.030	16

[a] Reprinted with permission from Gerstle and Kemnitz.[70]

[b] Gaseous hydrocarbons expressed as methane.

[c] Expressed as acetic acid.

and pear have an especially high aldehyde content--13-18 mg/kg. Some species contain three carbonyl compounds, others as many as 20 (Table 5-15); 2-trans-hexenal (2TH) is the most common. Schauenstein et al. speculated that the unusually wide distribution of 2TH indicates that it probably is formed during the processing of the fruit. According to their interpretation, some aldehyde in apple and fruit juices is formed biogenically in the fruit, and some of the remainder is produced by enzymatic and nonenzymatic reactions during processing. Vegetables are not without their share of aldehydes.[167] Acetaldehyde, propionaldehyde, isobutyraldehyde, and butyraldehyde have been detected in beans, broccoli, brussels sprout, cabbage, carrot, cauliflower, celery, cucumber, lettuce, onion, potato, and soybean.

Woody species also contain aldehydes, but there is disagreement as to whether they occur in healthy, as well as in injured, tissue. According to Schauenstein et al.,[153] one report stated that 2TH was emitted by Robinia pseudoacacia in the absence of injury when the plant was enclosed within a plastic bag for 12 h. More numerous reports cite the capacity for formation of 2TH in injured trees, such as Ginkgo biloba, Albizia julibrissin, and Ailanthus glandulosa. It has been suggested that the biosynthesis of aldehydes is a defense against biologic attack, for example, in the resistance shown by ginkgo to fungi.[123]

In the course of inquiry into the possible etiologic factors of nasopharyngeal tumors among the Chinese and Kenyans, Gibbard and Schoenthal[71] made a semiquantitative measurement of sinapylaldehyde and related aldehydes in the wood of eight angiosperms and two gymnosperms. The aldehyde content varied according to species, with Eucalyptus sp. and Fagus sylvatica having the highest and Juniperus procera and Larix decidua the lowest content (Table 5-16).

There is some information on the location of aldehydes in plant tissue.[153] A report by Lamberton and Redcliffe established that the long-chain aldehydes occur in cuticular plant waxes. On measuring the aldehyde content as a percentage of total lipids, they found a range from 0.2% in purple loosestrife (Lythrum salicaria) to 14.3% in cranberry (Table 5-17). Apparently, the distribution pattern of aldehydes is so characteristic for a species that it may have value in taxonomic studies.

Aldehydes emanating from vegetation have been detected in ambient air.[73] Thirty-six plant volatiles including aliphatic, olefinic, aromatic, and cyclic aldehydes have been cited by Graedel (Table 5-18).

When plant material is burned either deliberately for disposal of agricultural waste or unintentionally as in forest fires, the increase in aldehyde emission may become significant. Because the nature of the plant material and the conditions of the burning can vary widely, it is difficult to characterize the emission. In a special experiment with a tower that simulated field conditions, Darley et al.[51] compared emission from the three principal types of agricultural waste in the San Francisco Bay area. The number of pounds of total hydrocarbon emitted per ton of plant material was 9.7 for fruit prunings, 14.5 for barley straw, and 4.4 for native brush. Aldehydes

TABLE 5-15

Carbonyl Compounds in Various Fruits[a]

Pineapple

formaldehyde
acetaldehyde
furfural

Apple

formaldehyde
acetaldehyde
propanal
1-butanal
pentanal
hexanal
2-hexenal
furfural
C_{24-30} aldehydes

Grapefruit

acetaldehyde
citral
C_{7-11} aldehydes

Lime

octanal
nonanal
citral
dodecanal
furfural

Banana

acetaldehyde
1-pentanal
2-hexenal
C_{24}, C_{26}, C_{28},
 C_{29}, C_{30}, C_{31},
 C_{32} aldehydes

Peach

acetaldehyde
benzaldehyde
furfural
C_{24}, C_{26},
 C_{30} aldehydes

Pear

acetaldehyde
propanal
2-hexenal

Lemon

heptanal
octanal
nonanal
decanal
undecanal
dodecanal
C_{13-17} aldehydes
citral
neral
geranial
citronellal

Blackcurrant

acetaldehyde
butanal
pentanal
hexanal
2-hexenal
benzaldehyde

Strawberry

acetaldehyde
propanal
2-propenal
2-butenal
2-pentenal
hexanal
3-cis-hexenal
heptanal
benzaldehyde
furfural
methylfurfural

Orange

acetaldehyde
pentanal
hexanal
2-hexenal
heptanal
octanal
octenal
nonanal
decanal
undecanal
citral
neral
geranial
dodecanal
α, β-substituted acroleins
C_{24}, C_{26}, C_{28}, C_{30} aldehydes

Grape

acetaldehyde
butanal
hexanal
2-hexenal
benzaldehyde

Raspberry

acetaldehyde
propanal
2-propenal
2-methylpropenal
2-pentenal
2-hexenal
3-cis-hexenal
benzaldehyde
furfural
methylfurfural

---

[a]Reprinted with permission from Schauenstein et al.[153] Unless otherwise stated, alkenals have 2-trans configuration.

TABLE 5-16

Yields of Aldehydes in Wood of Various Tree Species[a]

Tree	Aldehyde Yield, µg/g			
	Sinapyl	Syringic	Coniferyl	Vanillin
Eucalyptus sp.	3,000	3,000	150	100
Fagus sylvatica L. (beech)	800	800	250	250
Tectona grandis (teak)	700	500	600	450
Santalum album (sandalwood)	600	500	300	200
Quercus robur L. (oak)	500	600	250	200
Chinese incense	500	600	250	250
Indian incense	100	200	100	50
Cocos nucifera L. (coconut)	300	500	300	300
Juniperus procera Hochst	0	0	450	600
Larix decidua (larch)	0	0	600	600

[a]Reprinted with permission from Gibbard and Schoental.[71]

TABLE 5-17

Aldehydes in Surface Waxes of Leaves and Fruits[a]

Source	Aldehydes (% of total lipids)	Distribution of aldehydes (%)								
		C_{24}	C_{25}	C_{26}	C_{27}	C_{28}	C_{29}	C_{30}	C_{31}	C_{32}
Pea (*Pisum sativum*)	5·0	1·4		55·7	1·0	40·9		1·0		
Dogbane (*Apocynum androsaemifolium* L.)	2·5			2·5		15·5	7·2	74·8		
Indian hemp (*Apocynum cannabinum* L.)	1·1	4·2		5·6		18·1	6·6	68·5		
Cypress spurge (*Euphorbia cyparissias* L.)	2·8		4·1	8·5		75·6		11·8		
Spurge (*Euphorbia esula* L.)	0·5			11·3		73·5		15·2		
Toadflax (*Linaria vulgaris* L.)	3·3	6·1	1·7	20·8	6·8	8·2	11·6	22·5		22·3
Purple loosestrife (*Lythrum salicaria* L.)	0·2	2·3	3·9	16·4	8·6	21·8	12·0	32·9	2·1	
Fat hen (*Chenopodium album*)	11·6	2·9	0·6	13·6	1·2	46·0	0·3	14·6		9·3
Ryegrass (*Lolium perenne* L.)	9·1		1·6	55·7	1·2	22·3		15·8		1·7
Grape		12·4	2·8	41·7	2·5	21·8	1·1	7·5	0·5	2·8
Cranberry	14·3			1·1	1·9	10·2	5·2	76·2	0·8	3·9

[a]Reprinted with permission from Schauenstein et al.[153]

TABLE 5-18

Aldehydes Emitted from Natural Plant Sources[a]

Aliphatic

Formaldehyde
Acetaldehyde
Propanal
n-Butanal
Isobutanal
n-Pentanal
Isopentanal
Hexanal
Heptanal
Octanal
Nonanal
Decanal
Undecanal
Tetradecanal

Aromatic

Benzaldehyde
Cuminaldehyde
Dihydrocuminaldehyde--
 Phenylpropanal
Cinnamaldehyde
p-Hydroxybenzaldehyde
Anisaldehyde
o-Methoxycinnamaldehyde
Piperonal
Vanillin
Veratraldehyde
Coniferaldehyde
Evernicaldehyde

Olefinic

1-Hexen-2-al
trans-2-Hexenal
3,7-Dimethyl-2,6-octadien-1-al
3,7-Dimethyl-6-octen-1-al
4-Hydroxy-3,7-dimethyl-6-octen-1-al
2,6-Nonaldien-1-al

Cyclic

Furfural
Safranal

---

[a]Data from Graedel.[73]

were not mentioned specifically. In the burning of landscape refuse (lawn clippings, leaves, and tree branches), Gerstle and Kemnitz[70] reported 30 lb of hydrocarbons and only 0.005 lb of formaldehyde per ton of material. Combustion of the same amounts of municipal refuse and automobile components yielded about the same amount (30 lb) of hydrocarbon, but an increase in formaldehyde from 0.005 lb/ton to 0.095 and 0.030 lb/ton, respectively.

INDOOR SOURCES OF ALDEHYDES

Aldehydes enter the indoor environment through infiltration of outdoor air and from a variety of sources within the indoor environment itself. Indoor sources include aldehyde-containing building materials, combustion appliances, tobacco smoke, and a large variety of consumer products. Measurements of aldehydes in the indoor environment have focused almost exclusively on formaldehyde. In general, indoor formaldehyde concentrations exceed outdoor concentrations. The contribution of formaldehyde in outdoor air to indoor formaldehyde concentration appears to be minor. This section considers some of the important indoor sources.

Building Materials

The low cost and superior bonding properties of formaldehyde polymers make them excellent choices as resins for the production of various building materials, especially plywood and particleboard. Resins used for building materials include urea-formaldehyde (UF), phenol-formaldehyde, and melamine-formaldehyde.

Urea-formaldehyde resin is the most common adhesive used for the production of indoor plywood and particleboard. It is also used in protective coatings and for treating paper and textiles. UF resin contains some free formaldehyde and decomposes and releases formaldehyde gas at high temperature and high humidity.

Phenol-formaldehyde resin, which does not release formaldehyde as readily as UF resin, is used as adhesive for wood products requiring greater moisture resistance (i.e., outdoor plywood). Phenol-formaldehyde resin, however, is not generally used for most indoor wood products, because of its higher cost.

Plywood is composed of several sheets of thin wood glued together. Particleboard is made by saturating small wood shavings with a resin (usually UF resin) and pressing the resulting mixture, usually at a high temperature, into the final form. Particleboard continuously emits formaldehyde, but at a steadily decreasing rate over a period of several years; in dwellings where it is used for furniture, partitions, etc., the emission may become large and even exceed the OSHA time-weighted average of 3 ppm. The emission rate varies as a function of several conditions, such as the original manufacturing process, the nature of the wood used, the quantity of catalyst used in curing the resin, quality control of fabrication, porosity, humidity, cutting of the board for final use, rate of infiltration, and ventilation.

The problems with plywood and particleboard are especially severe in mobile homes.[19] Within the last few years, there has been a trend to make mobile homes more airtight in an effort to conserve heat in the winter and minimize cooling demands in the summer. Hence, there is less turnover of the air in a mobile trailer, and formaldehyde emission from plywood and particleboard has become much more obvious and of increased concern. Because air-exchange rates affect indoor air quality, the rate of release of formaldehyde from these building products and the air-exchange rates in the design of mobile homes are especially important for the control of pollution.

Insulation

UF foam is used as thermal insulation in the side walls of existing buildings,[116] mainly single-family residential buildings. It is convenient and inexpensive to inject the foam through small holes that can be sealed after insulation is completed.

Installation involves mixing partially polymerized UF resin with a surfactant (foaming agent) and an acid catalyst under pressure that forces air into the mixture to create a foam. The foam hardens within minutes and cures and dries completely within a few days. Building codes in the United States, concerned with the fire-safety aspects of UF-foam insulation, rate UF foam as a combustible material. The codes require that, when used on the inside of buildings, the UF foam must be protected by a thermal barrier of fire-resistant material. In England and Holland, UF insulation materials are certified for use only in masonry cavities of buildings.

If the foam insulation is improperly mixed, or if improperly formulated UF resin is used,[159,160] the insulation may release formaldehyde into the building. Specific factors that have been identified as contributors to formaldehyde release include excessive formaldehyde in the resin concentrate, excessive acid catalyst in the foaming agent (especially important), excessive foaming agent (surfactant), foaming during periods of high humidity and temperature, foaming with cold chemicals (optimal temperature, 50-80°F), improper use of vapor barriers, improper use of foams (in ceilings, etc.), and excessive resin.[2,116]

Combustion Appliances

Several recent studies have reported on combustion-generated indoor air pollutants, namely air contaminants from gas stoves and heating systems in residential buildings. Laboratory studies have shown that gas stoves emit substantial aldehyde; formaldehyde has been identified as the major component of the aldehydes measured (Schmidt and Gotz;[154] G. Traynor, personal communication). Formaldehyde emission rates for a gas stove have been measured at approximately

25,000 µg/h and 15,000 µg/h for the oven and each top burner, respectively (Traynor, personal communication).

Tobacco Smoke

Tobacco smoke is a source of several chemical pollutants, including aldehydes (Table 5-19), that can reach high concentrations in the indoor environment. The smoker's exposure to the chemical pollutants results principally from smoke inhaled directly into the lungs (mainstream smoke). The smoke that is not inhaled directly into the lungs enters the space surrounding the smoker (sidestream smoke). It is the sidestream smoke that is the major contributor to indoor pollution. The inhalation of tobacco smoke involuntarily, commonly referred to as "passive smoking," has only recently been the subject of investigation. Analysis by Hobbs et al.[146] indicated acrolein to be an important component of tobacco smoke. Weber[95] used a smoking machine in an environmental chamber and identified substantial amounts of acrolein. Data on formaldehyde, acetaldehyde, and acrolein in cigarette smoke are presented in Table 5-20.

Harke et al.[80] measured concentrations of nicotine, carbon monoxide, acrolein, and aldehydes (expressed as acetaldehyde) in the air of an unventilated room in which a series of experiments with a smoking machine were performed. Important concentrations of all four of these compounds were observed in these experiments; however, the number of cigarettes per unit time was unusually high.

It has been demonstrated that the quality of smoke from Burley tobacco depends on the potassium and magnesium composition of the leaves.[111] When potassium was applied to the soil at 224 kg/hectare, the aldehyde content of tobacco smoke increased from 0.41 to 0.55 mg/cigarette. At the same time, the total particulate material in cigarette smoke decreased. Thus, the researchers were faced with both harmful and beneficial effects on smoke quality and therefore recommended bioassays to evaluate the potential consequence for health.

AGRICULTURAL AND DISINFECTANT PRODUCTS

Commercially grown plants require fertilizers for optimal growth and pesticides for disease control. Both may involve the use of aldehydes and theoretically could contribute to the aldehyde content of indoor air. Urea-formaldehyde polymers represent one of several groups of fertilizers and are used not only to obtain a more uniform release rate than is possible with soluble nitrogen, but also to minimize the hazards of water pollution by nitrates leached out of the soil.[76] Fertilizers with aldehyde compounds as a source of slow-release nitrogen have been used on field crops,[76] turfgrass,[186] pine seedlings,[19] and geranium.[176]

Formaldehyde has been used in a wide variety of agricultural operations to disinfect seeds, bulbs, roots, soil, and contaminated

TABLE 5-19

Aldehydes Identified in Tobacco Smoke[a]

Aromatic

Formaldehyde
Acetaldehyde
Glyoxylic Acid
Propanal
2-Oxopropanal
n-Butanal
Isobutanal
Galactose

Olefinic

Acrolein
Crotonaldehyde

Aromatic

Benzaldehyde

Cyclic

Furfural
5-Hydroxymethylfurfural

---

[a]Data from Graedel.[73]

TABLE 5-20

Quantities of Some Aldehydes in Cigarette Smoke

Aldehyde	Amount in Cigarette Smoke, mg/cigarette	Reference
Acetaldehyde	0.18-1.44	181
Acrolein	0.7	95
Formaldehyde	0.02-0.04	181

equipment, such as pots, tools, storage bins, and greenhouses. Walker[187] described its use to disinfect wheat and barley by steeping in a formaldehyde solution (1 pint of formalin in 40 gal of water) for 5 min and then holding in a covered container for 2 h. Leafspot in beets is prevented by dipping in a solution of 1 pint of formalin in 8 gal of water. Bacterial blight in celery is combatted by soaking seeds for 15-30 min in a solution of 1 pint of formalin in 32 gal of water. Williams and Siegel[194] found bactericidal concentrations of formaldehyde on the shells of eggs exposed to formalin at 1.2 ml/ft^3 of incubator space. Infected laboratory animal housing can be decontaminated with paraformaldehyde at 10 g/m^3 heated to 232°C to release formaldehyde.[136]

Glutaraldehyde in 2% alkaline solution has a germicidal spectrum similar to that of formaldehyde, although it is more expensive and less stable.[163]

There are at least 60 registered pesticides containing formaldehyde and 75 containing paraformaldehyde as active ingredients. At prescribed rates, they can be used on some vegetable, field, and ornamental crops. Formaldehyde can also be used on equipment used in the culture of mushrooms, potatoes, and other crops.

Formaldehyde is an effective disinfectant against bacteria, fungi, and viruses. It kills bacteria in 6-12 h in concentrations of 1:200 and bacterial spores in 2-4 d. It is effective against tubercle bacilli. It is used in dilute solutions as a disinfectant and preservative in cosmetics (see Chapter 7). Formaldehyde is used in a variety of applications as a preservative and tissue fixative for biologic and histologic specimens and in embalming.[163]

OTHER CONSUMER PRODUCTS

Urea-formaldehyde resin is used by the paper industry to give increased wet strength to various grades of paper. Typical paper products treated with UF resin include grocery bags, waxed paper, facial tissues, napkins, paper towels, and disposable sanitary products. Formaldehyde polymers are used extensively in the manufacture of floor coverings and as carpet backing. UF resin is used in binders in the textile industry to improve the adherence of pigments, fire retardants, or other material to cloth. It is also used to impart stiffness, wrinkle resistance, and water repellency to fabrics.

THE MECHANISM OF ALDEHYDE GENERATION IN THE ATMOSPHERE

THE UNPOLLUTED, NATURAL ATMOSPHERE

There are natural precursors of formaldehyde even in the atmosphere that is unpolluted by man. It contains methane, CH_4, at about 1.6 ppm and smaller amounts of various other hydrocarbons that are emitted from the earth through natural processes--escape of gases from the earth, tree and plant emission, etc. The reaction of these naturally occurring hydrocarbons with photochemically generated HO radicals is the major natural source of formaldehyde in the clean lower troposphere. The HO-radical is formed through a variety of reactions. One important reaction sequence is initiated by the photodissociation of ozone, O_3, at the short wavelengths present in sunlight:

$$O_3 + h\nu(\lambda < 3200 \text{ A}) \rightarrow O(^1D) + O_2(^1\Sigma_g^+, ^1\Delta_g, \text{ or } ^3\Sigma_g^-) \quad (1)$$

The $O(^1D)$ atom is an electronically excited species that may be deactivated to a normal ground-state atom, $O(^3P)$, by collisions with O_2 and N_2 in the air (Reaction 2), or it may, on encountering a water molecule, form HO radicals (Reaction 3):

$$O(^1D) + N_2(\text{or } O_2) \rightarrow O(^3P) + N_2(\text{or } O_2) \quad (2)$$

$$O(^1D) + H_2O \rightarrow 2HO \quad (3)$$

The HO radicals react in part with hydrocarbons present in the atmosphere. In the case of reaction with methane, the following reaction sequence (somewhat abbreviated) may occur and lead to formaldehyde:

$$HO + CH_4 \rightarrow H_2O + CH_3 \quad (4)$$

$$CH_3 + O_2 \; (+ N_2 \text{ or } O_2) \rightarrow CH_3O_2 \; (+ N_2 \text{ or } O_2) \quad (5)$$

$$CH_3O_2 + NO \rightarrow CH_3O + NO_2 \quad (6)$$

$$CH_3O_2 + CH_3O_2 \rightarrow CH_3O + CH_3O + O_2 \quad (7)$$

$$\rightarrow CH_3OH + HCHO + O_2 \quad (8)$$

$$CH_3O + O_2 \rightarrow HCHO + HO_2 \quad (9)$$

Formaldehyde absorbs the short wavelengths of sunlight ($\lambda < 3700$ A) and undergoes photodecomposition. It is also destroyed by reactions with the HO radical and other reactive atmospheric species. Levy used a somewhat incomplete reaction mechanism involving these various formaldehyde formation and decay processes with the rate-constant estimates then available to estimate the theoretical formaldehyde concentration-versus-altitude profile shown in Figure 5-1. A more complete reaction scheme and updated rate and photochemical data lead to somewhat higher formaldehyde concentrations than those predicted by Levy.[37]

Thus, one anticipates in theory that, in the clean atmosphere near ground level during the daylight hours, the formaldehyde concentration will be around 1.4×10^{10} molecules/cm^3 (about 0.0006 ppm), owing to the chemistry involving only the naturally occurring components of the atmosphere. Indeed, concentrations of this magnitude are observed even in the remote and seemingly uncontaminated regions of the lower atmosphere. If one were to include formaldehyde source terms from the naturally occurring nonmethane hydrocarbons, a somewhat higher ground-level formaldehyde concentration is anticipated.

The Mechanism of Aldehyde Generation within the Polluted Lower Atmosphere

In addition to the clean-air mechanism of formaldehyde generation outlined briefly in the preceding section, many other reactions occur within the polluted troposphere that lead to the formation of formaldehyde and the higher aldehydes. The major sources are the reactions of the anthropogenic and natural nonmethane hydrocarbons (alkanes, alkenes, and aromatic hydrocarbons) with HO radicals and ozone present in the atmosphere. It will be instructive to consider here some examples of these important reaction mechanisms.

The Aldehyde-Generating Reactions of the HO Radical with the Alkanes

There is now an abundance of both direct experimental and theoretical evidence that the reactive HO radical is present in the sunlight-irradiated lower atmosphere; for examples, see Calvert,[35] Wang et al.,[188] Davis et al.,[52,53] Calvert and McQuigg,[39] and Crutzen and Fishman.[49] These HO radicals formed within the atmosphere react by H-atom abstraction with all the impurity-alkane molecules present in the air. The rate constants for these reactions are very much larger for the higher-molecular-weight hydrocarbons than for methane, and all the reactions lead to aldehyde formation at least in part. As an example, consider the reactions initiated by the attack of HO on n-butane, n-C_4H_{10}, a typical alkane impurity found in the urban atmosphere. Both secondary and primary H atoms may be abstracted in this case:

$$CH_3CH_2CH_2CH_3 + HO \rightarrow CH_3CH_2CH_2CH_2\cdot + H_2O \qquad (10)$$

$$\rightarrow CH_3\dot{C}HCH_2CH_3 + H_2O \qquad (11)$$

The rate of Reaction 11 is about 3.5 times that of Reaction 10 at 25°C. Even the slower of these two reactions has a rate constant about 200 times larger than that of HO radical with methane (Reaction 4). To illustrate the mechanism in which the aldehydes are formed following Reactions 10 and 11, the sequence of reactions of the n-butyl radical, C_4H_9, product of Reaction 10 may be considered in Figure 5-7; the aldehyde products are highlighted by enclosing them in boxes. Note that, during the course of these reactions, every possible straight-chain aldehyde of four or fewer carbon atoms is formed. Some other reactions of the alkylperoxy, RO_2, and alkoxy, RO, radicals not shown in Figure 5-7 compete with those given here,

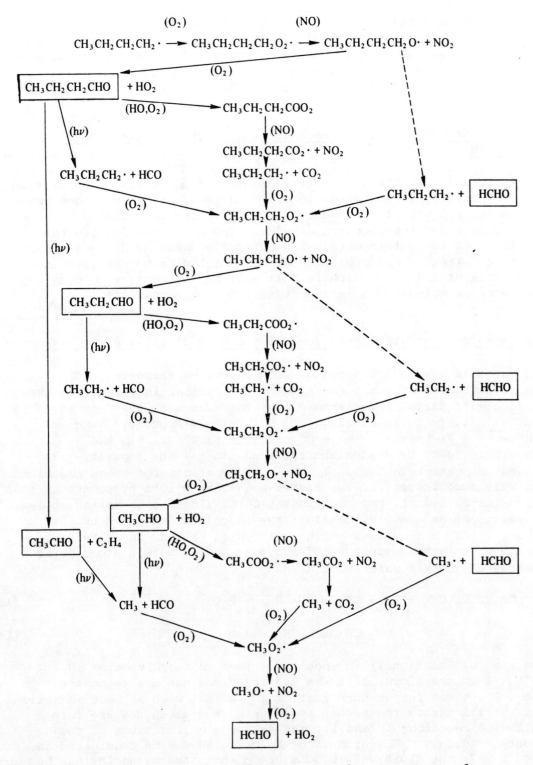

FIGURE 5-7 Example of aldehyde-forming reaction sequences after n-butyl radical generation from n-butane in polluted troposphere; pathways shown by dashed arrows are much less important than those shown by solid arrows.

but the aldehyde-forming reactions are expected to dominate in the polluted atmosphere.

A similar set of reactions occurs following Reaction 11 in which HCHO, CH_3CHO, CH_3CH_2CHO, and methyl ethyl ketone are the expected major products. Indeed, all the impurity-alkane molecules present in the polluted atmosphere are potential sources of the aldehydes through similar reaction sequences. For further examples and a consideration of the detailed reaction mechanisms of hydrocarbon photooxidation, see Demerjian et al.[55] Present evidence suggests that the major atmospheric loss mechanism for the alkanes involves HO attack on these species. If one assumes an HO-radical concentration for the polluted troposphere that is consistent with theory and experiment, about 3×10^{-7} ppm, then the half-life of n-butane may be estimated from the sum of the known rate constants, $k_{10} + k_{11}$, to be about 10 h.[10] Other representative alkanes, such as isobutane and isopentane, have similar half-lives--about 10 and 8 h, respectively. During this rather short period in which the typical alkane decays, the photooxidation reactions commonly lead to more aldehyde molecules than molecules of hydrocarbon that have reacted.

The Aldehyde-Generating Reactions of the HO Radical with Alkenes

The most reactive class of hydrocarbons, the alkenes, also are major sources of aldehydes. The HO radical is only one of the reactants that stimulate aldehyde formation in this case. The mechanism of the reactions can be illustrated with the simple alkene, propylene, C_3H_6. The complete mechanism of the HO-alkene reactions is not entirely clear, but it now appears probable that the dominant primary reaction is HO addition to the carbon-carbon double bond of the alkene; presumably, both terminal and internal additions may occur with propylene:

$$CH_3CH = CH_2 + HO \rightarrow CH_3\dot{C}HCH_2OH \tag{12}$$

$$\rightarrow CH_3\overset{\overset{\displaystyle OH}{|}}{C}HCH_2^{\cdot} \tag{13}$$

The radical product of Reaction 12 may react by the following possible steps:

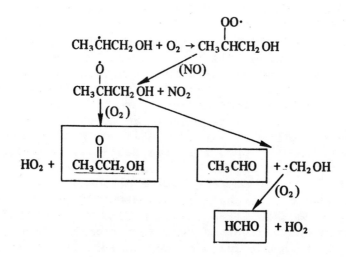

In a similar fashion, the radical product of Reaction 13 may form formaldehyde and acetaldehyde, among other products. The rate constants for the HO-radical reaction with the alkenes are in general larger than those for the alkanes.[10] For the typical HO-radical concentration in the sunlight-irradiated, polluted troposphere, [HO] ≃ 3 × 10^{-7} ppm, and propylene, isobutene, and trans-2-butene have half-lives of only 1.0, 0.5, and 0.4 h, respectively. Because the aldehydes are major products of this rapid interaction, the HO-alkene reactions are expected to be major sources of aldehydes in the usual hydrocarbon-polluted atmosphere.

Aldehyde Generation through the Ozone-Alkene Reactions

As ozone builds up in a sunlight-irradiated, polluted atmosphere, the interaction of ozone with the impurity-alkene molecules can become important, and reactions between these molecules are an efficient source of aldehydes. In illustration, consider the attack of ozone on propylene. The primary reaction leads to an unstable, energy-rich ozonide (Reaction 14). Both theory and experiment suggest that, in the atmosphere, this species will react rapidly, in part to form aldehydes (Reactions 15 and 16):

$$O_3 + CH_3CH=CH_2 \rightarrow CH_3CH\underset{\underset{}{}}{-}CH_2 \text{ (with O-O-O bridge)} \quad (14)$$

$$\xrightarrow{(15)} CH_3CHO + CH_2OO$$
$$\xrightarrow{(16)} HCHO + CH_3CHOO$$
$$\xrightarrow{(17)} \text{other products}$$

The intermediate CH_2O_2 and CH_3CHO_2 species formed in Reactions 15 and 16 may fragment by a variety of reaction paths, but they may also lead to aldehydes through Reactions 18-21 when easily oxidized compounds, such as NO and SO_2, are present:[41]

$$CH_2O_2 + NO \rightarrow \boxed{HCHO} + NO_2 \qquad (18)$$

$$CH_2O_2 + SO_2 \rightarrow \boxed{HCHO} + SO_3 \qquad (19)$$

$$CH_3CHO_2 + NO \rightarrow \boxed{CH_3CHO} + NO_2 \qquad (20)$$

$$CH_3CHO_2 + SO_2 \rightarrow \boxed{CH_3CHO} + SO_3 \qquad (21)$$

Present evidence[117,124,141,170] suggests that a significant fraction (greater than 20%) of the gas-phase ozonolysis of the simple alkenes proceeds through the so-called Criegee mechanism, of which Reactions 15 and 16 are critical parts. Again, aldehydes are among the major products formed.

The half-lives of the impurities of propylene, isobutene, and trans-2-butene for reaction with ozone in a highly polluted atmosphere, where the concentration of ozone may be about 0.2 ppm, are 3.7, 3.3, and 0.2 h, respectively.[83,90] Because aldehydes are major products of this system, it is evident that the ozone-alkene reactions may be an important source of aldehydes in the polluted atmosphere.

In most urban areas, the total amount of aldehydes from direct emission (autos, refuse burning, chemical plants, power plants, etc.) is usually below that of the nonmethane reactive hydrocarbons. As we have seen, the atmospheric chemistry results in the formation of at least one molecule of aldehyde from each molecule of hydrocarbon within a relatively short period (a few hours to a few days). Thus, it appears that the largest share of the total aldehyde content of urban air is created in the atmosphere from hydrocarbon precursors and that control of the direct emission of hydrocarbon, as well as aldehydes, will be a necessary part of any newly developed strategy to control ambient concentrations of the aldehydes.

ALDEHYDE REMOVAL PROCESSES OPERATIVE IN THE ENVIRONMENT

The accumulation of aldehydes in the atmosphere is suppressed by several natural removal processes. Many of the chemical steps are seemingly well understood; other chemical and physical processes remain speculative. The action of sunlight on the aldehydes results in their decomposition. The reaction of the reactive molecular fragments that are present in the atmosphere--HO, HO_2, $O(^3P)$, and NO_3--may also result in chemical degradation or transformation of the aldehydes. These and other important natural removal processes are considered in this section.

THE PHOTODECOMPOSITION OF THE ALDEHYDES

There is a substantial overlap between the ultraviolet-wavelength region of the light absorbed by the simple aldehydes and the solar spectral distribution incident on the earth's surface. This can be seen in Figure 5-8 for formaldehyde, acetaldehyde, and propionaldehyde. The rather weak absorption bands in the near-ultraviolet region for the aldehydes originate from a weakly allowed n → π* electronic transition, which involves largely the promotion of an electron in a nonbonding (n) orbital on oxygen to the antibonding π* orbital associated with the carbon-oxygen double bond in the aldehyde.

The initial electronically excited states of the aldehydes that are formed in this process are short-lived, and a large fraction of the excited molecules undergo molecular fragmentation or rearrangement very quickly. In the case of formaldehyde, the decay of the excited molecules occurs efficiently through either of two primary processes:

$$HCHO + h\nu \rightarrow HCHO^* \rightarrow H + HCO \tag{I}$$

$$\rightarrow H_2 + CO \tag{II}$$

Many measurements of the quantum efficiencies of these processes--i.e., the fraction of the excited molecules that decay by a given path--have been made in recent years; for a review of this extensive literature, see Calvert.[37] The results derived from two of these studies that should be most applicable to the reactions in the lower atmosphere at 25°C are summarized in Figures 5-9 and 5-10. It is apparent from these data that the quantum yield of fragmentation of excited formaldehyde into the reactive free-radical fragments, H and HCO, in process I (ϕ_I) increases from near zero at 3380 A to near 0.8 at 3000 A. Process II, forming molecular hydrogen and carbon monoxide, has a longer wavelength onset, and ϕ_{II} maximizes near 3350 A.

After process I in air at 1 atm, the radicals formed react largely through Reactions 22 and 23 to generate HO_2 radicals and carbon monoxide:

$$H + O_2 (+ N_2 \text{ or } O_2) \rightarrow HO_2 (+ N_2 \text{ or } O_2) \tag{22}$$

$$HCO + O_2 \rightarrow HO_2 + CO \tag{23}$$

The photolysis of formaldehyde in air can be a major source of the HO_2 radical.

If one couples the formaldehyde-absorption data,[16] the primary quantum-yield estimates for processes I and II (Figures 5-9 and 5-10), and the actinic-flux data for various solar zenith angles,[56] the apparent first-order rate constants J_I and J_{II} for the occurrence of processes I and II, respectively, in air can be calculated. These and the total decay constant for formaldehyde photodecomposition ($J_I + J_{II}$) are shown in Figure 5-11; here, the rate of process I (or II)

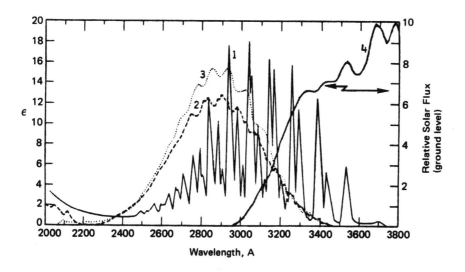

FIGURE 5-8 Absorption spectra for (1) formaldehyde, 75°C; (2) acetaldehyde, 25°C; (3) propionaldehyde, 25°C (reprinted with permission from Calvert and Pitts[40]). Curve 4, actinic flux received at ground level for typical atmospheric conditions during the day (reprinted with permission from Demerjian et al.[56]). $\varepsilon = \log(I_o/I)/[\text{aldehyde}]\ell$, L/mol-cm.

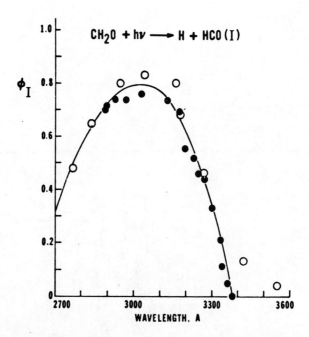

FIGURE 5-9 Wavelength dependence of primary quantum yield of process I in formaldehyde photolysis. Closed circles, data of Horowitz and Calvert.[87,88] Open circles, data of Moortgat and Warneck.[129]

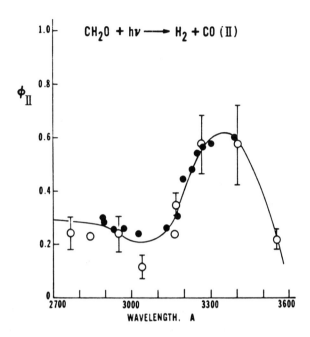

FIGURE 5-10 Wavelength dependence of primary quantum yield of process II in formaldehyde photolysis. Closed circles, data of Horowitz and Calvert.[87,88] Open circles, data of Moortgat and Warneck.[129]

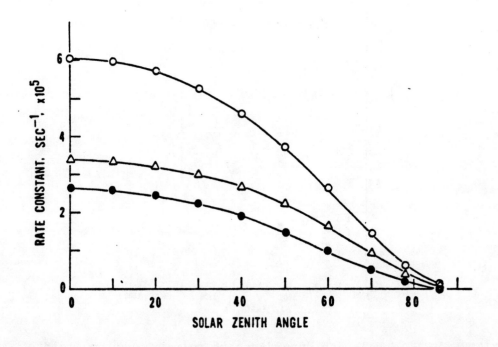

FIGURE 5-11 Theoretical first-order decay constants for photo-decomposition of formaldehyde by primary processes I and II in lower troposphere as function of solar zenith angle; J_I, solid circles; J_{II}, triangles; $J_I + J_{II}$, open circles. Reprinted from Calvert.[37]

is given by R_I (or R_{II}) = J_I (or J_{II})[HCHO]. With a solar zenith angle (angle between the sun and the vertical line perpendicular to the earth's surface at the point of observation) of 0, 20, or 40°, the half-life of formaldehyde decay by photodecomposition in the atmosphere near sea level is expected to be 3.2, 3.4, or 4.2 h, respectively. The rates of HO_2-radical generation through the occurrence of process I (R_{HO_2} = $2J_I$[HCHO]) can be reasonably large and may influence the timing of the chemistry that controls ozone formation.

The nature of the photochemical decay paths and their quantum efficiencies in air are less well established for the higher aliphatic aldehydes, acrolein, and the aromatic aldehydes. However, present evidence shows that both free-radical and intramolecular primary processes occur; the chemical nature of these processes for the first few members of the aliphatic aldehyde series are as follows:[40]

$$CH_3CHO + h\nu \rightarrow (CH_3CHO)^* \rightarrow CH_3 + HCO \quad (III)$$
$$\xrightarrow{(O_2)} CH_4 + CO \quad (IV)$$
$$\rightarrow \text{products?}$$

$$CH_3CH_2CHO + h\nu \rightarrow (CH_3CH_2CHO)^* \rightarrow C_2H_5 + HCO \quad (V)$$
$$\xrightarrow{(O_2)} C_2H_6 + CO \quad (VI)$$
$$\rightarrow \text{products?}$$

$$CH_3CH_2CH_2CHO + h\nu \rightarrow (CH_3CH_2CH_2CHO)^* \rightarrow n\text{-}C_3H_7 + HCO \quad (VII)$$
$$\rightarrow C_3H_8 + CO \quad (IX)$$
$$\xrightarrow{(O_2)} C_2H_4 + CH_2=CHOH \quad (X)$$
$$\downarrow$$
$$CH_3CHO$$
$$\rightarrow \text{products?}$$

Although many studies related to these processes have been made, the quantum efficiency of each for molecules in air at 1 atm remains unclear. Thus, Demerjian et al.[56] have reviewed the present information on acetaldehyde primary quantum yields, and they could suggest only a large range of values that may apply for processes III and IV. From the data of Table 5-21, it can be seen that the

TABLE 5-21

Estimated First-Order Rate Constants for Photodecomposition of Acetaldehyde as Function of Solar Zenith Angle (χ) in Lower Atmosphere[a]

χ, deg.	Rate Constant, s^{-1}			
	Process III		Process IV	
	Upper Limit	Lower Limit	Upper Limit	Lower Limit
0	3.75×10^{-5}	7.22×10^{-6}	1.48×10^{-6}	3.42×10^{-7}
10	3.69×10^{-5}	7.10×10^{-6}	1.45×10^{-6}	3.34×10^{-7}
20	3.51×10^{-5}	6.65×10^{-6}	1.32×10^{-6}	3.04×10^{-7}
30	3.19×10^{-5}	5.90×10^{-6}	1.13×10^{-6}	2.58×10^{-7}
40	2.75×10^{-5}	4.88×10^{-6}	0.87×10^{-6}	1.98×10^{-7}
50	2.17×10^{-5}	3.64×10^{-6}	0.59×10^{-6}	1.32×10^{-7}
60	1.48×10^{-5}	2.28×10^{-6}	0.31×10^{-6}	0.70×10^{-7}
70	0.76×10^{-5}	1.02×10^{-6}	0.11×10^{-6}	0.24×10^{-7}
78	0.29×10^{-5}	0.33×10^{-6}	0.03×10^{-6}	0.06×10^{-7}
86	0.05×10^{-5}	0.05×10^{-6}	0.003×10^{-6}	0.006×10^{-7}

[a] Reprinted with permission from Demerjian et al.[56]

theoretical half-life of acetaldehyde from photodecomposition in the lower atmosphere ($\chi = 0$) is 4.9-25 h.

Estimates of the range of photodecomposition rate constants for propionaldehyde and butyraldehyde decay in the lower atmosphere ($\chi = 40°$) have been made by Demerjian et al.[55] with the older estimates of actinic irradiance given by Leighton;[112] these are summarized in Table 5-22. Theoretical photodecomposition half-lives of these aldehydes in air ($\chi = 40°$) are in the range of those estimated for the other simple aldehydes (4-9 h).

All the photodecomposition data on the simple aldehydes suggest that the photodecomposition reactions are major loss reactions and that these decay paths can be an important source of free radicals in the atmosphere. The occurrence of processes III, V, and VII in the lower atmosphere will always be followed by the formation of an alkylperoxy radical (CH_3O_2, $C_2H_5O_2$, or $\underline{n}$-$C_3H_7O_2$) and a hydroperoxy radical (HO_2):

$$CH_3 + O_2 \rightarrow CH_3O_2 \tag{5}$$

$$C_2H_5 + O_2 \rightarrow C_2H_5O_2 \tag{24}$$

$$\underline{n}\text{-}C_3H_7 + O_2 \rightarrow \underline{n}\text{-}C_3H_7O_2 \tag{25}$$

$$HCO + O_2 \rightarrow HO_2 + CO \tag{23}$$

These radicals act to initiate the chain oxidation of NO to NO_2 and in turn can influence the concentration of ozone reached in the polluted atmosphere.

REACTIONS OF THE ALDEHYDES WITH REACTIVE INTERMEDIATES IN THE ATMOSPHERE

Several of the reactive species that are present in a sunlight-irradiated, NO_x- and hydrocarbon-polluted atmosphere react measurably with the aldehydes. These include HO, $O(^3P)$, HO_2, NO_3, and O_3. Bimolecular rate constants for these reactions with some of the aldehydes have been determined and are summarized in Table 5-23. Typical concentrations of the reactive intermediates in highly polluted air, as estimated theoretically by computer simulation (J.G. Calvert and W.R. Stockwell, personal communication), and the approximate relative rates of attack of these species on formaldehyde are summarized in Table 5-24.

The transient species whose rates of reaction with formaldehyde appear to be of particular importance are those for the HO and HO_2 radicals. The reaction with NO_3 may contribute a small amount, and it may be the dominant loss reaction for nighttime conditions for which the NO_3 concentration may remain high as the NO_2-O_3 reaction continues to generate this species. In the case of the HO and NO_3 radicals, the reactions are those of H-atom abstraction from

TABLE 5-22

Theoretical Estimates of First-Order Decay Constants for Propionaldehyde and $\underline{n}$-Butyraldehyde in Lower Atmosphere (X = 40°)[a]

Process			Rate Constant, s^{-1}
$C_2H_5CHO + h\nu$	$\longrightarrow C_2H_5 + HCO$	(V)	$(4.2-2.0) \times 10^{-5}$
	$\longrightarrow C_2H_6 + CO$	(VI)	1.0×10^{-6}
$\underline{n}\text{-}C_3H_7CHO + h\nu$	$\longrightarrow \underline{n}\text{-}C_3H_7 + HCO$	(VII)	$(3.2-2.2) \times 10^{-5}$
	$\longrightarrow C_3H_8 + CO$	(IX)	1.0×10^{-6}
	$\longrightarrow C_2H_4 + CH_3CHO$	(X)	1.0×10^{-5}

[a]Reprinted with permission from Demerjian et al.[55]

TABLE 5-23

Bimolecular Rate Constants for Reactions of Various Reactive Atmospheric Species with Aldehydes

Reactive Species	Aldehyde	Rate Constant at 25°C, cc·molec^{-1}s^{-1}	Reference
HO	HCHO	$(1.4 \pm 0.35) \times 10^{-11}$	130
		$(1.5 \pm 0.1) \times 10^{-11}$	142
		$(0.65 \pm 0.15) \times 10^{-11}$	165
		$(0.94 \pm 0.10) \times 10^{-11}$	11
		$(0.99 \pm 0.11) \times 10^{-11}$	169
	CH_3CHO	$(1.5 \pm 0.38) \times 10^{-11}$	132
		$(1.60 \pm 0.16) \times 10^{-11}$	11
		$(1.6 \pm 0.2) \times 10^{-11}$	142
		$> 2.0 \times 10^{-11}$	48
	C_2H_5CHO	$(2.1 \pm 0.1) \times 10^{-11}$	142
	C_6H_6CHO	$(1.3 \pm 0.1) \times 10^{-11}$	142
$O(^3P)$	HCHO	$(1.5 \pm 0.5) \times 10^{-13}$	84
		$(1.5 \pm 0.2) \times 10^{-13}$	121
		1.64×10^{-13}	138
		$(1.50 \pm 0.10) \times 10^{-13}$	106
		$(1.61 \pm 0.17) \times 10^{-13}$	105
		$(1.9 \pm 0.4) \times 10^{-13}$	44
	CH_3CHO	4.3×10^{-13}	162
		4.8×10^{-13}	122
		4.5×10^{-13}	31
		5.0×10^{-13}	50
	C_2H_5CHO	7.0×10^{-13}	162
		2.3×10^{-13}	30
	$\underline{n}$-C_3H_7CHO	9.5×10^{-13}	162
		2.5×10^{-13}	92
	$\underline{iso}$-C_3H_7CHO	1.2×10^{-12}	162
	$CH_2=CHCHO$	2.7×10^{-13}	30
		4.9×10^{-13}	65
	$CH_3CH=CHCHO$	0.83×10^{-12}	30
		1.09×10^{-12}	65
NO_3	CH_3CHO	1.2×10^{-15}	131
HO_2	HCHO	1.0×10^{-14}	172
O_3	HCHO	$< 2.1 \times 10^{-24}$	24

TABLE 5-24

Typical Theoretical Concentrations of Reactive Intermediates
in Sunlight-Irradiated, NO_x- and RH-Polluted Atmosphere, Approximate
Rate Constants, and Relative Rate of Attack of These Species
on Formaldehyde (25°C, 1 atm)

Species	Typical Concentrations molec/cc	ppm	Approximate Rate Constant, $cc \cdot molec^{-1} s^{-1}$	Relative Rate Reaction with HCHO
HO	7.4×10^6	3×10^{-7}	1.1×10^{-11}	1.00
HO_2	4.9×10^9	2×10^{-4}	1.0×10^{-14}	0.61[a]
NO_3	2.5×10^9	1×10^{-4}	1.2×10^{-15}[b]	0.036
$O(^3P)$	1.7×10^5	7×10^{-9}	1.6×10^{-13}	0.00034
O_3	4.9×10^{12}	2×10^{-1}	$<2.1 \times 10^{-24}$	1.3×10^{-7}

[a] Rate of addition; net rate is lower as result of reverse reaction.

[b] Taken as equal to that for $NO_3 + CH_3CHO$ measured by Morris and Niki.[131]

formaldehyde; the CHO radical formed here will react primarily to form HO_2 and carbon monoxide:

$$HO + HCHO \rightarrow H_2O + HCO \qquad (26)$$

$$NO_3 + HCHO \rightarrow HONO_2 + HCO \qquad (27)$$

$$HCO + O_2 \rightarrow HO_2 + CO \qquad (23)$$

For the HO_2-radical reaction, recent studies show that the addition of the radical to formaldehyde, rather than H-atom abstraction, is the major step:[171][172]

$$HO_2 + HCHO \rightarrow (HO_2CH_2O) \rightarrow O_2CH_2OH \qquad (28)$$

However, the reverse of this reaction does occur with $k_{29} \simeq 1.5$ s^{-1} (25°C), and the removal of formaldehyde does not result with each occurrence of Reaction 28:

$$O_2CH_2OH \rightarrow (HO_2CH_2O) \rightarrow HO_2 + HCHO \qquad (29)$$

In laboratory studies, the O_2CH_2OH radical has been shown to react either by dissociation (Reaction 29), by disproportionation with HO_2 radicals (Reaction 30), or by disproportionation with other O_2CH_2OH radicals (Reaction 31):

$$HO_2 + O_2CH_2OH \rightarrow HO_2CH_2OH + O_2 \qquad (30)$$

$$2O_2CH_2OH \rightarrow 2OCH_2OH + O_2 \qquad (31)$$

The unusual, newly identified compound, HO_2CH_2OH, forms formic acid in laboratory experiments through the overall reaction:

$$HO_2CH_2OH \rightarrow HCO_2H + H_2O \qquad (32)$$

The OCH_2OH radical product of Reaction 31 reacts rapidly to form HCO_2H:

$$OCH_2OH + O_2 \rightarrow HCO_2H + HO_2 \qquad (33)$$

It has been estimated that the rate of HO_2CH_2OH generation in a typical, highly polluted atmosphere in which [HCHO] $\simeq 0.02$ ppm and $[HO_2] \simeq 2 \times 10^{-4}$ ppm will be about 0.4 ppt/min. Conceivably, these or related reactions account for a portion of the HCO_2H that is generated in highly polluted atmospheres.

It is instructive to compare the relative rates of removal of formaldehyde by the various chemical and photochemical pathways that have been described. In making the estimates in Table 5-25, the theoretical concentrations of the reactive species shown in Table 5-24 were used. It is seen that the attack on formaldehyde by the HO radical and the photodecomposition of formaldehyde are the two

TABLE 5-25

Theoretical Relative Rates of Major Chemical and Photochemical HCHO Removal Reactions for Highly Polluted, Sunlight-Irradiated ($\chi = 0$) Lower Atmosphere

Reaction		Relative Rate (approximate)
$HCHO + h\nu \longrightarrow H + HCO$	(I)	0.24 ⎱ 0.57
$\longrightarrow H_2 + CO$	(II)	0.33 ⎰
$HCHO + HO \longrightarrow HCO + H_2O$	(26)	1.00
$HCHO + HO_2 \rightleftharpoons O_2CH_2OH$	(28,29) ⎱	>0.0082
$HO_2 + O_2CH_2OH \longrightarrow HO_2CH_2OH + O_2$	(30) ⎰	
$NO_3 + HCHO \longrightarrow HONO_2 + HCO$	(27)	0.037

dominant homogeneous pathways for formaldehyde removal in the polluted atmosphere. The relative rate of removal as a result of the reversible HO_2-radical addition reaction and the later reaction of the O_2CH_2OH radical is shown as a lower limit, because other radical reactions of this species (possible with CH_3O_2, RO_2, etc.) will probably act as a permanent sink as well and also compete with the dissociation reaction (Reaction 29).

If the processes considered here alone describe the removal of formaldehyde in the lower atmosphere, then the half-life of formaldehyde for these conditions, typical of the highly polluted atmosphere, would be somewhat less than 2.6 h. Ill-defined heterogeneous reaction pathways involving rainout of formaldehyde and removal by surface water, rock, and soil must also occur and shorten the lifetime of formaldehyde. Thus, O.C. Zafiriou and A.M. Thompson (personal communication, 1979) estimated that in the vicinity of Woods Hole, Massachusetts, formaldehyde enters the ocean from the atmosphere at the rate of 6 μg/cm^2 per year. The flux of gaseous formaldehyde into the sea at a remote, marine site in the equatorial Pacific was measured by Zafiriou et al.[197] at 5 μg/cm^2 per year; for these same conditions, the rainout and washout of formaldehyde amounted to about 1 μg/cm^2 per year. These various processes restrict the formaldehyde buildup in the atmosphere.

The atmospheric transport of the aldehydes over long distances is probably not very important, because of their short lifetimes. It is probably less important as a source of aldehydes in remote areas than the local generation from transported, longer-lived precursors, such as the less reactive hydrocarbons. The lifetime of formaldehyde in aqueous media may be somewhat greater, because the hydrated form of formaldehyde ($HOCH_2OH$) dominates in these conditions, and it does not absorb sunlight appreciably. In this case, microorganisms in the water appear to play an important role in the degradation process, which may take 30-72 h under natural conditions commonly encountered.

All available evidence at hand suggests that the removal paths for acetaldehyde, propionaldehyde, etc., are very similar to those outlined for formaldehyde. The accuracy of the data on these compounds does not warrant a detailed analysis now.

The commonly observed unsaturated aldehyde, acrolein, is comparatively stable toward photodecomposition.[145] In view of this, it has been suggested that there may be a higher persistence for acrolein than the other aldehydes in photochemical smog--a conclusion of special interest, in light of the high degree of eye irritation attributed to acrolein. The HO attack on acrolein is expected in theory to be the dominant removal mechanism, although estimates of the rate constant for this reaction have been made only by theoretical methods. Acrolein and crotonaldehyde appear to be as reactive as the aliphatic aldehydes in photooxidation in NO_x-containing mixtures, and it is likely that their lifetimes in the atmosphere are determined largely by the rate of HO-radical attack.[4,57]

The photochemistry of benzaldehyde and the higher homologues of the aromatic aldehydes is marked by the relatively high photochemical stability of the excited states toward decomposition. In solution-phase studies, photoreduction and electronic energy-transfer processes are commonly observed with these compounds.[40] Benzaldehyde, 2-methylbenzaldehyde, and 3-methylbenzaldehyde show very low reactivity when photooxidized in dilute $NO-NO_2$ mixtures in air in smog-chamber experiments.[57] In contrast, 1-methylbenzaldehyde shows a high reactivity characteristic of the aliphatic aldehydes. Present data do not allow quantitative estimates of the half-lives of the aromatic aldehydes toward photodecomposition or other possible light-induced reactions, but they appear to be somewhat longer than those observed for the aliphatic aldehydes in most cases.

REMOVAL PROCESSES IN AQUEOUS SYSTEMS

Very little information is available on the factors that affect the stability of aldehydes in aqueous systems. This section addresses reactions that could occur in the aquatic environment with the carbonyl group. It should be noted that some aldehydes may have other functional groups that contribute to or dominate their chemistry in aqueous systems.

A reaction that many aldehydes undergo in water is hydration at the carbonyl group(s) to produce gem-diols (gem = geminal, with both hydroxyl groups on the same carbon atom):

$$RCHO + H_2O \rightarrow RCH(OH)_2$$

The extent of hydration depends on the nature of the R group (or substituent); electron-withdrawing substituents favor a greater degree of hydration.[17] The degrees of hydration at equilibrium, calculated from the hydration-rate data of Bell and McDougall[18] and Smith[164] for formaldehyde, chloral, acrolein, and acetaldehyde are 99.9, 99.8, 95.0, and 60.0%, respectively, at 25°C.

At a given temperature, the ratio of the nonhydrated to the hydrated form of an aldehyde in water is constant. Determining chemical and biologic transformation and transport processes of aldehydes can be difficult, because the hydration equilibrium will shift to replenish the form removed by these processes. Because the hydration reaction is associated with a complex kinetic expression that entails both kinetic and equilibrium processes, it is difficult to estimate the simple half-life of an aldehyde in water.

Biotransformation is perhaps the most important process that will remove aldehydes from water. It has been shown that both aliphatic and aromatic aldehydes are biotransformed in the aquatic environment. In their review of formaldehyde as an environmental contaminant, Kitchens and co-workers[104] reported evidence that some bacteria in sewage sludge can use formaldehyde as a sole carbon source and that complete degradation can be achieved in 48-72 h if the temperatures and nutrient conditions are maintained. They also cited a study showing that microorganisms in stagnant lake water could completely degrade formaldehyde in 30 h at 20°C under aerobic conditions and in

48 h under anaerobic conditions. That study also showed no detectable loss of formaldehyde when incubated in sterilized lake water for 48 h.

Bowmer and Higgins[23] reported that acrolein introduced into water samples from an agricultural area had a half-life of 29 h. When they reduced microbiologic activity by adding thymol to the water, the half-life increased to 43 h. Acrolein has been reported to be effectively biotransformed in activated sewage sludge and in the biotreatment systems that process the water used by refineries.[34]

Keith[99] compared the concentration of several organic compounds (including four aldehydes) in an effluent from a kraft paper mill before and after effluent treatment by biodegradation. The treatment process completely removed benzaldehyde from the effluent and removed 73, 54, and 43% of the vanillin, salicylaldehyde, and syringaldehyde, respectively.

Although there is ample evidence that aldehydes are oxidizable, oxidation in the aquatic environment by the alkylperoxyl radical ($RO_2\cdot$) is very slow. The rate constant for this radical in abstracting the H atom from the acyl carbon is $0.1\ M^{-1}\ s^{-1}$, and Mill[127] estimated the RO_2 concentration in the aquatic environment to be 10^{-9} M. Assuming that RO_2 addition to the aldehydes is not important in the liquid phase (although it is important in the gas phase), these values indicate that the half-life of aldehydes through oxidation by the RO_2 radical will probably be several years. However, aldehyde reactions with hydroxyl and alkoxyl radicals and other oxidizing agents are much faster, and these species may account for additional pathways that should be included. No information is available to indicate that oxidation of the diol form of the aldehydes would occur more rapidly or to suggest what other chemical oxidation processes might affect the persistence of the aldehydes in the natural waters.

The effect of light on aldehydes in aqueous systems is unknown. It is likely that aldehydes undergo photolysis in water, but probably at a lower rate than in the atmosphere, because light is scattered and diffracted in water, and color and turbidity limit the intensity and depth of penetration. Hydration of an aldehyde should substantially retard its photolysis, because hydration will completely destroy the carbonyl chromophore responsible for light absorption and the potential for photodecomposition.

SOME IMPORTANT SECONDARY EFFECTS OF ALDEHYDES IN THE CHEMISTRY OF THE POLLUTED ATMOSPHERE

INFLUENCE OF ALDEHYDES IN PHOTOCHEMICAL SMOG FORMATION

Bufalini and Brubaker[28] showed many years ago that the irradiation of the simplest aldehyde, formaldehyde, in dilute $NO-NO_2$-air mixtures could induce the NO-to-NO_2 conversion and ozone formation characteristic of photochemical smog. Altshuller et al.[5] found that the ultraviolet-irradiated aliphatic aldehydes in the parts-per-million range in NO- and NO_2-free, dilute mixtures of the olefinic and aromatic hydrocarbons in air induced the photooxidation of the hydrocarbons. They expressed concern that these results could modify current considerations of whether control of the nitrogen oxides would effectively reduce photochemical air pollution.

Using dilute NO-NO$_2$-aldehyde and/or -hydrocarbon mixtures in air, Dimitriades and Wesson[57] studied the smog-forming reactivities of several aldehydes--formaldehyde, acetaldehyde, propionaldehyde, n-butyraldehyde, acrolein, crotonaldehyde, benzaldehyde, o-tolualdehyde, m-tolualdehyde, and p-tolualdehyde. Several criteria were used to establish the reactivity of the aldehyde or olefin used: rate of NO$_2$ formation; maximal concentrations of ozone, peroxyacetylnitrate, peroxybenzoylnitrate, and formaldehyde; and the time-weighted exposures (ppm x min) for these four products. These workers concluded that the aldehydes present in auto exhaust as a group should be classified among the reactive exhaust components. The specific reactivity (reactivity per part per million) of formaldehyde, as measured by the rate of NO-to-NO$_2$ conversion, was comparable with that of the average exhaust alkene. However, with respect to oxidant yield, the specific reactivity of formaldehyde was considerably lower than that of the average exhaust hydrocarbon. The specific reactivity of the higher aldehydes was in every respect comparable with that of the average exhaust alkene. When tested individually, benzaldehyde and m- and p-tolualdehyde were unreactive, and o-tolualdehyde was reactive. In mixtures, benzaldehyde and presumably all the aromatic aldehydes manifested reactivity as precursors of the strong eye irritants, the peroxybenzoylnitrates.

Dimitriades and Wesson[57] observed another important effect of formaldehyde: mixtures containing formaldehyde appeared to have higher oxidant-yield reactivity than expected from the sum of the individual effects observed from the specific reactivity and compositional data alone; the difference increased with increasing formaldehyde content.

Kopczynski et al.[108] found that the photooxidation of dilute mixtures of formaldehyde, acetaldehyde, and propionaldehyde in the presence of nitrogen oxides produces the same products and biologic effects (eye irritation and plant damage) as does the hydrocarbon photooxidation. Propionaldehyde was found to be the most reactive, with respect to highest product yields, eye irritation, and plant damage. These workers concluded that, inasmuch as aldehydes are both primary (directly emitted) and secondary (photochemically formed) products, their substantial reactivities are of special importance. They may be expected to contribute to photochemical air pollution problems, not only in the central city, but in the urban, suburban, and rural areas downwind.

Computer modeling of the complex chemical changes expected to occur in simulated, sunlight-irradiated, NO-, NO$_2$-, hydrocarbon-, and aldehyde-polluted atmospheres has confirmed the observed aldehyde effects and pointed to the specific chemistry responsible for these effects.[14,38,39,42,55,58,74,103,120,139] It has shown that the aldehydes (formaldehyde and acetaldehyde) present initially in polluted air will decrease the induction period observed for ozone, peroxyacetylnitrate, and other products formed and their final concentrations, which increase in simulated smog mixtures.

Jeffries and Kamens[94] have demonstrated this aldehyde effect in experiments in a large outdoor smog chamber (Figure 5-12). In matched experiments carried out in sunlight simultaneously in two equivalent, isolated portions of the chamber, nearly equivalent amounts of a typical pollutant composition, hydrocarbon mixture (urban mix) and nitrogen oxides, were injected. In only one side, additional acetaldehyde was added initially (about 10% of the nonmethane hydrocarbon). The photochemical reactions forming ozone proceeded faster and significantly higher ozone concentrations developed in the experiment with additional added acetaldehyde.

Pitts et al.[149] have observed a similar effect in smog-chamber photooxidation experiments with a surrogate mixture of hydrocarbons with and without added formaldehyde (Figure 5-13). The initial rate of ozone formation and the final ozone concentration reached during the experiment both were increased greatly by the addition of small amounts of formaldehyde.

The reactions that determine the influence of the aldehydes in these simulated smog mixtures are largely those already described: radical formation through photodecomposition of the aldehydes and the reactions of the HO radical with the aldehydes. The ozone concentration developed in the NO_x- and RH-polluted, sunlight-irradiated atmosphere is related to the NO_2-to-NO ratio, as a result of the following rapid reactions involving NO, NO_2, and ozone:

$$NO_2 + h\nu \rightarrow NO + O \quad (34)$$

$$O + O_2 \; (+N_2, O_2) \rightarrow O_3 \; (+N_2, O_2) \quad (35)$$

$$O_3 + NO \rightarrow O_2 + NO_2 \quad (36)$$

For the usual conditions in these highly polluted atmosphere, one expects Equation 37 to hold approximately:[35,36,112]

$$[O_3] \simeq ([NO_2]/[NO])(k_{34}/k_{36}) \quad (37)$$

The presence of the aldehydes can provide an additional source of the hydroperoxy (HO_2) and alkylperoxy radicals (CH_3O_2, $C_2H_5O_2$, RO_2, etc.), which may pump NO to NO_2 and hence increase the ozone concentration through its close relation to the $[NO_2]/[NO]$ ratio:

$$RCHO + h\nu \rightarrow R + HCO \quad (38)$$

$$R + O_2 \rightarrow RO_2 \quad (39)$$

$$HCO + O_2 \rightarrow HO_2 + CO \quad (23)$$

$$RO_2 + NO \rightarrow RO + NO_2 \quad (40)$$

$$HO_2 + NO \rightarrow HO + NO_2 \quad (41)$$

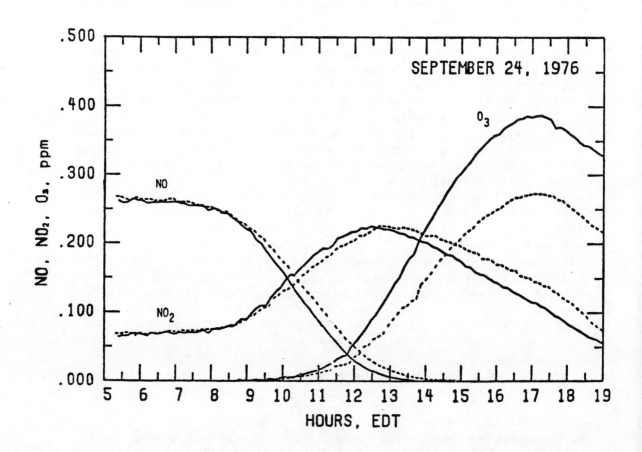

FIGURE 5-12 Comparison of nitric oxide, nitrogen dioxide, and ozone profiles in urban mix with (solid lines) and without (dashed lines) additional added acetaldehyde (10% of initial nonmethane hydrocarbon concentration). Initial conditions: with added acetaldehyde, NO_x, 0.325 ppm, urban mix, 2.32 ppmC; without added acetaldehyde, NO_x, 0.332 ppm, urban mix, 2.45 ppmC. Reprinted with permission from Jeffries and Kamens.[94]

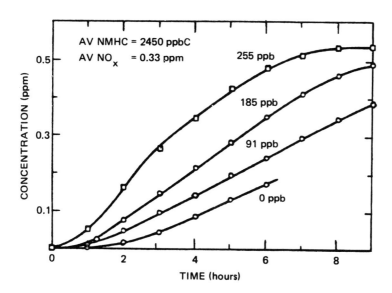

FIGURE 5-13 Effect of added formaldehyde on ozone formation in irradiation of surrogate mixtures of hydrocarbons and nitrogen oxides. Reprinted with permission from Pitts et al.[149]

The alkoxy (RO) and hydroxyl radicals formed in Reactions 40 and 41 can regenerate HO_2 and RO_2 radicals in further reactions with the impurity aldehydes, as well as the hydrocarbons present:

$$HO + HCHO \rightarrow H_2O + HCO \tag{26}$$

$$HCO + O_2 \rightarrow HO_2 + CO \tag{23}$$

$$HO + RCHO \rightarrow RCO + H_2O \tag{42}$$

$$RCO + O_2 \rightarrow RCOO_2 \tag{43}$$

$$RCOO_2 + NO \rightarrow RCO_2 + NO_2 \tag{44}$$

$$RCO_2 \rightarrow R + CO_2 \tag{45}$$

$$R + O_2 \rightarrow RO_2 \tag{39}$$

Obviously, the aldehydes provide a new route for a chain reaction driving NO to NO_2 in these systems, and hence they can influence the generation of ozone in photochemical smog. It is clear that the control of aldehyde emission, as well as hydrocarbon emission, is important in the strategy for ozone control.

THE INFLUENCE OF ALDEHYDES ON THE FORMATION OF PEROXYACYLNITRATES IN THE POLLUTED ATMOSPHERE

The aliphatic and aromatic aldehydes are important precursors of the notorious peroxyacylnitrates and peroxybenzoylnitrates. For example, all the experimental evidence and theoretical considerations support the view that acetaldehyde is a direct precursor of peroxyacetylnitrate (PAN) in the real atmosphere:[55]

$$CH_3CHO + HO \rightarrow CH_3CO + H_2O \tag{46}$$

$$CH_3CO + O_2 \rightarrow CH_3COO_2 \tag{47}$$

$$CH_3COO_2 + NO_2 \rightarrow CH_3COO_2NO_2 \text{ (PAN)} \tag{48}$$

$$CH_3COO_2 + NO \rightarrow CH_3CO_2 + NO_2 \tag{49}$$

Similar reactions presumably lead to the formation of the higher homologues of PAN in the case of the higher aliphatic aldehydes (propionaldehyde, etc.)[68] and the aromatic aldehydes (benzaldehyde, etc.).[85] (Gay et al.[68] cited references to the earlier studies of the peroxyacylnitrates.) The simplest member of the aldehyde family, formaldehyde, does not form the analogous peroxyformylnitrate, $HCOO_2NO_2$, in substantial amounts; presumably, this is a consequence of the unique disproportionation of the HCO radical with

O_2, which dominates the association reaction (Reaction 50) in this case:

$$HCO + O_2 \rightarrow HO_2 + CO \qquad (23)$$

$$HCO + O_2 \ (+N_2 \text{ or } O_2) \rightarrow HCOO_2 \ (+N_2 \text{ or } O_2) \qquad (50)$$

However, in the photooxidation of formaldehyde in NO_2-containing mixtures, the less stable peroxynitric acid, HO_2NO_2, results from the HO_2 reaction with NO_2:[78][140]

$$HO_2 + NO_2 \rightleftharpoons HO_2NO_2 \qquad (51)$$

We may conclude that the presence of the aldehydes or their precursors (hydrocarbons) in the polluted atmosphere is directly involved in the formation of the important class of highly oxidizing, eye-irritating secondary pollutants, the peroxyacylnitrates and the peroxybenzoylnitrates.

THE POTENTIAL ROLE OF FORMALDEHYDE IN THE ORIGIN OF FORMIC ACID IN THE POLLUTED ATMOSPHERE

The gas-phase photooxidation of formaldehyde at low concentrations in air has been shown to lead to the products H_2O_2, CO, CO_2, H_2, and HCO_2H.[28][89][140][147] Recently, HO_2CH_2OH has been identified as an intermediate product of this system,[172] and its reaction to form formic acid was noted. The kinetic results suggest that the reaction of HO_2 addition to formaldehyde leads to this new product. Other recent experiments with C_2H_4, O_3, and formaldehyde mixtures at parts-per-million concentrations in air showed that the reactive CH_2O_2 intermediate product from the C_2H_4-O_3 reaction may lead to formic acid as well.[170] In this case, an unidentified intermediate is formed first by CH_2O_2-formaldehyde reaction; this leads to formic acid anhydride and hence to formic acid; a possible reaction scheme consistent with the kinetics is the following:

$$C_2H_2 + O_3 \rightarrow CH_2O_2 + HCHO \qquad (52)$$

$$CH_2O_2 + HCHO \rightarrow CH_2OOCH_2O \rightarrow \underset{O}{\overset{O-O}{CH_2CH_2}} \rightarrow \qquad (53)$$

$$OCH_2OCH_2O \rightarrow HOCH_2OCHO$$

$$HOCH_2OCHO \rightarrow H_2 + H\overset{O}{\overset{\|}{C}}O\overset{O}{\overset{\|}{C}}H \qquad (54)$$

$$\overset{O}{\underset{\|}{H C}}\overset{O}{\underset{\|}{O C H}} + H_2O(\text{aerosol}) \rightarrow 2HCO_2H \tag{55}$$

These newly discovered reaction routes to formic acid may explain, at least in part, the large amount of formic acid identified in aged, highly polluted atmospheres.[178] Conceivably, the apparent correlation of formaldehyde content of smog mixtures with eye irritation[3,85,195] is related in part to the HCO_2H formation, which would follow roughly the formaldehyde concentration in these systems through the reactions outlined.

THE POTENTIAL ROLE OF FORMALDEHYDE IN THE GENERATION OF BIS(CHLOROMETHYL) ETHER IN HYDROGEN CHLORIDE-FORMALDEHYDE-POLLUTED ATMOSPHERES

It has been observed that bis(chloromethyl) ether (BCME) is formed from moist air containing formaldehyde and hydrogen chloride gases.[63,98,157] The overall reaction is:

$$2HCl + 2HCHO \rightleftharpoons ClCH_2OCH_2Cl + H_2O \tag{56}$$

Studies by Drew et al.,[59] Laskin et al.[110] and Kuschner et al.[109] have shown that the chloromethyl ethers are respiratory tract carcinogens, and epidemiologic studies have indicated that they are human carcinogens.[1] In a recent study,[157] it was demonstrated that Reaction 56 occurred under dynamic conditions at room temperature with relatively high formaldehyde and HCl concentrations in the gas phase--about 1,000 and 6,500 ppm, respectively. Chronic exposure of rats to dilute HCl-formaldehyde-chloromethyl ether mixtures-- bis(chloromethyl) ether at about 2.8 ppb, HCl at 10.7 ppm, and formaldehyde at 14.6 ppm--caused a markedly increased incidence of squamous metaplasia of the nasal cavity and squamous cell carcinoma of the nasal epithelium after 136-390 d.

From the very limited data at hand, it is impossible to extrapolate with great confidence to the lower, more representative concentrations of bis(chloromethyl) ether that would be formed with the HCl and formaldehyde concentrations commonly encountered in the atmosphere. However, we can derive present "best" estimates from both experimental and theoretical data on the $HCHO-HCl-ClCH_2OCH_2Cl$ system. The reaction kinetics of the chloromethyl ether formation has not been determined. But, for a worst-case estimate, we may assume that an equilibrium concentration of the ether is formed in the atmosphere; this will overestimate the actual concentrations somewhat. We may make the reasonable assumption that equilibrium was achieved at the longest reaction times used in the experiments of Sellakumar et al.[157] and Frankel et al.[63] These data give: $K_{56} \approx 600 \pm 300 \text{ atm}^{-2}$, where $K_{56} = (P_{H_2O})(P_{BCME})/(P_{HCl})^2(P_{HCHO})^2$, and P_{BCME} is the pressure of the bis(chloromethyl) ether.

A rough check may be made on the reasonableness of the K_{56} estimate derived from the experiments. We may use the estimated enthalpy and entropy changes for Reaction 56, $\Delta H°_{56}$ and $\Delta S°_{56}$, to derive a theoretical estimate of K_{56}: $K_{56} = e^{\Delta S°/R_e - \Delta H°/RT}$. Benson's[20] approximate thermochemical methods may be used to derive the unknown thermodynamic quantities for the bis(chloromethyl) ether: $\Delta H°_f \simeq -58.4 \pm 3$ kcal/mol, $S° \simeq 86.4 \pm 2$ cal mol^{-1} deg^{-1} (25°C, 1 atm). Coupling these quantities with the measured experimental enthalpies of formation and absolute entropies of the other reactants and products, we calculate the theoretical range of values, which should include K_{56}: $6,580 > K_{56} > 0.036$ atm^{-2}. As one anticipates if the approach in estimating K_{56} is reasonable, the experimental value is within this range. Thus, one might use these data to obtain reasonable, order-of-magnitude results for the concentrations of bis(chloromethyl) ether in polluted atmospheres.

We have used the maximal concentrations of formaldehyde observed in the urban atmosphere (about 0.10 ppm) and a seemingly reasonable maximum for hydrogen chloride (about 10 ppb) in air at 50% relative humidity at 25°C.* Using the upper limit estimated for K_{56}, 6,580 atm^{-2}, we estimate the maximal equilibrium concentration of bis(chloromethyl) ether for these conditions at about 4×10^{-16} ppb. Thus, we may conclude tentatively that there is probably little impact on human health from the generation of bis(chloromethyl) ether from formaldehyde and HCl in the urban atmosphere.

We must be cognizant of the potential hazard under conditions more favorable to bis(chloromethyl) ether formation. Thus, in principle, this compound could be formed in HCl-rich plumes from the incineration of polyvinyl chloride or other HCl-producing processes in which formaldehyde may be present at a high concentration. The potential for bis(chloromethyl) ether generation exists if fairly concentrated HCl solutions are brought into contact with formaldehyde-containing particleboard or other formaldehyde-copolymer materials. Such polymers may contain free formaldehyde or they may hydrolyze to form formaldehyde and then interact with HCl to lead to $ClCH_2OCH_2Cl$. There is no evidence of which the Committee is aware that allows an evaluation of these potential problems.

Further direct tests for bis(chloromethyl) ether in the ambient air and water and new and more precise measurements of K_{56} and the rate-determining reactions that control its rates of formation and decay are required, in order to evaluate quantitatively the potential extent of human exposure to and the influence of bis(chloromethyl) ether.

---

*This estimate of HCl concentration is about 10 times the number estimated theoretically for the "clean" lower troposphere by the Livermore Kinetic-Transport Model, from which [HCl] = 0.9 ppb (D.J. Wuebbles, personal communication, 1979). It is also somewhat greater than the highest concentrations observed in ambient air near the ground.[61 69 96 97]

REFERENCES

1. Albert, R., B. Pasternak, R. Shore, M. Lippmann, N. Nelson, and B. Ferris. Mortality patterns among workers exposed to chloromethyl ethers--A preliminary report. Environ. Health Perspect. 2:209-214, 1975.
2. Allan, G. G., J. Dutkiewicz, and E. J. Gilmartin. Long-term stability of urea-formaldehyde foam insulation. Environ. Sci. Technol. 14:1235-1240, 1980.
3. Altshuller, A. P. Assessment of the contribution of chemical species to the eye irritation potential of photochemical smog. J. Air Pollut. Control Assoc. 28:594-598, 1978.
4. Altshuller, A. P., and I. R. Cohen. Photo-oxidation of acrolein-nitrogen oxide mixtures in air. Int. J. Air Water Pollut. 7:1043-1049, 1963.
5. Altshuller, A. P., I. R. Cohen, and T. C. Purcell. Photooxidation of hydrocarbons in the presence of aliphatic aldehydes. Science 156:937-939, 1967.
6. Altshuller, A. P., L. J. Leng, and A. F. Wartburg. Source and atmospheric analyses for formaldehyde by chromotropic acid procedures. Int. J. Air Water Pollut. 6:381-385, 1962.
7. Altshuller, A. P., and S. P. McPherson. Spectrophotometric analysis of aldehydes in the Los Angeles atmosphere. J. Air Pollut. Control Assoc. 13:109-111, 1963.
8. Andersen, I., G. R. Lundqvist, and L. Mølhave. Formaldehyde in the atmosphere in Danish homes. Ugeskr. Laeg. 136:2133-2139, 1974. (in Danish; English summary)
9. Andersen, I., G. R. Lundqvist, and L. Molhave. Indoor air pollution due to chipboard used as construction material. Atmos. Environ. 9:1121-1127, 1975.
10. Atkinson, R., K. R. Darnall, A. C. Lloyd, A. M. Winer, and J. N. Pitts, Jr. Kinetics and mechanisms of the reactions of the hydroxyl radical with organic compounds in the gas phase. Adv. Photochem. 11:375-487, 1979.
11. Atkinson, R., and J. N. Pitts, Jr. Kinetics of the reactions of the OH radical with HCHO and CH_3CHO over the temperature range 299-426°K. J. Chem. Phys. 68:3581-3584, 1978.
12. Bachman, K. C., and E. L. Kayle. Reactive exhaust emissions from a stratified-charge engine vehicle. Presented at American Chemical Society Meeting, Philadelphia, Pa., April 6-11, 1975. Am. Chem. Soc. Div. Pet. Chem. Preprints 20(3):652-658, 1975.
13. Baker, R. E., J. A. Harrington, J. H. Baudino, L. C. Copeland, W. J. Koehl, L. J. McCabe, and W. T. Wotring. Utilization of methanol as an automotive fuel. A report from IIEC-2, the Inter-Industry Emission Control Program, pp. 2.7.1.-2.7.8. In Proceedings of the International Symposium on Alcohol Fuel Technology: Methanol and Ethanol, Wolfsburg, F.R. Germany, November 21-23, 1977. Washington, D.C.: U.S. Department of Energy, 1978.

14. Baldwin, A. C., J. R. Barker, D. M. Golden, and D. G. Hendry. Photochemical smog. Rate parameter estimates and computer simulations. J. Phys. Chem. 81:2483-2492, 1977.
15. Barbe, A., P. Marché, C. Secroun, and P. Jouve. Measurements of tropospheric and stratospheric H_2CO by an infrared high resolution technique. Geophys. Res. Lett. 6:463-465, 1979.
16. Bass, A. M., L. C. Glasgow, C. Miller, J. P. Jesson, and D. L. Filkin. Temperature dependent absorption cross-sections for formaldehyde (CH_2O): The effect of formaldehyde on stratospheric chlorine chemistry, pp. 467-477. In Proceedings of the NATO Advanced Study Institute on Atmospheric Ozone. Its Variation and Human Influences. Washington, D.C.: U.S. Department of Transportation, Federal Aviation Administration, 1980.
17. Bell, R. P. The reversible hydration of carbonyl compounds. Adv. Phys. Org. Chem. 4:1-29, 1966.
18. Bell, R. P., and A. O. McDougall. Hydration equilibria of some aldehydes and ketones. Trans. Faraday Soc. 56:1281-1285, 1960.
19. Bengtson, G. W., and G. K. Voigt. A greenhouse study of relations between nutrient movement and conversion in a sandy soil and the nutrition of slash pine seedlings. Soil Sci. Soc. Am. Proc. 26:609-612, 1962.
20. Benson, S. W. Thermochemical Kinetics. Methods for the Estimation of Thermochemical Data and Rate Parameters. 2nd ed. New York: John Wiley & Sons, 1976. 320 pp.
21. Berger, V., and M. Tomas. Urea-formaldehyde resins with a low content of free formaldehyde. Drevo 19:211-214, 1964. (in Czech)
22. Blejer, H. P., and B. H. Miller. Occupational Health Report of Formaldehyde Concentrations and Effects on Workers at the Bayly Manufacturing Company, Visalia. Study Report S-1806. Los Angeles: State of California Health and Welfare Agency, Dept. of Public Health, Bureau of Occupational Health, 1966. 6 pp.
23. Bowmer, K. H., and M. L. Higgins. Some aspects of the persistence and fate of acrolein herbicide in water. Arch. Environ. Contam. Toxicol. 5:87-96, 1976.
24. Braslavsky, S., and J. Heicklen. The gas-phase reaction of O_3 with H_2CO. Int. J. Chem. Kinet. 8:801-808, 1976.
25. Breeding, R. J., J. P. Lodge, Jr., J. B. Pate, D. C. Sheesley, H. B. Klonis, B. Fogle, J. A. Anderson, T. R. Englert, P. L. Haagenson, R. B. McBeth, A. L. Morris, R. Pogue, and A. F. Wartburg. Background trace gas concentration in the central United States. J. Geophys. Res. 78:7057-7064, 1973.
26. Breysse, P. A. The environmental problems of urea-formaldehyde structures. Formaldehyde exposure in mobile homes. Presented at American Medical Association Congress on Occupational Health, 26 Oct. 1979.
27. Brinkman, N. D. Effect of Compression Ratio on Exhaust Emissions and Performance of a Methanol-Fueled Single-Cylinder Engine. SAE Paper No. 770791. Warrendale, Pa.: Society of Automotive Engineers, Inc. 16 pp.

28. Bufalini, J. J., and K. L. Brubaker. The photooxidation of formaldehyde at low partial pressures, pp. 225-240. In C. S. Tuesday, Ed. Chemical Reactions in Urban Atmospheres. Symposium Held at the General Motors Research Laboratories, Warren, Michigan, 1969. New York: American Elsevier Publishing Company, Inc., 1971.
29. Bykowski, B. B. Gasohol, TBA, MTBE Effects on Light-Duty Emissions. Final Report of Task Force 6, Contract 68-03-2377. San Antonio, Texas: Southwest Research Institute, October 1979.
30. Cadle, R. D., S. S. Lin, and R. F. Hausman, Jr. The reaction of $O(^3P)$ with propionaldehyde and acrolein in a fast-flow system. Chemosphere 1:15-20, 1972.
31. Cadle, R. D., and J. W. Powers. The reaction of $O(^3P)$ with acetaldehyde in a fast-flow system. J. Phys. Chem. 71:1702-1706, 1967.
32. Cadle, S. H., G. J. Nebel, and R. L. Williams. Measurements of Unregulated Emissions from General Motors' Light-Duty Vehicles. SAE Paper No. 790694. Warrendale, Pa.: Society of Automotive Engineers, Inc., 1979. 21 pp.
33. California Air Resources Board. Staff Report. 1972.
34. Callahan, M. A., M. W. Slimak, N. W. Gabel, I. P. May, C. F. Fowler, J. R. Freed, P. Jennings, R. L. Durfee, F. C. Whitmore, B. Maestri, W. R. Mabey, B. R. Holt, and C. Gould. Water-Related Environmental Fate of 129 Priority Pollutants. Vol. II. Halogenated Aliphatic Hydrocarbons, Halogenated Ethers, Monocyclic Aromatics, Phthalate Esters, Polycyclic Aromatic Hydrocarbons, Nitrosamines, and Miscellaneous Compounds. U.S. Environmental Protection Agency Report No. EPA-440/4-79-029b. Washington, D.C.: U.S. Environmental Protection Agency, Office of Water Planning and Standards, 1979. [714] pp.
35. Calvert, J. G. Hydrocarbon involvement in photochemical smog formation in Los Angeles atmosphere. Environ. Sci. Technol. 10:256-262, 1976.
36. Calvert, J. G. Test of the theory of ozone generation in Los Angeles atmosphere. Environ. Sci. Technol. 10:248-256, 1976.
37. Calvert, J. G. The homogeneous chemistry of formaldehyde generation and destruction within the atmosphere, pp. 153-190. In Proceedings of the NATO Advanced Study Institute on Atmospheric Ozone: Its Variation and Human Influences. Washington, D.C.: U.S. Department of Transportation, Federal Aviation Administration, 1980.
38. Calvert, J. G., K. L. Demerjian, J. A. Kerr. Computer simulation of the chemistry of a simple analogue to the sunlight-irradiated auto-exhaust polluted atmosphere. Environ. Lett. 4:123-140, 1973.
39. Calvert, J. G., and R. D. McQuigg. The computer simulation of the rates and mechanisms of photochemical smog formation, pp. 113-154. In S. W. Benson, D. M. Golden, and J. R. Barker, Eds. Proceedings of the Symposium on Chemical Kinetics Data for the Upper and Lower Atmosphere. Held at Warrenton, Virginia,

40. Calvert, J. G., and J. N. Pitts, Jr. Photochemistry. New York. John Wiley & Sons, Inc., 1966. 899 pp.
41. Calvert, J. G., F. Su, J. W. Bottenheim, and O. P. Strausz. Mechanism of the homogeneous oxidation of sulfur dioxide in the troposphere. Atmos. Environ. 12:197-226, 1978.
42. Carter, W. P. L., A. C. Lloyd, J. L. Sprung, and J. N. Pitts, Jr. Computer modeling of smog chamber data: Progress in validation of a detailed mechanism for the photooxidation of propene and n-butane in photochemical smog. Int. J. Chem. Kinet. 11:45-101, 1979.
43. Cauler, H. Some problems of atmospheric chemistry. In Compendium of Meteorology. Baltimore: Waverly Press, Inc., 1951.
44. Chang, J. S., and J. R. Barker. Reaction rate and products for the reaction $O(^3P) + H_2CO$. J. Phys. Chem. 83:3059-3064, 1979.
45. Chui, G. K., R. D. Anderson, R. E. Baker, and F. B. P. Pinto. Brazilian vehicle calibration for ethanol fuels. Paper II-17. Proceedings of Third International Symposium on Alcohol Fuels Technology, Asilomar, Cal., May 29, 1979. 10 pp. Washington, D.C.: U.S. Department of Energy, 1980.
46. Cleveland, W. S., T. E. Graedel, and B. Kleiner. Urban Formaldehyde: Observed correlation with source emissions and photochemistry. Atmos. Environ. 11:357-360, 1977.
47. Coleman, W. E., R. D. Lingg, R. G. Melton, and F. C. Kopfler. The occurrence of volatile organics in five drinking water supplies using gas chromatography/mass spectrometry, pp. 305-327. In L. H. Keith, Ed. Identification and Analysis of Organic Pollutants in Water. Ann Arbor, Mich.: Ann Arbor Science Publishers, Inc., 1976.
48. Cox, R. A., R. G. Derwent, P. M. Holt, and J. A. Kerr. Photolysis of nitrous acid in the presence of acetaldehyde. J. Chem. Soc., Faraday Trans. I 72:2061-2075, 1976.
49. Crutzen, P. J., and J. Fishman. Average concentrations of OH in the troposphere, and the budgets of CH_4, CO, H_2, and CH_3CCl_3. Geophys. Res. Lett. 4:321-324, 1977.
50. Cvetanović, R. J. Reaction of oxygen atoms with acetaldehyde. Can. J. Chem. 34:775-784, 1956.
51. Darley, E. F., J. T. Middleton, and M. J. Garber. Plant damage and eye irritation from ozone-hydrocarbon reactions. Agric. Food Chem. 8:483-485, 1960.
52. Davis, D. D. OH Radical Measurements: Impact on Power Plant Plume Chemistry. Electric Power Research Institute Report No. EA-465. Palo Alto: Electric Power Research Institute, 1977. 60 pp.
53. Davis, D. D., W. Heaps, and T. McGee. Direct measurements of natural tropospheric levels of OH via an aircraft borne tunable dye laser. Geophys. Res. Lett. 3:331-333, 1976.

54. Deimel, M. Erfahrungen über Formaldehyd--Raumluft--Konzentrationen in Schulneubauten. Presented at the Conference on Organic Pollution in Indoor and Outdoor Air, Berlin, 1976.
55. Demerjian, K. L., J. A. Kerr, and J. G. Calvert. The mechanism of photochemical smog formation. Adv. Environ. Sci. Technol. 4:1-262, 1974.
56. Demerjian, K. L., K. L. Schere, and J. T. Peterson. Theoretical estimates of actinic (spherically integrated) flux and photolytic rate constants of atmospheric species in the lower troposphere. Adv. Environ. Sci. Technol. 10:369-459, 1980.
57. Dimitriades, B., and T. C. Wesson. Reactivities of exhaust aldehydes. J. Air Pollut. Control Assoc. 22:33-38, 1972.
58. Dodge, M. C., and T. A. Hecht. Rate constant measurements needed to improve a general kinetic mechanism for photochemical smog, pp. 155-163. In S. W. Benson, D.M. Golden, and J. R. Barker, Eds. Proceedings of the Symposium on Chemical Kinetics Data for the Upper and Lower Atmosphere. Held at Warrenton, Virginia, September 15-18, 1974. Published as a book supplement to Int. J. Chem. Kinet. 7. New York: John Wiley & Sons, Inc., 1975.
59. Drew, R. T., S. Laskin, M. Kuschner, and N. Nelson. The inhalation carcinogenicity of alpha halo ethers. I. The acute inhalation toxicity of chloromethyl methyl ether and bis(chloromethyl) ether. Arch. Environ. Health 30:61-69, 1975.
59a. Ewing, B. B., E. S. K. Chian, J. C. Cook, C. A. Evans, and P. K. Hopke. Monitoring to Detect Previously Unrecognized Pollutants in Surface Waters--Appendix: Organic Analysis Data. U.S. Environmental Protection Agency Report No. EPA-560/6-77-015a. Washington, D.C.: U.S. Environmental Protection Agency, 1977. 304 pp.
60. Fanning, L. Z. Formaldehyde in Office Trailers. Lawrence Berkeley Laboratory Report No. LBID-084. Berkeley, California: Lawrence Berkeley Laboratory, Energy and Environment Division, 26 October 1979. 7 pp.
61. Farmer, C. B., O. F. Raper, B. D. Robbins, R. A. Toth, and C. Muller. Simultaneous spectroscopic measurements of stratospheric species: O_3, CH_4, CO, CO_2, N_2O, H_2O, HCl, and HF at northern and southern mid-latitudes. J. Geophys. Res. 85:1621-1632, 1980.
62. Fox, M. E. Fate of selected organic compounds in the discharge of Kraft paper mills into Lake Superior, pp. 641-659. In L. H. Keith, Ed. Identification and Analysis of Organic Pollutants in Water. Ann Arbor, Mich.: Ann Arbor Science Publishers, Inc., 1976.
63. Frankel, L. S., K. S. McCallum, and L. Collier. Formation of bis(chloromethyl)ether from formaldehyde and hydrogen chloride. Environ. Sci. Technol. 8:356-359, 1974.
64. Freeman, H. G., and W. C. Grendon. Formaldehyde detection and control in the wood industry. Forest Prod. J. 21(9):54-57, 1971.
65. Gaffney, J. S., R. Atkinson, and J. N. Pitts, Jr. Relative rate constants for the reaction of $O(^3P)$ atoms with selected

olefins, monoterpenes, and unsaturated aldehydes. J. Amer. Chem. Soc. 97:5049-5051, 1975.

66. Gamble, J. F., A. J. McMichael, T. Williams, and M. Battigelli. Respiratory function and symptoms: An environmental epidemiological study of rubber workers exposed to a phenol-formaldehyde type resin. Am. Ind. Hyg. Assoc. J. 37:499-513, 1976.

67. Garry, V. F., L. Oatman, R. Pleus, and D. Gray. Formaldehyde in the home. Some environmental disease perspectives. Minn. Med. 63:107-111, 1980.

68. Gay, B. W., Jr., R. C. Noonan, J. J. Bufalini, and P. L. Hanst. Photochemical synthesis of peroxyacyl nitrates in gas phase via chlorine-aldehyde reaction. Environ. Sci. Technol. 10:82-85, 1976.

69. Georgii, H.-W. Untersuchungen über Atmosphärische Spurenstoffe und ihre Bedeutung für die Chemie der Niederschläge. Geofis. Pura Appl. 47:155-171, 1960.

70. Gerstle, R. W., and D. A. Kemnitz. Atmospheric emissions from open burning. J. Air Pollut. Control Assoc. 17:324-327, 1967.

71. Gibbard, S., and R. Schoental. Simple semi-quantitative estimation of sinapyl and certain related aldehydes in wood and in other materials. J. Chromat. 44:396-398, 1969.

72. Giger, W., M. Reinhard, C. Schaffner, and F. Zürcher. Analyses of organic constituents in water by high-resolution gas chromatography in combination with specific detection and computer-assisted mass spectrometry, pp. 433-452. In L. H. Keith, Ed. Identification and Analysis of Organic Pollutants in Water. Ann Arbor, Mich.: Ann Arbor Science Publishers, Inc., 1976.

73. Graedel, T. E. Chemical Compounds in the Atmosphere. New York: Academic Press, Inc., 1979. 440 pp.

74. Graedel, T. E., L. A. Farrow, and T. A. Weber. The influence of aerosols on the chemistry of the troposphere, pp. 581-594. In S. W. Benson, D. M. Golden, and J.R. Barker, Eds. Proceedings of the Symposium on Chemical Kinetics Data for the Upper and Lower Atmosphere. Held at Warrenton, Virginia, September 15-18, 1974. Published as a book supplement to Int. J. Chem. Kinet. 7. New York: John Wiley & Sons, Inc., 1975.

75. Grob, K. Organic substances in potable water and in its precursor. Part I. methods for their determination by gas-liquid chromatography. J. Chromat. 84:255-273, 1973.

76. Hagin, J., and L. Cohen. Nitrogen fertilizer potential of an experimental urea formaldehyde. Agron. J. 68:518-520, 1967.

77. Hangebrauck, R. P., D. J. von Lehmden, and J. E. Meeker. Emissions of polynuclear hydrocarbons and other pollutants from heat-generation and incineration processes. J. Air Pollut. Control Assoc. 14:267-278, 1964.

78. Hanst, P. L., and B. W. Gay, Jr. Photochemical reactions among formaldehyde, chlorine, and nitrogen dioxide in air. Environ. Sci. Technol. 11:1105-1109, 1977.

79. Hanst, P. L., W. E. Wilson, R. K. Patterson, B. W. Gay, Jr., L. W. Chaney, and C. S. Burton. A spectroscopic study of California smog, pp. 17-70. In Proceedings of the 6th Annual Symposium. Trace Analysis and Detection in the Environment, 29 April-1 May, 1975. Edgewood Arsenal Special Report EO-SP-76001. Aberdeen Proving Ground, Md.: Edgewood Arsenal 1975.

80. Harke, H.-P., A. Baars, B. Frahm, H. Peters, and C. Schultz. The problem of passive smoking. Concentration of smoke constituents in the air of large and small rooms as a function of number of cigarettes smoked and time. Int. Arch. Arbeitsmed. 29:323-339, 1972. (in German; English summary)

81. Harrington, J. A., D. D. Brehob, and E. H. Schanerberger. Evaluation of methyl tertiary-butyl ether as a gasoline blending component. Paper III-53, pp. 1-13. In Proceedings of Third International Symposium on Alcohol Fuels Technology, Asilomar, Cal., May 29, 1979. Washington, D.C.: U.S. Department of Energy, 1980.

82. Harris, D. K. Health problems in the manufacture and use of plastics. Br. J. Ind. Med. 10:255-268, 1953.

83. Herron, J. T., and R. E. Huie. Rate constants for the reactions of ozone with ethene and propene, from 235.0 to 362.0 K. J. Phys. Chem. 78:2085-2088, 1974.

84. Herron, J. T., and R. D. Penzhorn. Mass spectrometric study of the reactions of atomic oxygen with ethylene and formaldehyde. J. Phys. Chem. 73:191-196, 1969.

85. Heuss, J. M., and W. A. Glasson. Hydrocarbon Reactivity and Eye Irritation. Environ. Sci. Technol. 2:1109-1116. 1968.

86. Hollowell, C. D., J. V. Berk, M. L. Boegel, R. R. Miksch, W. W. Nazaroff, and G. W. Traynor. Building Ventilation and Indoor Air Quality. Lawrence Berkeley Laboratory Report LBL-10391. Berkeley, Cal.: Lawrence Berkeley Laboratory, for U.S. Department of Energy, 1980. 12 pp.

87. Horowitz, A., and J. G. Calvert. The quantum efficiency of the primary processes in formaldehyde photolysis at 3130 Å and 25°C. Int. J. Chem. Kinet. 10:713-732, 1978.

88. Horowitz, A., and J. G. Calvert. Wavelength dependence of the quantum efficiencies of the primary processes in formaldehyde photolysis at 25°C. Int. J. Chem. Kinet. 10:805-819, 1978.

89. Horowitz, A., F. Su, and J. G. Calvert. Unusual H_2-forming chain reaction in the 3130 Å photolysis of formaldehyde-oxygen mixtures at 25°C. Int. J. Chem. Kinet. 10:1099-1117, 1978.

90. Huie, R. E., and J. T. Herron. Temperature dependence of the rate constants for reactions of ozone with some olefins. Int. J. Chem. Kinet. 7 (Suppl.):165-181, 1975.

91. Jackson, M.W. Effect of catalytic emission control on exhaust hydrocarbon composition and reactivity. SAE Paper No. 780624. Warrendale, Pa.: Society of Automotive Engineers, Inc. 1978. 24 pp.

92. Jaffee, S., and E. Wan. Thermal and photochemical reactions of NO_2 with butyraldehyde in gas phase. Environ. Sci. Technol. 8:1024-1025, 1974.
93. Jeffries, H. E., and R. M. Kamens. A Critical Review of Ambient Air Aldehyde Measurement Methods and an Analysis of Houston Aldehyde Data. Part II. An Analysis of Houston Aldehyde Data and Comparison with Cincinnati and St. Paul Data. Prepared for Houston Area Oxidant Study. Chapel Hill, N.C.: University of North Carolina School of Public Health, 1977.
94. Jeffries, H. E., and R. M. Kamens. Outdoor Simulation of Air Pollution Control Strategies. Progress Report for 1975-1976. U.S. Environmental Protection Agency Grant 800916. Chapel Hill, N.C.: University of North Carolina, School of Public Health, 1977.
95. Jermini, C., A. Weber, and E. Grandjean. Quantitative determination of various gas-phase components of the side-stream smoke of cigarettes in the room air as a contribution to the problem of passive smoking. Int. Arch. Occup. Environ. Health 36:169-181, 1976. (in German)
96. Junge, C. E. Chemical analysis of aerosol particles and gas traces on the island of Hawaii. Tellus 9:528-537, 1957.
97. Junge, C. E. Recent investigations in air chemistry. Tellus 8:127-139, 1956.
98. Kallos, G. J., and R. A. Solomon. Investigations of the formation of bis(chloromethyl)ether in simulated hydrogen chloride-formaldehyde atmospheric environments. Amer. Ind. Hyg. Assoc. J. 34:469-473, 1973.
98a. Katou, T. Research of the smog materials in the atmosphere by CG method, pp. 419-425. In Proceedings. International Symposium on Air Pollution 1972 Tokyo. The Status of Air Pollution and the Progress of the Preventive Technology. Tokyo: Union of Japanese Scientists and Engineers, 1972.
99. Keith, L. H. GC/MS analysis of organic compounds in treated Kraft paper mill wastewaters, pp. 671-707. In L. H. Keith, Ed. Identification and Analysis of Organic Pollutants in Water. Ann Arbor, Mich.: Ann Arbor Sciences Publishers, Inc., 1976.
100. Keith, L. H., A. W. Garrison, F. R. Allen, M. H. Carter, T. L. Floyd, J. D. Pope, and A. D. Thruston, Jr. Identification of organic compounds in drinking water from thirteen U.S. cities, pp. 329-373. In L. H. Keith, Ed. Identification and Analysis of Organic Pollutants in Water. Ann Arbor, Mich.: Ann Arbor Science Publishers, Inc., 1976.
101. Kelly, M. W. Critical Literature Review of Relationships between Processing Parameters and Physical Properties of Particleboard. General Technical Report FPL-10. Madison, Wisc.: U.S. Department of Agriculture, Forest Service, Forest Products Laboratory, 1977. 69 pp.
102. Kerfoot, E. J., and T. F. Mooney, Jr. Formaldehyde and paraformaldehyde study in funeral homes. Am. Ind. Hyg. Assoc. J. 36:533-535, 1975.

103. Kerr, J. A., J. G. Calvert, and K. L. Demerjian. The mechanism of photochemical smog formation. Chem. Britain 8:252-257, 1972.
104. Kitchens, J. E., R. E. Casner, G. S. Edwards, W. E. Harward III, and B. J. Macri. Investigation of Selected Potential Environmental Contaminants: Formaldehyde. U.S. Environmental Protection Agency Report No. EPA-560/2-76-009. Washington, D.C.: U.S. Environmental Protection Agency, Office of Toxic Substances, 1976. 217 pp.
105. Klemm, R. B., Absolute rate parameters for the reactions of formaldehdye with O-atoms and H-atoms over the temperature range, 250-500°K. J. Chem. Phys. 71:1987-1993, 1979.
106. Klemm, R. B., E. G. Skolnik, and J. V. Michael. Absolute rate parameters for the reaction of $O(^3P)$ with H_2CO over the temperature range 250 to 750 K. J. Chem. Phys. 72:1256-1264, 1980.
107. Kleopfer, R. D. Analysis of drinking water for organic compounds, pp. 399-416. In L. H. Keith, Ed. Identification and Analysis of Organic Pollutants in Water. Ann Arbor, Mich.: Ann Arbor Science Publishers, Inc., 1976.
108. Kopczynski, S. L., A. P. Altshuller, and F. D. Sutterfield. Photochemical reactivities of aldehyde-nitrogen oxide systems. Environ. Sci. Technol. 8:909-918, 1974.
109. Kuschner, M., S. Laskin, R. T. Drew, V. Cappiello, and N. Nelson. The inhalation carcinogenicity of alpha-halo ethers. III. Lifetime and limited period inhalation studies with bis(chloromethyl)ether at 0.1 ppm. Arch. Environ. Health 30: 73-77, 1975.
110. Laskin, S., R. T. Drew, V. Cappiello, M. Kuschner, and N. Nelson. Inhalation carcinogenicity of alpha-halo ethers. II. Chronic inhalation studies with chloromethyl methyl ether. Arch. Environ. Health 30: 70-72, 1975.
111. Leggett, J. E., J. L. Sims, D. R. Gossett, U. R. Pal, and J. F. Benner. Potassium and magnesium nutrition effects on yield and chemical composition of Burley tobacco leaves and smoke. Can. J. Plant Sci. 57:159-166, 1977.
112. Leighton, P. A. Photochemistry of Air Pollution. New York: Academic Press, Inc., 1961. 300 pp.
113. Levy, H., II. Tropospheric budgets for methane, carbon monoxide, and related species. J. Geophys. Res. 78:5325-5332, 1973.
114. Lin, C., R. N. Anaclerio, D. W. Anthon, L. Z. Fanning, and C. D. Hollowell. Indoor/outdoor measurements of formaldehyde and total aldehydes. Presented at American Chemical Society Meeting, Washington, D.C., Sept. 9-14, 1979.
115. Lloyd, A. C. Tropospheric chemistry of aldehydes, pp. 27-48. In J. T. Herron, R. E. Huie, and J. A. Hodgeson, Eds. Chemical Kinetic Data Needs for Modeling the Lower Troposphere. National Bureau of Standards Special Publication 557. Washington, D.C.: U.S. Government Printing Office, 1979.
116. Long, K. R., D. A. Pierson, S. T. Brennan, C. W. Frank, and R. A. Hahne. Problems Associated with the Use of Urea-Formaldehyde Foam for Residential Insulation. Part I. The Effects of

Temperature and Humidity on Formaldehyde Release from Urea-Formaldehyde Foam Insulation. Oak Ridge National Laboratory Report ORNL/SUB-7559/I. Washington, D.C.: U.S. Government Printing Office, 1979. 89 pp.

117. Lovas, F. J., and R. D. Suenram. Identification of dioxirane (H_2COO) in ozone-olefin reactions via microwave spectroscopy. Chem. Phys. Lett. 51:453-456, 1977.

118. Lovell, R. J. Emissions Control Options for the Synthetic Organic Chemicals Manufacturing Industry: Formaldehyde Product Report. EPA Contract No. 68-02-2577. Knoxville, Tenn.: Hydroscience, Inc., for U.S. Environmental Protection Agency, Office of Air Quality Planning and Standards, 1979. [150] pp.

119. Lyles, G. R., F. B. Dowling, and V. J. Blanchard. Quantitative determination of formaldehyde in the parts per hundred million concentration level. J. Air Pollut. Control Assoc. 15:106-108, 1965.

120. MacCracken, M. C., and G. D. Sauter, Eds. Development of an Air Pollution Model for the San Francisco Bay Area. Final Report to the National Science Foundation. Vol. 2. Appendixes. Livermore, Calif.: Lawrence Livermore Laboratory, 1975. 948 pp. (abstract in Energy Res. Abstr. 2:14714, 1977)

121. Mack, G. P. R., and B. A. Thrush. Reaction of oxygen atoms with carbonyl compounds. Part 1. Formaldehyde. J. Chem. Soc., Faraday Trans. I 69:208-215, 1973.

122. Mack, G. P. R., and B. A. Thrush. Reaction of oxygen atoms with carbonyl compounds. Part 2. Acetaldehyde. J. Chem. Soc., Faraday Trans. I 70:178-186, 1974.

123. Major, R. T. The Ginkgo, the most ancient living tree. Science 157(3794):1270-1273, 1967.

124. Martinez, R. I., R. E. Huie, and J. T. Herron. Mass spectrometric detection of dioxirane, H_2COO, and its decomposition products, H_2 and CO, from the reaction of ozone with ethylene. Chem. Phys. Lett. 51:457-459, 1977.

125. McKee, D. E., F. C. Ferris, and R. E. Goeboro. Unregulated emissions from PROCO engine powered vehicles. SAE Report No. 780592. Warrendale, Pa.: Society of Automotive Engineers, Inc., 1978. 23 pp.

126. McKinney, J. D., R. R. Maurer, J. R. Hass, and R. O. Thomas. Possible factors in the drinking water of laboratory animals causing reproductive failure, pp. 417-432. In L. H. Keith, Ed. Identification and Analysis of Organic Pollutants in Water. Ann Arbor: Ann Arbor Science Publishers, Inc., 1976.

127. Mill, T. Structure Reactivity Correlations for Environmental Reactions. U.S. Environmental Protection Agency Report No. EPA-560/11-79-012. Menlo Park, Calif.: SRI International, for U.S. Environmental Protection Agency, 1979. 66 pp.

128. Miller, B. H., and H. P. Blejer. Report of an Occupational Health Study of Formaldehyde Concentrations at Maximes, 400 E. Colorado Street, Pasadena, California. Study No. S-1838. Los Angeles: State of California Health and Welfare Agency,

Department of Public Health, Bureau of Occupational Health, 1966. 5 pp.

129. Moortgat, G. K., and P. Warneck. CO and H_2 quantum yields in the photodecomposition of formaldehyde in air. J. Chem. Phys. 70:3639-3651, 1979.

130. Morris, E. D., Jr., and H. Niki. Mass spectrometric study of the reaction of hydroxyl radical with formaldehyde. J. Chem. Phys. 55:1991-1992, 1971.

131. Morris, E. D., Jr., and H. Niki. Reaction of the nitrate radical with acetaldehyde and propylene. J. Phys. Chem. 78:1337-1338, 1974.

132. Morris, E. D., Jr., D. H. Stedman, and H. Niki. Mass spectrometric study of the reactions of the hydroxyl radical with ethylene, propylene, and acetaldehyde in a discharge-flow system. J. Am. Chem. Soc. 93:3570-3572, 1971.

133. Morris, R. B., F. B. Higgins, Jr., J. A. Lee, R. Newirth, and J. W. Pervier. Engineering and Cost Study of Air Pollution Control for the Petrochemical Industry. Vol. 4. Formaldehyde Manufacture with the Silver Catalyst Process. U.S. Environmental Protection Agency Report No. EPA-450/3--73-006-d. Research Triangle Park, N.C.: U.S. Environmental Protection Agency, Office of Air Quality Planning and Standards, 1975. 94 pp.

134. Morris, R. B., F. B. Higgins, Jr., J. A. Lee, R. Newirth, and J. W. Pervier. Engineering and Cost Study of Air Pollution Control for the Petrochemical Industry. Vol. 5. Formaldehyde Manufacture with Mixed Oxide Catalyst Process. U.S. Environmental Protection Agency Report No. EPA-450/3-73-006-e. Research Triangle Park, N.C.: U.S. Environmental Protection Agency, Office of Air Quality Planning and Standards, 1975. 82 pp.

135. Moschandreas, D. J., J. W. C. Stark, J. E. McFadden, and S. S. Morse. Indoor Air Pollution in the Residential Environment. Vol. I. Data Collection, Analysis and Interpretation. U.S. Environmental Protection Agency Report No. EPA-600/7-78-229a. Research Triangle Park, N.C.: U.S. Environmental Protection Agency, Office of Research and Development, Environmental Monitoring and Support Laboratory, 1978. 201 pp.

136. National Research Council, Institute of Laboratory Animal Resources. Laboratory Animal Housing. Proceedings of a Symposium Held at Hunt Valley, Maryland, September 22-23, 1976, p. 113. Washington, D.C.: National Academy of Sciences, 1978.

137. Natusch, D. F. S. Potentially carcinogenic species emitted to the atmosphere by fossil-fueled power plants. Environ. Health Persp. 22:79-90, 1978.

138. Niki, H., E. E. Daby, and B. Weinstock. Mass spectrometric study of the kinetics and mechanism of the ethylene-atomic oxygen reaction by the discharge flow technique at 300 K, p. 277. In 12th International Symposium on Combustion, Poitiers, France, July 14-20, 1968. Pittsburgh, Pa.: The Combustion Institute, 1969.

139. Niki, H., E. E. Daby, and B. Weinstock. Mechanisms of smog reactions. In Photochemical Smog and Ozone Reactions. Two Symposia, 1971. Adv. Chem. Series 113:16-57, 1972.
140. Niki, H., P. D. Maker, C. M. Savage, and L. P. Breitenbach. Fourier transform IR spectroscopic observation of pernitric acid formed via $HOO + NO_2 \rightarrow HOONO_2$. Chem. Phys. Lett. 45: 564-566, 1977.
141. Niki, H., P. D. Maker, C. M. Savage, and L. P. Breitenbach. Fourier transform IR spectroscopic observation of propylene ozonide in the gas phase reaction of ozone-cis-2-butene-formaldehyde. Chem. Phys. Lett. 46:327-330, 1977.
142. Niki, H., P. D. Maker, C. M. Savage, and L. P. Breitenbach. Relative rate constants for the reaction of hydroxyl radical with aldehydes. J. Phys. Chem. 82:132-134, 1978.
143. Ninomiya, J. S., and A. Golovoy. Effects of Air-Fuel Ratio on Composition of Hydrocarbon Exhaust from Isooctane, Diiosbutylene Toluene, and Toluene-n-Heptane Mixture. SAE Paper No. 690504. New York: Society of Automotive Engineers, Inc., 1969. 11 pp.
144. Novák, J., J. Žlutický, V. Kubelka, and J. Mostecký. Analysis of organic constituents present in drinking water. J. Chromat. 76:45-50, 1973.
145. Osborne, A. D., J. N. Pitts, Jr., and E. F. Darley. On the stability of acrolein towards photooxidation in the near ultra-violet. Int. J. Air Water Pollut. 6:1-3, 1962.
146. Osborne, J. S., S. Adamek, and M. E. Hobbs. Some components of gas phase of cigarette smoke. Anal. Chem. 28:211-215, 1956.
147. Osif, T. L., and J. Heicklen. Oxidation of HCO radicals. J. Phys. Chem. 80:1526-1531, 1976.
148. Pefley, R. K., M. A. Saad, M. A. Sweeney, and J. D. Kilgroe. Performance and emission characteristics using blends of methanol and dissociated methanol as an automotive fuel. Paper No. 719008, pp. 36-46. In Proceedings of the SAE Intersociety Energy Conversion Engineers Conference, Boston, Mass., August 3-5, 1971. New York: Society of Automotive Engineers, Inc.
149. Pitts, J. N., Jr., A. M. Winer, K. R. Darnall, G. J. Doyle, and J. M. McAfee. Chemical Consequences of Air Quality Standards and of Control Implementation Programs. Roles of Hydrocarbons, Oxides of Nitrogen and Aged Smog in the Production of Photochemical Oxidant. Final Report to the California Air Resources Board. Contract No. 4-214. Riverside: University of California, Statewide Air Pollution Research Center, 1976. 444 pp.
150. Rasmussen, R. A. What do the hydrocarbons from trees contribute to air pollution? J. Air Pollut. Control Assoc. 22:537-543, 1972.
151. Renzetti, N. A., and R. J. Bryan. Atmospheric sampling for aldehydes and eye irritation in Los Angeles smog--1960. J. Air Pollut. Control. Assoc. 11:421-424, 427, 1961.
152. Santodonato, J., D. Basu, and P. Howard. Health Effects Associated with Diesel Exhaust Emissions. Literature Review and Evaluation. U.S. Environmental Protection Agency Report No.

EPA-600/1-78-063. Research Triangle Park, N.C.: U.S. Environmental Protection Agency, Health Effects Research Laboratory, November 1978. 163 pp.

153. Schauenstein, E., H. Esterbauer, and H. Zollner. Aldehydes in Biological Systems. Their Natural Occurrence and Biological Activities, London: Pion Limited, 1977. 205 pp.

154. Schmidt, A., and H. Götz. The formation of formaldehyde during the burning of natural gas in household devices. Gas-Wasserfach, Gas-Erdgas 118:112-115, 1977. (in German)

155. Schuck, E. A., E. R. Stephens, and J. T. Middleton. Eye irritation response at low concentrations of irritants. Arch. Environ. Health 13:570-575, 1966.

156. Scott Research Laboratories, Inc. Atmospheric Reaction Studies in the Los Angeles Basin. Phase I. Vol. II. pp. 143-149, 310-314. Washington, D.C.: U.S. Department of Health, Education, and Welfare, Public Health Service, National Air Pollution Control Administration, 1969.

157. Sellakumar, A. R., R. E. Albert, G. M. Rusch, G. V. Katz, N. Nelson, and M. Kuschner. Inhalation carcinogenicity of formaldehyde and hydrogen chloride in rats. Abstract No. 424. Proc. Am. Assoc. Cancer Res. 21:106, 1980.

158. Shackleford, W. M., and L. H. Keith. Frequency of Organic Compounds Identified in Water. U.S. Environmental Protection Agency Report No. EPA-600/4-76-062. Athens, Georgia: U.S. Environmental Protection Agency, Office of Research and Development, Environmental Research Laboratory, 1976. 626 pp.

159. Sheldrick, J. E., and T. R. Steadman. Product/Industry Profile and Related Analysis on Formaldehyde and Formaldehyde-Containing Consumer Products. Part II. Products/Industry Profile on Urea Formaldehyde. Columbus, Ohio: Batttelle Columbus Division, for U.S. Consumer Product Safety Commission, 1979. [24] pp.

160. Sheldrick, J. E., and T. R. Steadman. Product/Industry Profile and Related Analysis on Formaldehyde and Formaldehyde-Containing Consumer Products. Part III. Consumer Products Containing Formaldehyde. Columbus, Ohio: Battelle Columbus Division, for U.S. Consumer Product Safety Commission, 1979. [39] pp.

161. Shipkovitz, H. D. Formaldehyde Vapor Emissions in the Permanent-Press Fabrics Industry. Report No. TR-52. Cincinnati: U.S. Department of Health, Education, and Welfare, Public Health Service, Consumer Protection and Environmental Health Service, Environmental Control Administration, 1968. 18 pp.

162. Singleton, D. L., R. S. Irwin, and R. J. Cvetanović. Arrhenius parameters for the reactions of $O(^3P)$ atoms with several aldehydes and the trend in aldehydic C-H bond dissociation energies. Can. J. Chem. 55:3321-3327, 1977.

163. Smith, C. R., and E. H. Spaulding. Myobactericidal agents, p. 1039. In L. S. Goodman and A. Gilman, Eds. Pharmacological Basis of Therapeutics. 4th ed. New York: Macmillan Publishing Co., Inc., 1970.

164. Smith, C. W., Ed. Acrolein. New York: John Wiley & Sons, Inc., 1962. 273 pp.
165. Smith, R. H. Rate constants and activation energy for the gaseous reaction between hydroxyl and formaldehyde. Int. J. Chem. Kinet. 10:519-527, 1978.
166. Springer, K. J., and R. C. Stahman. Diesel Car Emissions--Emphasis on Particulate and Sulfate. SAE Paper No. 770254. Warrendale, Pa.: Society of Automotive Engineers, Inc., 1977. 32 pp.
167. SRI International. Class Study Report. Aldehydes. Menlo Park, Calif.: SRI International, for National Cancer Institute, Chemical Selection Working Group, 1978. 19 pp. + data package.
168. Stahl, Q. R. Preliminary Air Pollution Survey of Aldehydes. A Literature Review. National Air Pollution Control Administration Publication No. APTD 69-24. Raleigh, N.C.: U.S. Department of Health, Education, and Welfare, Public Health Service, National Air Pollution Control Administration, 1969.
169. Stief, L. J., D. F. Nava, W. A. Payne, and J. V. Michael. Rate constant for the reaction of hydroxyl radical with formaldehyde over the temperature range 228-362 K, pp. 479-481. In Proceedings of the NATO Advanced Study Institute on Atmospheric Ozone: Its Variation and Human Influences. Washington, D.C.: U.S. Department of Transportation, Federal Aviation Administration, 1980.
170. Su, F., J. G. Calvert, and J. H. Shaw. An FT IR spectroscopic study of the ozone-ethene reaction mechanism in O_2-rich mixtures. J. Phys. Chem. 84:239-246, 1980.
171. Su, F., J. G. Calvert, and J. H. Shaw. The Mechanism of the photooxidation of gaseous formaldehyde. J. Phys. Chem. 83: 3185-3190, 1979.
172. Su, F., J. G. Calvert, J. H. Shaw, H. Niki, P. D. Maker, C. M. Savage, and L. D. Breitenbach. Spectroscopic and kinetic studies of a new metastable species in the photooxidation of gaseous formaldehyde. Chem. Phys. Lett. 65:221-225, 1979.
173. Suta, B. E. Population Exposures to Atmospheric Formaldehyde inside Residences. CRESS Report No. 113. Menlo Park, Calif.: SRI International, for U.S. Environmental Protection Agency, Office of Research and Development, 1980. [78] pp.
174. Suta, B. E. Production and Use of 13 Aldehyde Compounds. Menlo Park, California: SRI International, for U.S. Environmental Protection Agency, Office of Research and Development, 1979. [32] pp.
175. Tabershaw, I.R., H.N. Doyle, L. Gaudette, S. H. Lamm, and O. Wong. A Review of the Formaldehyde Problems in Mobile Homes. Rockville, Maryland: Tabershaw Occupational Medical Associates, P.A., for National Particleboard Institute, 1979. 19 pp.
176. Tija, B. and T. J. Sheehan. Effect of urea formaldehyde slow release fertilizers on growth and leaf pigmentation of _Euphorbia pulcherrima_ Willd. HortScience 12:383, 1977. (abstract)
177. Tomas, M. Liberation of formaldehyde during the hardening of urea-formaldehyde resins at high temperatures. Holztechnot.

5:89-91, 1964. (in German; English abstract in Abstr. Bull. Inst. Paper Chem. 35:1118, 1965)

178. Tuazon, E. C., R. A. Graham, A. M. Winer, R. R. Easton, J. N. Pitts, Jr., and P. L. Hanst. A kilometer pathlength Fourier-transform infrared system for the study of trace pollutants in ambient and synthetic atmospheres. Atm. Environ. 12:865-875, 1978.

179. Tuazon, E. C., A. M. Winer, R. A. Graham, and J. N. Pitts, Jr. Application of a kilometer pathlength FT-IR spectrometer to analysis of trace pollutants in ambient and simulated atmospheres, pp. 798-802. In Proceedings of the 4th Joint Conference on Sensing Environmental Pollutants, 1977. Washington, D.C.: American Chemical Society, 1978.

180. U.S. Department of Health, Education, and Welfare, National Air Pollution Control Administration. Air Quality Criteria for Hydrocarbons. DHEW Publication No. AP-64. Washington, D.C.: U.S. Government Printing Office, 1970. 118 pp.

181. U.S. Department of Health, Education, and Welfare, Public Health Service, Health Services and Mental Health Administration. The Health Consequences of Smoking. A Report of the Surgeon General: 1972. DHEW Publication No. (HSM)72-7516. Washington, D.C.: U.S. Government Printing Office, 1972. 158 pp.

182. U.S. Department of Labor, Occupational Safety and Health Administration. Table Z-2, p. 594. In Occupational safety and health standards. Subpart Z. Toxic and hazardous substances. 29 CFR 1901.1000, July 1, 1980.

183. U.S. Environmental Protection Agency, Office of Toxic Substances. Preliminary Assessment of Suspected Carcinogens in Drinking Water. Interim Report to Congress. Report No. EPA 560/4-75/003. Washington, D.C.: U.S. Environmental Protection Agency, 1975. 39 pp.

184. U.S. Environmental Protection Agency, Region VI. Analytical Report: New Orleans Area Water Supply Study. Report No. EPA 906/9-75-003. Dallas: U.S. Environmental Protection Agency, 1975. 90 pp.

185. United States of America Standards Institute. USA Standard. Acceptable Concentrations of Formaldehyde. New York: United States of America Standards Institute, 1967. 8 pp.

186. Waddington, D. V., E. L. Moberg, J. M. Duich, and T. L. Watschke. Long-term evaluation of slow-release nitrogen sources on turfgrass. Soil Sci. Soc. Am. J. 40:593-597, 1976.

187. Walker, J. F. Formaldehyde. American Chemical Society Monograph No. 120. 2nd ed. New York: Reinhold Publishing Corp., 1953. 575 pp.

188. Wang, C. C., L. I. Davis, Jr., C. H. Wu, S. Japar, H. Niki, and B. Weinstock. Hydroxyl radical concentrations measured in ambient air. Science 189:797-800, 1975.

189. Wanner, H. U., A. Deuber, J. Satish, M. Meier, and H. Sommer. Air pollution in the vicinity of streets, pp. 99-107. In M. W. Benarie, Ed. Proceedings of the 12th International Colloquium on Atmospheric Pollution, 1976. Amsterdam: Elsevier, 1976.

190. Weber-Tschopp, A., T. Fischer, and E. Grandjean. Irritating effects of formaldehyde on men. Int. Arch. Occup. Environ. Health 39:207-218, 1977. (in German)
191. Wigg, E. E. Reactive Exhaust Emissions from Current and Future Emission Control Systems. SAE Paper No. 730196. New York: Society of Automotive Engineers, Inc., 1973. 21 pp.
192. Wigg, E. E., R. J. Campion, and W. L. Petersen. The Effect of Fuel Hydrocarbon Composition on Exhaust Hydrocarbon and Oxygenate Emissions. SAE Paper No. 72051. New York: Society of Automotive Engineers, Inc., 1972. 17 pp.
193. Wild, H. The liberation of formaldehyde during the hardening of urea-formaldehyde resins. Holztechnot. 5:92-95, 1964. (in German; English abstract in Abstr. Bull. Inst. Paper Chem. 35: 1120-1121, 1965)
194. Williams, J. E., and H. S. Siegel. Formaldehyde levels on and in chicken eggs following preincubation fumigation. Poultry Sci. 48:552-558, 1969.
195. Wilson, W. E., Jr., A. Levy, and E. H. McDonald. Role of SO_2 and photochemical aerosol in eye irritation from photochemical smog. Environ. Sci. Technol. 6:423-427, 1972.
196. World Health Organization Working Group. Health Aspects Related to Indoor Air Quality. Copenhagen, Denmark: World Health Organization, 1979. 34 pp.
197. Zafiriou, O. C., J. Alford, M. Herrera, E. T. Peltzer, R. B. Gagosian, and S. C. Liu. Formaldehyde in remote marine air and rain: flux measurements and estimates. Geophys. Res. Lett. 7, 341-344, 1980.

CHAPTER 6

ANALYTICAL METHODS FOR THE DETERMINATION OF ALDEHYDES

Air-quality standards and pollution-control legislation are generally based on the assumption that exceeding some concentration of any given pollutant will have harmful effects on human health that outweigh any economic disadvantage of imposing regulatory standards. Accurate determination of such "threshold concentrations" demands accurate methods of analysis.

This chapter discusses analytical methods currently used for aldehydes, including techniques of sampling and calibration, and other available or potentially available methods. In general, the analytical methods for aldehydes are difficult, and much developmental work is needed. Where possible, estimates of the accuracy, precision, and applicability of the various measurement methods are presented. An assessment of the state of the art is given in Chapter 2, and some recommendations for future action are presented in Chapter 3.

METHODS OF GENERATING STANDARDS

All methods of analysis have in common the need for calibration. Calibration is performed by applying the chosen method of analysis to a standard. The standard can be prepared by weighing (a primary standard) or measured by an independent primary reference method of analysis (a secondary standard). In the case of aldehydes, the standard is usually a liquid solution or a gas-phase mixture of one or more aldehydes. Liquid solutions are static; gas-phase mixtures can be static or dynamic (i.e., generated continuously). This section discusses the preparation of standards and their application to calibration.

STATIC METHODS

Aqueous solutions of aldehydes can be used as standards for calibration. The solutions are usually obtained by dissolving an appropriate amount of the desired aldehyde in water. Ordinary reagent-grade aldehydes are often used without purification, although for accurate work it is imperative to distill the aldehyde before use, because oxidation and polymerization occur on standing.

Primary standardization can be achieved by straightforward application of gravimetric or volumetric methods. It is also possible to prepare a secondary-standard solution of aldehyde by oxidative titration. Two methods described by Walker[142] are suited to the analysis of aldehydes other than formaldehyde: the alkaline peroxide and iodometric methods, which rely on the oxidation of an aldehyde to its corresponding carboxylic acid. Once oxidized, the acidic solution can be titrated. These reactions are characteristic of all aldehydes, so there should be no problems in applying the methods to the preparation of a secondary-standard solution of any (pure) aldehyde.

It is difficult to prepare a primary-standard solution of formaldehyde, because pure formaldehyde is not readily available. There are, however, two ways to prepare formaldehyde solutions for standardization by a primary reference method. The easier (but less desirable) is to dilute commercial formalin (37% formaldehyde w/w) to the approximate desired concentration. Unfortunately, solutions so obtained will contain methanol, which is added to formalin as a stabilizer, as an impurity. A methanol-free formaldehyde solution can be obtained by refluxing an appropriate amount of pure paraformaldehyde in water and filtering the resulting solution.

For standardizing formaldehyde solutions prepared by these methods, Walker[142] described several methods. A simple and accurate primary reference method involves the addition of an aliquot of formaldehyde solution to a neutral solution of sodium sulfite to form a bisulfite addition product and sodium hydroxide. The hydroxide released can be neutralized with a primary acid standard to standardize the solution. The neutralization can be monitored with a pH meter.

A second method is the bisulfite-iodine titration procedure.[10,139] Excess sodium bisulfite is added to the formaldehyde solution to form a bisulfite-formaldehyde adduct at neutral pH. The unreacted bisulfite is then destroyed with iodine. Addition of a carbonate buffer releases the bisulfite from the bisulfite-formaldehyde adduct, and the freed bisulfite is titrated with iodine (starch is used an an indicator). The iodine solution itself must be standardized with sodium thiosulfate. Furthermore, one may encounter problems associated with the stability of the iodine reagent. In summary, the method is complex and has several sources of possible error.

Standardization methods based on bisulfite are recommended for use only with formaldehyde, because the formation of the bisulfite-aldehyde adduct with other aldehydes may be less than quantitative.[142]

DYNAMIC METHODS

Aldehydes are reactive compounds, so it is difficult to make calibration gases that are stable for any useful period. This precludes the use of gas-tank standards, unless the concentration of aldehyde is very high (several percent). Recent advances in rendering gas-tank surfaces inert may alter this situation, but no data are available. For most applications, it is currently necessary to use dynamic methods to generate gas-phase-aldehyde standards.

Permeation tubes have been used to generate dynamic gas standards for many different types of compounds and can be used for aldehydes as well. These tubes contain pure compound in a length of Teflon tubing capped at both ends. Over time, material diffuses through the Teflon wall at a low and constant rate, provided that the temperature is held constant.[119] Tubes for acetaldehyde, propionaldehyde, and benzaldehyde are commercially available, and tubes could undoubtedly be constructed for other aldehydes. These tubes are calibrated gravimetrically (and are thus classified as primary reference standards) and can be used with a constant-flow system to generate primary gas standards in the concentration range of parts per billion to parts per million.

Permeation tubes containing pure formaldehyde do not exist. The vapor pressure of pure formaldehyde would be too high to permit the construction of permeation tubes, if it were not already prone to polymerization at room temperature. Construction of a permeation tube for formaldehyde has been attempted with paraformaldehyde. At 80°C, the decomposition rate of the polymer is great enough that a usable permeation rate can be obtained. However, the gas in equilibrium with paraformaldehyde is not pure formaldehyde; it contains substantial amounts of methylal, methyl formate, orthoformate, and water.[142] Thus, co-emission of these gases with formaldehyde from the paraformaldehyde permeation tube may make gravimetric calibration impossible.

One of the simplest methods for generating a gaseous aldehyde is to use the headspace vapor of an aqueous solution of the aldehyde. This method has been used to generate acrolein for use in assessing molecular sieves as aldehyde adsorbents.[48] The method is especially applicable to the generation of gaseous formaldehyde standards. It must be noted that, because formaldehyde is almost entirely hydrated to methylene glycol, $CH_2(OH)_2$, in aqueous solution, it has a much lower vapor pressure than would otherwise be expected. The apparent Henry's law constant (2.77 torr/mol-fraction) for formaldehyde was determined in 1925 by Ledbury and Blair.[76]

Use of aqueous headspace vapor does not provide a primary standard directly. The gas must be standardized in a secondary manner—usually by measuring the amount of aldehyde lost from the solution. It is possible to assess the efficiency of a collection device by comparing the amount lost from a source solution with the amount of aldehyde trapped.

A second, related method for generating gas-phase aldehyde standards involves the slow addition of a dilute aqueous solution of an aldehyde to a gas stream in such a way that the aldehyde solution evaporates entirely. By knowing the rate at which the aldehyde is being added to the gas stream and the flow rate of the dilution gas, one can determine the aldehyde concentration in the gas stream. A device implementing this method was constructed with a syringe pump to inject the aldehyde solutions into a heated section of tubing through which the dilution gas was flowing.[77,146] The purity of the gas standards generated by this method depends on the purity of the liquid solutions. In the case of formaldehyde, again, it is desirable to use methanol-free formaldehyde solutions. Gases made this way will contain a great deal of water (as occurs with the headspace technique), which is undesirable in some cases. There may also be some decomposition of aldehyde.[77] As a secondary reference method, the technique must be used with caution.

A promising, although relatively unused, technique that has been used to generate low concentrations of aldehydes involves the thermal or catalytic decomposition of precursor compounds. In one study, formaldehyde was generated through the decomposition of a gas stream of S-trioxane (the cyclic trimer of formaldehyde) as it passed over a phosphoric acid-coated substrate (A. Gold, personal communication). In a second study, olefinic alcohols were thermally decomposed into a mixture of an aldehyde and an olefin (e.g., 3-methyl-3-butene-1-ol gives formaldehyde, 4-pentene-2-ol gives acetaldehyde, and 5-methyl-1,5-hexadiene-3-ol gives acrolein). The olefinic alcohol was introduced into the gas phase with a diffusion or permeation tube and is decomposed in a heated gold tube. Decomposition of the parent olefinic alcohol is virtually quantitative, so the technique generates a primary standard. It is also possible to use gas chromatography as a secondary reference method to analyze for the olefin produced in the reaction. When this method is used to generate standards for gas-chromatographic analytic techniques, the olefin can be used as an internal standard. One advantage of this method is that undesirable compounds are never handled in bulk, inasmuch as they are generated only in small amounts as they are used. This method has been used to generate standards of formaldehyde, acetaldehyde, and acrolein as low as a few parts per million.[136] Other thermal decompositions of precursor compounds have been used to obtain vinyl chloride and acrylonitrile.[43]

SAMPLING

An essential aspect of any analytic technique is the method of sampling. Choice of a method of sampling must be consistent with the information desired. Techniques that take an integrated sample over a long period can concentrate pollutants and simplify analysis. Such techniques are applicable when the determination of mean exposure is desired. Techniques that provide real-time measurements usually require sophisticated equipment, but may be required when it is desirable to observe concentration fluctuation during a short period. In the monitoring of compliance of pollutant concentrations with specific values set by a government agency, high precision is needed; in the study of trends, it is more important to have a reproducible method.

AIR

In the analysis of air pollutants, both direct and indirect sampling methods may be used. The direct method uses such instruments as infrared and microwave spectrophotometers, which are capable of measuring the concentrations of compounds in situ. Direct sampling techniques and direct investigative methods are discussed later in this chapter. When the compounds of interest are present in extremely low concentrations, thus precluding direct measurement, or when sampling sites are inaccessible to sophisticated instruments, indirect sampling techniques are commonly used.

Indirect sampling can consist merely of taking a representative grab sample. Air to be sampled is admitted into a previously evacuated vessel or pumped into a deflated bag. Inert materials--such as Teflon, Tedlar, and stainless steel--are used to construct grab-sampling containers. The sample is returned to a central laboratory and analyzed as though the measurement were being made in situ.

Grab sampling suffers from two defects. Because no preconcentration has been effected, the laboratory measurement technique must be sensitive enough to determine ambient concentrations directly. A more serious problem arises from the relatively long time that the low concentrations of the pollutants to be measured are in contact with the high surface area of the grab-sampling container. Nonspecific site adsorption occurs often, and a substantial fraction of the sample is lost. The container may develop a "memory" and give rise to spuriously high determinations on successive samples. Careful calibration and scrupulous analytic technique may minimize this latter defect.[31,98,122,146]

Preconcentration Sampling with Subsequent Analysis

A common indirect sampling technique involves preconcentrating the sample at the sampling site, e.g., by passing air through an absorbing liquid. There are two advantages. Preconcentration makes analysis in a laboratory easier, inasmuch as a higher detectability limit can be tolerated. And preconcentration often stabilizes the sample. In sampling for aldehydes, preconcentration techniques are almost always used.

As noted previously, preconcentration devices are generally used in sampling aldehydes in ambient air. Impingers are used most often for trapping low-molecular-weight aldehydes. Many types of impingers have been constructed to accommodate different sampling applications (Figure 6-1).

If the collection efficiency of the trapping solution is less than 100%, it is desirable to use more than one impinger in series. A typical arrangement for the sampling of formaldehyde (as recommended by NIOSH)[139] consists of two midget impingers in series, each containing 10 ml of water. The sample is collected at a flow rate of 1

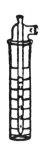

a. Midget Impinger.
   Ace Glass Co.

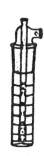

b. Midget Gas Bubbler
   (coarse frit).
   Ace Glass Co.

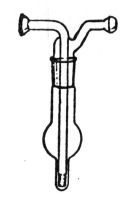

c. Nitrogen Dioxide Gas Bubbler.
   Ace Glass Co.

d. Spiral Type Absorber.
   American Society for Testing Materials:
   Tentative Methods of Sampling Atmospheres for Analysis of Gases and Vapors, Philadelphia, PA, July 24, 1956.

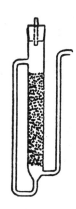

e. Packed Glass-Bead Column.
   American Society for Testing Materials:
   Tentative Methods of Sampling Atmospheres for Analysis of Gases and Vapors, Philadelphia, PA, July 24, 1956.

f. Midget Impinger.
   Lawrence Berkeley Laboratory.

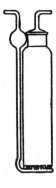

g. Bubbler Absorber with Diffuser.
   American Society for Testing Materials:
   Tentative Methods of Sampling Atmospheres for Analysis of Gases and Vapors, Philadelphia, PA, July 24, 1956.

FIGURE 6-1 Various types of impingers used to sample air. a-e and g reprinted with permission from Pagnotto;[98] f from C. D. Hollowell (personal communication).

standard liter per minute (slpm). The final solution is analyzed colorimetrically.

It is desirable to use an ice bath or a refrigerated sampler with impingers. Otherwise, low relative humidity or high ambient temperature may cause the impinger solution to evaporate, thus limiting the sampling time. The solubility and stability of the aldehyde in the trapping solution may also be adversely affected if impingers are not kept cold.

Figure 6-2 shows two designs for aldehyde samplers used by R.R. Miksch et al. (unpublished manuscript). The impinger sampling trains are contained in a small refrigerator. One sampler has a separate flow-control system that can sample air at a constant mass flow rate even when the pressure drop across the sampling train varies. The second sampler uses a critical orifice for flow control.

The absorbing solution used in the impinger depends on the aldehyde to be analyzed. In many cases, the solution contains a trapping reagent that is a constituent of the analytical procedure, thus simplifying operations. In general, there are two categories of trapping solutions for aldehydes. The first "category" is simply water. Formaldehyde reacts rapidly with water to form the relatively nonvolatile hydrate, methylene glycol. Methylene glycol does have a finite vapor pressure, however, and saturation may occur if sampling times are excessively long. This problem can be minimized by using two impingers in series. The collection efficiency of a single impinger containing water will decrease with time, but two impingers in series will maintain a collection efficiency of over than 95% for sampling times of over 48 h (Miksch et al., unpublished manuscript).

Water does not appear to be an especially good reagent for trapping higher-molecular-weight aldehydes, because the equilibria do not favor the formation of the hydrates.[16] To use aqueous bubblers to trap higher-molecular-weight aldehydes, an additional carbonyl scavenger must be present in the trapping solution. Carbonyl scavenger compounds constitute the second category of aldehyde-trapping solutions. The scavengers are chosen for their ability to react rapidly and quantitatively with carbonyl-containing compounds to form nonvolatile adducts. The reagents selected have included bisulfite, hydroxylamine, semicarbazone, and several phenylhydrazines, all of which have been shown to react extremely rapidly with aldehydes.[17] Table 6-1 shows the collection efficiency for different aldehydes of various trapping solutions that contain scavengers. The data are compiled from a number of sources and not always consistent, owing to the different experimental conditions used. The choice of a carbonyl-scavenger trapping agent depends on the analytical method to be used.

Higher-molecular-weight aldehydes also have been detected by means of solid adsorbents. The most widely used solid adsorbent is the porous polymer Tenax-GC, which has been used extensively to measure atmospheric organic compounds, including aldehydes, at low concentrations. In practice, the procedure is best suited for organics in the range C_6 to C_{12}. Pellizzari[103][104] has reported finding several aldehydes in ambient air with this method.

TABLE 6-1

Collection Efficiencies of Various Trapping Solutions for Aldehydes

Trapping Solution	Aldehyde	Collection Efficiency, %	Calibration Method[a]	Iced	Reference
Water	Formaldehyde	84	A	No	3, 8
		85	D	Yes	[b]
		80	E	No	139
1% aqueous bisulfite	Formaldehyde	100	B	Yes	146
		94	B	Yes	77
	Acetaldehyde	~100	B	Yes	77
	Propionaldehyde	100	B	Yes	146
		96	B	Yes	77
	n-Butanal	100	B	Yes	146
		~100	B	Yes	77
	i-Butanal	100	B	Yes	146
		98	B	Yes	77
	n-Pentanal	97	B	Yes	77
3-Methyl-2-benzothi- azolone hydrazine (MBTH)	Crotonaldehyde	92	B	Yes	77
	Formaldehyde	88	B	No	31, 115
		84	C	No	53
		92	E	No	92
	CH_2CHO	75	C	No	31
	Propionaldehyde	65	C	No	31
Chromotropic acid in concentrated sulfuric acid	Formaldehyde	99	A	No	8

139

Table 6-1 (continued)

Trapping Solution	Aldehyde	Collection Efficiency, %	Calibration Method[a]	Iced	Reference
Concentrated sulfuric acid	Formaldehyde	~99	A	No	8
Hydroxylamine	Mixed	>90	A	No	141
Dinitrophenylhydrazine (DNPH)	Mixed	~90	A	No	52
Girard-T	Mixed	Good	E	No	141
Ethanol	Acrolein	80-90	A	Yes	31
Ethanol and 4-hexyl-resorcinol	Acrolein	70-80	A	Yes	31

[a] Calibration methods: A, ratio of first impinger to total; B, syringe-pump evaporation of aldehyde solution; C, evaporation of aldehyde solution; D, use of headspace vapor over aldehyde solution; E, unknown.

[b] R. R. Miksch et al., unpublished manuscript.

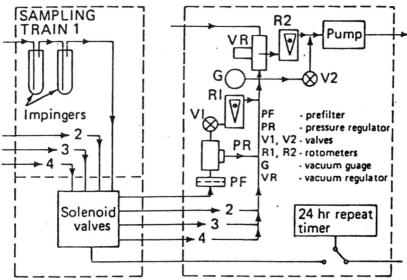

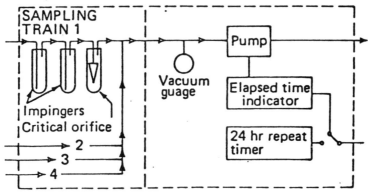

FIGURE 6-2 Sampling systems for sequential sampling of formaldehyde/aldehydes. Reprinted with permission from R. R. Miksch.

Other solid adsorbents also have been used. Molecular sieves have been used to capture low-molecular-weight aldehydes by physical entrapment. Samples can be desorbed with water for analysis by gas-chromatographic or colorimetric techniques. Formaldehyde, acetaldehyde, and acrolein have all been detected with molecular sieves, but quantitative data are available only on acrolein.[48] The solid adsorbents, charcoal and silica gel, also have been investigated, but the results have not been promising. It has been difficult to effect quantitative desorption of collected aldehydes.

It should be noted that a standard source of aldehyde gas is not required to estimate the collection efficiency of a given sampling device. Several devices can be placed in series and the fraction of the total sample collected in each device determined. This technique has been used to obtain collection efficiencies,[8,31] but the method is not necessarily reliable. It can be determined that a sampling device is unsatisfactory; in that event, the sample will be distributed throughout the system. However, the observation that no sample has gotten past the first trap does not guarantee that the collection has been quantitative, inasmuch as the sample may have decomposed or have been otherwise lost. The only reliable means for determining the collection efficiency of a sampling device is the use of a gas standard.

A sampling device of recent development that has not been applied to aldehydes is the passive monitor. The monitor consists of a diffusion tube containing a trapping agent at one end. The device is inexpensive and easy to use, expediting large-scale sampling. Palmes et al.[99] have been instrumental in developing the theory of passive monitors and successfully constructing a passive monitor for nitrogen dioxide.

Continuous Samplers

As stated earlier, there are direct investigative methods for determining the concentrations of compounds in situ, e.g., infrared and microwave spectroscopy. None of these methods has been rendered sufficiently portable to be used in field studies. The details of these methods and their potential future applications are discussed later in this chapter.

Several continuous analyzers based on wet chemical methods have been constructed.[27,84,150] These analyzers are intended to combine the best elements of direct and indirect sampling. Air is sampled via an impinger apparatus to generate a preconcentrated sample that, instead of being transported to a central laboratory for analysis, is analyzed in the field.

The continuous analyzer described by Yunghans and Munroe[150] and Cantor[27] is manufactured by Combustion Engineering Associates (CEA). The instrument can use the pararosaniline method to analyze for formaldehyde, or it can use the Purpald method to measure total aldehydes. One problem with this instrument is that it is not thermostated. The color-development rate of pararosaniline is temperature-sensitive (Lahmann and Jander[73] and Miksch et al., unpublished manuscript), and this may lead to erratic results. The impinger absorber coil is also sensitive to temperature fluctuations, because the collection efficiency of the absorbing solution and the amount that evaporates into the air stream depend on temperature. The mercury reagents used with the pararosaniline procedure are toxic. Finally, the recommended color-development time is too short to allow full color development that ensures maximal sensitivity and stability.

WATER

Sampling of water for analysis of aldehydes entails obtaining one or more representative grab samples. Because industrial effluents and water from natural bodies of water are not homogeneous, some investigators prefer to collect several subsamples and combine them for analysis. These subsamples are usually collected at various times and from different locations.

The preferred sample container is a glass jar with a Teflon-lined cap. Both jar and cap should be thoroughly cleaned with detergent and water and rinsed well with distilled water and, if an organic solvent is used for extraction, with the organic solvent. The volume of sample collected depends on the desired detection. Usually 1-2 L is sufficient if the desired detection exceeds 2 ppb, and the GC/MS technique is used after proper sample extraction and concentration.

If the time between collection and analysis is expected to be fairly long, the samples should be stored at 4°C or, preferably, kept frozen, to prevent biologic or chemical degradation of the aldehydes.

PLANT MATERIAL

A literature review of the last two decades reveals many variations in the preparation of samples, methods of extraction, and analytic techniques for measuring aldehydes in plant tissue.

Sample preparation has involved several procedures. Free-run juice of apple and grape have been concentrated 100 times and used for analysis.[41,132] Tomato fruit has been cored, quartered, and reduced to a slurry in a stainless-steel sampler before analysis.[107] Winter and Sundt[147] crushed plant tissue under nitrogen because of their evidence that 2-hexenal content of plant tissue varies with the oxygen concentration in the atmosphere at the time of crushing. Teranishi et al.[135] avoided crushing and processing and sampled aromas from fresh fruit directly.

Extraction of aldehydes from plant products has been accomplished either with solvents or with distillation. Purified isopentane has been used to extract aldehydes from apple and fruit juices; after the extract is dried, it is washed with propylene glycol to remove alcohols; the remaining oil is ready for aldehyde analysis.[132] Alternatively, steam distillation has been used to recover aldehydes: Major et al.[85] steam-distilled fresh Ginkgo biloba leaves, collected the distillate, and extracted it with ether. Tomato was similarly distilled and extracted with diethyl ether and dried over anhydrous sodium sulfate. Winter distilled strawberry fruit under nitrogen to avoid the oxidation of unsaturated fatty acids that are the precursors of aldehydes.

WET-CHEMISTRY SPECTROPHOTOMETRIC ANALYSIS

Wet-chemistry spectrophotometric methods of analysis for aldehydes continue to be the most popular and widely used. The sensitivity associated with the formation of a dye chromophore and the ease of measurement with readily available spectrophotometers are not easily matched by other techniques. Field samples can usually be easily generated with simple equipment. However, spectrophotometric techniques are subject to error. The specificity and degree of completion of the chromophore-forming reaction must be considered, as well as the stability and standardization of reagents. In many cases, spectrophotometric techniques are slower than more direct measurement methods.

To sample air, wet-chemistry spectrophotometric methods are often applied to preconcentrated samples that are generated with impingers. It is often overlooked that the detection limit for aldehydes in air depends on both the sensitivity of the analytical method and the degree of preconcentration. If the time or flow rate is changed in sampling with impingers, the detection limit can be changed radically. Typically, aldehydes in air are sampled for 0.5-8 h at flow rates of 0.5-2.0 L/min.

FORMALDEHYDE

To date, only spectrophotometric techniques have been applied in field studies of formaldehyde. Table 6-2 lists a variety of spectrophotometric techniques that can be used to analyze formaldehyde. The most widely used methods have been based on chromotropic acid, as tentatively recommended both in NIOSH[139] and in American Public Health Association Intersociety Committee.[10] Pararosaniline has been the next most popular reagent and may have some advantages over chromotropic acid. The remaining reagents have not been widely used. Some are inappropriate for field sampling, and others have not been adequately tested.

Chromotropic Acid

Ever since Eegriwe[37] described the use of chromotropic acid in a spot-test method for the detection of formaldehyde, there has been widespread interest in using this reagent for spectrophotometric determination of formaldehyde. As stated above, a tentative method using this reagent has been suggested by both NIOSH[139] and the

TABLE 6-2

Summary of Spectrophotometric Methods for Formaldehyde Analysis in Ambient Air

Method	Minimal Detectable Concentration: μg/ml	Minimal Detectable Concentration: ppm (est.)[a]	λ_{max}, nm	Interferences	Reference
Chromotropic acid	0.25	0.1	580	Nitrogen dioxide, alkenes, acrolein, acetaldehyde, phenol, formaldehyde precursors	8,139
	0.1	0.04	580		
Pararosaniline	0.1	0.04	570	Sulfur dioxide, cyanide	b
	0.1	0.05	560	Virtually specific	84
Phenylhydrazine	1.1	0.4	520	Color not stable	88,135
J-acid (7-amino-4-hydroxy-2-naphthalene-sulfonic acid)	0.3	0.1	468	Formaldehyde precursors	116
Phenyl-J-acid	0.4	0.13	660	Formaldehyde precursors	116
p-Phenylenediamine	1.7	0.5	485	Sulfur dioxide, aliphatic aldehydes	12
Tryptophan	0.15	0.06	575	Virtually specific	30
MBTH	0.05	0.015	628	Higher aliphatic aldehydes	117
Purpald (4-amino-3-hydrazino-5-mercapto-1,2,4-triazole)	0.15	0.05	549	Higher aldehydes	62
Acetylacetone	1.4	0.5	412	Specific (?)	19

[a]Based on sampling at 1 slpm for 1 h into 25 ml of impinger solution and recording a difference of 0.05 absorben unit between blank and sample.

[b]R. R. Miksch *et al.*, unpublished manuscript.

Intersociety Committee[10] for determining formaldehyde concentration in occupational environments.

The chromotropic acid method suggested by NIOSH and the Intersociety Committee involves the collection of samples by passage of air through two midget impingers in series, each of which contains 20 ml of distilled water. When a suitable volume of air has been sampled (1 h of sampling at 1 slpm), the contents of the midget impingers are analyzed separately. For analysis, the contents of an impinger are diluted quantitatively to a known volume. With 1% chromotropic acid, an aliquot of the sample is brought up to 0.025% chromotropic acid. Concentrated sulfuric acid is then added at 3 parts acid to 2 parts sample. The heat of mixing develops the color; after cooling of the sample, the absorbance is read at 580 nm (extinction coefficient, ϵ, 8.9×10^3).

The reported sensitivity of the method is 0.1 μg/ml of color-developed solution, which corresponds to formaldehyde at approximately 0.04 ppm in the sampled air (see Table 6-2). Acrolein is reported to be a positive interference at few percent. Ethanol, higher-molecular-weight alcohols, and phenols can be negative interferences, but at concentrations not normally encountered in the atmosphere. Olefins in tenfold excess over formaldehyde can be negative interferences at approximately 10%. Aromatic hydrocarbons also constitute a negative interference. With the exception of olefins, the interferences listed are not likely to be encountered in substantial concentrations during atmospheric sampling. And even in the case of olefins, the interference is not serious.[129]

Early work by Altshuller et al.[8] indicated that nitrogen dioxide did not interfere. However, the same cannot be said for nitrite and nitrate. Indeed, there is a chromotropic acid assay for nitrate similar to the formaldehyde assay.[144] Cares[28] was the first to investigate systematically the nature of the nitrite and nitrate interference and methods of eliminating it. She found that both interfered with formaldehyde analysis--nitrite slightly more than nitrate. When they were present in tenfold molar excess, negative interferences of 60% and 30%, respectively were observed. Later work by Krug and Hirt[72] confirmed these findings. To overcome these interferences, Cares recommended a modified procedure that uses a solution of sodium bisulfite for sampling. This solution is neutralized and heated to reduce the oxides of nitrogen to nitric oxide, which outgasses from the solution. The sample is then analyzed as before, with chromotropic acid and sulfuric acid. This procedure has not been used in field studies, probably because of its complexity.

Oxides of nitrogen can probably interfere with analysis for formaldehyde with chromotropic acid. There is evidence that a major sink for NO_x in the atmosphere involves its transformation to nitric acid (or its subsequent transformation to nitrate-containing aerosols) by way of OH attached on nitrogen dioxide. Furthermore, nitrogen dioxide can be converted to nitrite and nitrate in the presence of water or sulfuric acid,[33] integral constituents of the analytic method.

It is not clear that the tentatively suggested method is optimized. Bricker and Johnson,[25] who originally developed a procedure using chromotropic acid, reported that full color development depended on heating of the reaction mixture for 30 min. Later work by West and Sen[145] and Altshuller et al.[8] suggested that the heat generated by the mixing of concentrated sulfuric acid with the sample solution was sufficient to drive the color-development reaction to completion. This conclusion is open to question, inasmuch as the peak temperature and duration of heating could be affected by the choice of reaction vessel and by the ambient temperature. Bricker and Johnson[25] also reported that the sulfuric acid concentration should be at least 67% for maximal color development. West and Sen,[145] however, reported that color development increased strongly with increasing sulfuric acid concentration until a value of 85% was reached, after which the dependence lessened. This finding was acknowledged by Altshuller et al.,[8] who went so far as to recommend that samples be collected with impingers containing chromotropic acid in concentrated sulfuric acid. Later simplex optimization work by Olansky and Deming[96] indicated that color development is maximal at 57% and declines at higher values.

In sum, it seems that the chromotropic acid method suffers from several deficiencies. It is not clear that the procedure is optimized for maximal sensitivity; the method suffers from interferences by a number of substances, some of which will undoubtedly be encountered during field sampling; and modifications designed to reduce these interferences introduce additional complexities.

Pararosaniline

A second reagent used for the measurement of formaldehyde concentrations is pararosaniline, which was first introduced in the form of a spot test by Schiff (1866).[120] In the classical Schiff test for aldehydes, the intense pink color of basic fuchsin is bleached with sulfur dioxide in basic solution. When an aldehyde is added to the solution, it reverses the bleaching process, and the basic fuchsin color returns. This spot test is neither quantitative nor formaldehyde-specific.

Lyles, Dowling, and Blanchard[84] were the first to develop a pararosaniline technique that produced a stable color and reproducible results. The technique is as follows. Samples are generated by passing air through a midget impinger containing distilled water. A reagent solution containing 0.05 M tetrachloromercurate II and 0.025% sodium sulfite is mixed with the sample in a ratio of 1 to 10. A second reagent solution, prepared by dissolving 0.16 g of pararosaniline and 24 ml of concentrated hydrochloric acid in water sufficient to total 0.1 L, is added to the sample in a ratio of 1 to 11. After 15 min, the absorbence is read at 560 nm.

Several aspects of this analysis require comment. Lyles et al. took note of earlier work[95][143] and were careful to use pure pararosaniline in place of basic fuchsin reagent. The latter is often contaminated with pararosaniline and is difficult to purify. Earlier problems with reagent stability and reproducibility may have resulted from insufficient purity.

The use of tetrachloromercurate II follows the work of West and Gaeke,[143] who used pararosaniline in conjunction with formaldehyde to determine sulfur dioxide. West and Gaeke sampled atmospheric sulfur dioxide by bubbling air through a solution of sodium tetrachloromercurate II. The sulfur dioxide was trapped and stabilized as a dichlorosulfitomercurate II complex, which then reacted with acidic pararosaniline and formaldehyde.

The pararosaniline method developed by Lyles et al.[84] is substantially the same as that used by the Combustion Engineering Associates (CEA) 555 continuous analyzer. The latter is used by many industrial hygienists to determine formaldehyde in workplace environments. Its primary virtue is its ability to give nearly real-time measurements.

Recent work has led to further refinements in the pararosaniline technique. Miksch et al. took note of the work of Lahmann and Jander,[73] German workers who investigated the dependence of the technique of Lyles et al. on each of the reagents used. In particular, the stability and sensitivity of the method could be markedly improved through a fivefold reduction in the sodium sulfite concentration. Substantial temperature effects on both stability and time of development of the color were also noted.

In the same study, Miksch et al. questioned the use of tetrachloromercurate II. Because the original role of this reagent had been to stabilize the sulfur dioxide collected in the procedure of West and Gaeke,[143] its function during formaldehyde determinations was not clear. Investigation revealed that reversing the order of addition of the reagents permitted the hazardous mercury reagent to be eliminated. No metal ion at all was found to be necessary.

The procedure developed by Miksch et al. is as follows. Samples are collected in impingers containing deionized distilled water. The samples are collected, shipped back, and stored at 5°C to enhance sample stability before analysis. In the laboratory, the contents of two impingers operated in series are pooled, and the solution is diluted to a known volume. A reagent solution, prepared by dissolving 0.16 g of pararosaniline and 20 ml of concentrated hydrochloric acid in water sufficient to total 100 ml, is added to an aliquot of the sample in a ratio of 1 to 10. After 10 min, a second addition of 0.1% sodium sulfite solution is added to the sample in a ratio of 1 to 11. The reaction vessels are capped (to prevent outgassing of sulfur dioxide), and the color is allowed to develop for 1 h. The absorbance is then determined at 570 nm (extinction coefficient, 1.88×10^4).

The procedure is specific for formaldehyde. Only sulfur dioxide, an integral part of the procedure in the form of sulfite, constitutes a potential interference. This interference can be largely removed by basifying the impinger solutions with 1 or 2 drops of 1 N sodium hydroxide before analysis to destroy any formaldehyde-sulfur dioxide adduct. This allows ambient concentrations of sulfur dioxide up to 500 ppb—higher than normally encountered—to be tolerated. Miksch et al. have reliably used the pararosaniline procedure in measuring several thousand indoor and outdoor air samples.

Acetylacetone

A very sensitive fluorimetric method for the determination of formaldehyde is based on the Hantzsch reaction between acetylacetone (2,4-pentanedione), ammonia, and formaldehyde to form 3,5-diacetyl-1,4-dihydrolutidine. The reagent was first used in a colorimetric procedure by Nash,[94] who also reported that the adduct fluoresced. Belman[19] developed a fluorimetric procedure based on this property.

The procedure of Belman[19] is as follows: Equal volumes of formaldehyde solution and a reagent consisting of 2 M ammonium acetate and 0.02 M acetylacetone (pH, 6) are mixed and incubated at 37°C for 1 h. After cooling to room temperature, the fluorescence is determined (λ_{excite} = 410 nm, λ_{emit} = 510 nm). The standard curve is linear with formaldehyde from 0.005 µg/ml to about 0.4 µg/ml and deviates slightly from linearity from 0.4 µg/ml to 1.0 µg/ml. Above 1.0 µg/ml, the formaldehyde can be determined colorimetrically.

This method has been particularly chosen by the wood industry in determining emission from particleboard and plywood.[20,90] Under controlled conditions in specially designed chambers, the formaldehyde content of headspace vapor over materials being examined is determined. This test is being considered for promulgation as a European standard.[113] Acetylacetone has not been used for sampling for formaldehyde in ambient air. In this application, possible interference by oxides of nitrogen, sulfur dioxide, and ozone must be considered.[90]

Other Methods

It has already been mentioned that there are a fairly large number of spectrophotometric methods for the determination of formaldehyde, in addition to the two discussed above. In general, these methods either have not been fully evaluated or suffer from major defects. Several alternative wet-chemistry spectrophotometric methods of analysis are listed in Table 6-2. Closely analogous methods, based on spectrofluorometry, have also been suggested, as shown in Table 6-3. One final analogous method deserving serious consideration is based on chemiluminescence. All these methods are discussed below.

An older reagent that has been considered as a candidate for the colorimetric determination of formaldehyde is phenylhydrazine. Reaction of this reagent with formaldehyde, followed by oxidation of the adduct with ferricyanide, leads to the formation of an anionic species absorbing at 512 nm.[88] The essential drawback encountered is that color is not stable and fades with time. Under some procedural conditions, aliphatic aldehydes interfere.[134] Other possible interferences have not been investigated.

TABLE 6-3

Summary of Spectrofluorometric Methods of Analysis for Formaldehyde

Method	Minimal Detectable Concentration µg/ml	ppm[a]	λ excite, nm	λ emission, nm	Interferences	Reference
1,3-Cyclohexanedione	0.2	0.07	395	460	Higher aldehydes	114
Dimedone (5,5-dimethyl-cyclohexanedione-1,3)	0.08	0.03	470	520	Higher aldehydes	114
J-acid (7-amino-4-hydroxy-2-naphthalene-sulfonic acid)	0.6	0.1	470	520	Aldehyde precursors, acrolein	114
J-acid	0.05	0.02	470	520	Aldehyde precursors, acrolein	116
Acetylacetone	1.2	0.4	410	510	Specific (?)	114
Acetylacetone	0.005	0.002	410	510	Specific (?)	19

[a]Based on sampling at 1 slpm for 1 h into 25 ml of impinger solution and recording a difference of 0.05 absorben unit between blank and sample.

A reagent similar to chromotropic acid in both its structure and its associated analytic technique is 7-amino-4-hydroxy-2-naphthalenesulfonic acid (J-acid).[116] The adduct formed is fluorescent, and a second, more sensitive, technique that takes advantage of this property has been developed.[118] Formaldehyde precursors interfere under the harsh conditions of both these techniques, and acrolein also interferes with the second technique. Other possible interferences have not been adequately investigated. The reagent phenyl-J-acid is a minor modification of J-acid.[116]

Two other reagents must be mentioned as potential candidates for the wet-chemistry determination of formaldehyde, although they have not been adequately tested. The reagent phenylenediamine may be oxidized by hydrogen peroxide to produce Bandrowski's base, 3,6-bis(4-aminophenylimino)cyclohexa-1,4-diene-1,4-diamine (λ_{max}, 485 nm). The reaction is catalyzed by formaldehyde[12] and may form the basis for an analytic procedure. At present, only sulfur dioxide in 100-fold excess is known to interfere. The second reagent is tryptophan, which reacts with formaldehyde in the presence of concentrated sulfuric acid and iron to give a colored species.[30] The reaction was found to be extremely sensitive and free of interference from a wide range of compounds, but its suitability as a field sampling method has not been tested. The instability of some of the reagents used may present a problem.

Two final reagents have occasionally been used to assay for formaldehyde: 3-methyl-2-benzothiazolone hydrazone (MBTH) and 4-amino-3-hydrazino-5-mercapto-1,2,4-triazole (Purpald). They are specific only for the class of aliphatic aldehydes as a whole, and precautions must be taken to ensure that separate formaldehyde is the only aldehyde present. These reagents are discussed more fully in the next section.

Several workers have attempted to develop fluorometric methods of analysis for the determination of formaldehyde. The better known examples are shown in Table 6-3. In general, the techniques are sensitive to the design of the instrument--note the different sensitivities reported for the same reagent at different times. The later work actually shows reduced sensitivity. Problems common to many fluorescence techniques are susceptibility to sample matrix variations and nonlinear standard curves. With the exception of acetylacetone, none of the reagents shown has been used in reported studies.

A final method deserving serious consideration is based on a chemiluminescent reaction of formaldehyde and gallic acid in the presence of alkaline peroxide.[128] In a flow system where the reagents can be mixed immediately before passage into an optical cell, formaldehyde concentrations as low as 3.0 ng/ml can be detected--an increase in sensitivity of more than an order of magnitude relative to the colorimetric procedures just described. A second distinct advantage is that the working linear range of response extends over five orders of magnitude.

The chemiluminescence method may not be completely formaldehyde-specific. Acetaldehyde was reported to give a response that was less than one-tenth that of formaldehyde. Other aldehydes were not tested. Two dicarbonyl compounds, glyoxal and methylglyoxal, gave a response equal in magnitude to that of formaldehyde.[128] These compounds would not normally be encountered, except perhaps in biologic samples.

Proper design of the flow system and optical cell are essential to the chemiluminescence method. With proper design, the apparatus can be inexpensive. The method is best suited to analyzing aqueous impinger solutions at a central laboratory or to continuous monitoring at selected stationary sites (C.D. Hollowell, personal communication).

TOTAL ALIPHATIC ALDEHYDES

Measurements of total aliphatic aldehydes are based on chemical reaction behavior imparted by the presence of the formyl group common to all aldehydes. As with formaldehyde, only wet-chemistry spectrophotometric techniques have been used for sampling total aliphatic aldehydes under field conditions. The application of more sophisticated instrumental techniques to the determination of total aliphatic aldehydes is inadvisable, because it is usually easier and more desirable to identify and measure each specific aldehyde separately.

3-Methyl-2-benzothiazolone Hydrazone

By far the most commonly used reagent for the determination of total aliphatic aldehydes is MBTH. First introduced by Sawicki et al.,[117] this reagent has been used for measuring lower-molecular-weight aliphatic aldehydes in auto exhaust and urban atmospheres (see Table 6-2).

A tentative method using MBTH for determining aldehydes in ambient air was given by the Intersociety Committee.[10] The method is as follows. Air to be sampled is bubbled through 0.05% aqueous MBTH contained in a midget impinger. After dilution to a known volume, an aliquot of an oxidizing reagent containing sulfamic acid and ferric chloride is added. After 12 min, the absorbence is read at 628 nm. At the recommended sampling rate of 0.5 slpm, assuming a minimal detectable absorbence change of 0.05 unit, a concentration of 0.03 ppm could be determined after sampling air for 1 h.

The original method of Sawicki et al.[117] used ferric chloride alone as the oxidizing reagent. Because of turbidity, acetone was incorporated into the dilution scheme. Hauser and Cummins[53] effectively eliminated the turbidity by adding sulfamic acid to the oxidizing reagent. The molar absorptivities of the aldehydic adducts formed vary between approximately 48,000 and 56,000. The formaldehyde adduct has a molar absorptivity of 65,000. Altshuller et al.[3] recommended that concentrations of aldehydes determined by MBTH should be multiplied by a factor of 1.25 to account for the difference in response between formaldehyde and the remaining aliphatic aldehydes. The recommendation has not been followed in reported uses of MBTH.

Many classes of compounds, particularly those containing nitrogen, react with MBTH to give colored products. Most of these compounds are not encountered during atmospheric sampling. Nitrogen dioxide has been reported to interfere through formation of nitrite and nitrate in water.

Purpald

A reagent recently developed for the determination of aliphatic aldehydes is Purpald. First described by Dickinson and Jacobsen,[35] the reagent can be used quantitatively as follows.[62] A basic solution of Purpald is added to aqueous samples containing formaldehyde. The mixture is aerated for 30 min to ensure oxidation, and the absorbence is determined at 549 nm. Assuming that impingers are used for sampling air at a rate of 1 slpm for 1 h and that the minimal detectable absorbence difference is 0.05 unit, a concentration of 0.04 ppm can be detected. Purpald suffers from the same drawback as MBTH: it gives different responses to different aldehydes. Potential interfering substances encountered in atmospheric sampling have not been completely examined, but no interference from a wide variety of test compounds was noted by the originators.[35]

Other Methods

2,4-Dinitrophenylhydrazine (DNPH) has received considerable attention as a reagent for determining aldehyde concentrations. The vast majority of DNPH techniques attempt to separate and identify the individual aldehydic adducts through the use of thin-layer chromatography, gas chromatography, or high-performance liquid chromatography. Wet-chemistry spectrophotometric procedures are based on the formation of a chromogen absorbing at 440 nm.[100][105] These procedures have been hampered by the interference of ketones and problems with reagent stability. The minimal detectable concentration of aldehydes with these procedures is about 0.2 ppm.

A method deserving mention is the bisulfite method published by the Los Angeles Air Pollution Control District.[74] Air is sampled with impingers containing aqueous bisulfite. The aldehydes react to form aldehyde- bisulfite adducts. The excess bisulfite is destroyed, and the solution is basified to liberate the bisulfite bound in the adducts. The freed sulfite is titrated with iodine and starch. The method is cumbersome, the adducts are not stable for long periods even if kept on ice, and the iodine reagent is sensitive to air and light.

ACROLEIN

Acrolein is a highly toxic aldehyde; a threshold limit value (TLV) of 0.1 ppm, has been established by the Occupational Health and Safety Administration (OSHA).[140] This standard is 30-fold lower than the corresponding TLV for formaldehyde. Acrolein is the only aldehyde other than formaldehyde for which there is a specific wet-chemistry spectrophotometric method of analysis.

4-Hexylresorcinol

The most popular method for determining acrolein in air uses 4-hexylresorcinol.[31,68,139] Air is typically drawn through two midget impingers at 1 slpm to collect the sample. The collecting solution can be either 1% sodium bisulfite or a reagent containing 4-hexylresorcinol, mercuric chloride, and trichloroacetic acid in ethanol. Samples collected in bisulfite are analyzed by adding 4-hexylresorcinol and mercuric chloride in ethanol and then a solution of trichloroacetic acid in ethanol. The solution is heated for 15 min at 60°C, and the resulting color is measured at 605 nm. Samples collected in 4-hexylresorcinol are analyzed simply by heating and measuring the color. For field sampling, the simplicity of the latter method is offset by the hazards of handling the toxic and corrosive reagent. In addition, the reagent and the samples collected are not very stable, and samples must be analyzed within a few hours. The bisulfite method is somewhat more complex, but it is safer to use. Besides using a less hazardous collecting solution, this method produces samples that are stable for up to 48 h if they are kept refrigerated, thus permitting later analysis at a central laboratory.

A recent paper by Hemenway et al.[54] pointed out a potential flaw in the 4-hexylresorcinol method given by NIOSH.[139] Apparently, the order of addition of reagents for analysis differs between field samples and calibrating solutions, and this may lead to underestimation by as much as 35%. The validity of this objection needs to be established.

Other Methods

A very sensitive procedure for the determination of acrolein is a fluorimetric method using m-aminophenol.[1] The procedure can detect acrolein at concentrations as low as 10 ng/ml in an aqueous test solution. No interference is reported for many alcohols, amines, amino acids, and polyamines. Other aldehydes do not interfere unless they have a double bond in conjugation with the formyl group analogous to acrolein (e.g., crotonaldehyde). The procedure shows promise, but has not been applied to environmental samples.

Chromotropic acid has also been suggested for acrolein determinations.[49] In the formaldehyde procedure using chromotropic acid, the response to acrolein is regarded as an interference;

however, the absorbance maximums are sufficiently different that it is possible to measure the two compounds separately. A method for the simultaneous determination of formaldehyde and acrolein in air has been proposed by Szelejewska.[133] Samples are collected in bisulfite at 1 slpm before the addition of chromotropic acid in sulfuric acid. After the addition of the chromotropic acid reagent, the absorbance is measured at both 420 and 575 nm. The two absorbances are fitted to a linear system of two equations that, when solved, gives the concentrations of the two aldehydes. This method has not been used in practice.

Two other reagents for the analysis of acrolein have been discussed by Cohen and Altshuller.[31] The first, phloroglucinol, reacts with acrolein to produce a red color. The reaction is subject to interference from formaldehyde and oxides of nitrogen and has not been used. The second, tryptophan, reacts with acrolein in acid solution to produce a purple color. The sensitivity of the tryptophan method is only one-fourth that of the 4-hexylresorcinol method. Furthermore, in view of the fact that a similar method has recently been used to determine formaldehyde,[30] the reagent may be subject to interference from formaldehyde.

ACETALDEHYDE

The only colorimetric method reported to be specific for acetaldehyde uses diazobenzene sulfonic acid.[142] Unfortunately, no data are available on the sensitivity or interferences associated with this method. Some efforts have been made to take advantage of the rather high volatility of acetaldehyde, in separating it by distillation from other aldehydes. Procedures that use this method are too cumbersome to be reliable. A method has recently been published for determining acetaldehyde in the presence of formaldehyde in biologic materials.[34] Acetylacetone reacts with the solution, which eliminates the formaldehyde, and then the acetaldehyde is analyzed with p-phenyl phenol. This method does not take into account interferences from higher aldehydes; it is not actually a procedure for acetaldehyde, but rather for nonformaldehyde aldehydes. The only methods available for determining acetaldehyde involve the separation of all the aldehydes that are present with gas or liquid chromatography.

OTHER ANALYTICAL METHODS

SPECTROSCOPIC METHODS

Microwave, infrared, and laser-fluorescence spectroscopy have all been used to measure aldehyde concentrations in ambient air directly. Each of the methods is prohibitively expensive for ordinary field sampling. The instrumentation required is often cumbersome and delicate, is seldom portable, and requires sophisticated maintenance and support facilities.

Microwave rotational spectroscopy can measure low concentrations of many compounds in gas-phase samples. Rotational resonances are very sharp at microwave frequencies and low partial pressures, so sample spectra can be easily resolved. Formaldehyde has been monitored continuously at concentrations as low as 10 ppb in air with a two-stage membrane separator for preconcentration.[57] Acetaldehyde has been detected directly at 15 ppm.[58] This is far above normal concentrations for ambient air, and the technique is not routinely applicable to ambient-air analysis. Microwave spectroscopy has also been used to determine acrolein, acetaldehyde, and formaldehyde in tobacco smoke.[66] The sensitivity of the technique was reported to be 2 ppm, but, again, this concentration is rather high and would not normally be encountered in ambient air. Furthermore, the response time of the instrument is long, rendering the technique insensitive to changes in concentrations.

Infrared spectroscopy is promising, owing to the sharpness of the rotational and vibrational peaks observed for gas-phase samples. Unfortunately, good spectral resolution (less than 0.1 cm^{-1}) and rapid measurements are hampered by the low power of infrared sources. To overcome this difficulty, Fourier-transform infrared (FTIR) methods have been developed in which conventional Fourier-transform methods are used to derive the absorption bands. FTIR instruments are commercially available, but are exceedingly expensive. In one application, formaldehyde was continuously monitored at ambient concentrations of less than 10 ppb with an FTIR system.[137] The system was used with a Michelson infrared interferometer with a sophisticated multiple-reflection optical cell whose pathlength was 2 km. Longer pathlengths could not be obtained, because of image overlap. Other aldehydes were not measured.

A fluorescence procedure based on the direct excitation of formaldehyde by a dye laser has been reported.[15] Formaldehyde as low as 50 ppb in air could be detected. The authors suggested that further refinements would increase the sensitivity. The application of this technique to other aldehydes is restricted by the weaker and less well-resolved absorption spectra in the accessible spectral region.

CHROMATOGRAPHIC METHODS

Three chromatographic techniques have been applied to the analysis of aldehydes: gas chromatography, liquid chromatography, and ion chromatography. Gas-chromatographic analysis of aldehydes generally takes one of two forms: direct analysis by gas or solution injection and derivatization followed by analysis.

Formaldehyde presents special problems with respect to direct analysis by injection. In a flame ionization detector (FID), a universal detector widely used for quantitative work, formaldehyde decomposes and gives a very small response. Thermal conductivity detectors (TCDs) are less sensitive and respond only to very high concentrations of formaldehyde. An electron capture detector (ECD) has a limited linear response range and is sensitive only to conjugated carbonyl groups. The photoionization detector (PID) is reported to be sensitive to formaldehyde (HNU Company, Newton Upper Falls, Mass.), but appears to have some drawbacks. Specifically, a high-energy lamp is required to detect formaldehyde; this drastically reduces both the selectivity and the lifetime of the detector.[36]

In principle, it is possible to circumvent the insensitivity of the FID to formaldehyde by catalytically reducing formaldehyde to easily detectable methane.[32] Because hydrogen is required for the operation of the FID, the reduction is easily achieved by passing a mixture of the column effluent and hydrogen gas over a short bed of catalyst before introduction into the FID. Deposits of nickel, thorium, and ruthenium on fine-mesh glass beads have all been used successfully to reduce formaldehyde to methane. The lack of success in applying the technique to routine analysis of formaldehyde can be attributed to problems in choosing proper gas-chromatographic conditions. Apparently, it is difficult to pass formaldehyde through any of a variety of column-packing materials quantitatively.[26]

With the exception of formaldehyde, aldehydes may be analyzed by direct gas injection if concentrations are high enough. By using a six-port valve equipped with a 1-ml gas-sampling loop, aldehydes can be routinely detected with an FID at concentrations as low as 0.03 ppm without preconcentration (Analytical Instrument Development, Inc., Avondale, Pa.). It is important to recall, however, that gas chromatography excels at separation, but provides minimal identification. Ambient-air samples often contain hydrocarbons, and their responses may overlap and obscure the aldehydic responses. Bellar and Sigsby[18] reported a complex automated gas chromatographic technique to analyze for C_2-C_5 aldehydes that avoided this problem. Hydrocarbons and aldehydes from an air sample flowed onto a polar cutter column, where the aldehydes were retained as the hydrocarbons were passed through and vented. The cutter column was then backflushed to a cryogenic trap, where the aldehydes were reconcentrated before introduction onto an analytic column. About an hour was required for a complete analysis. The method has not been used by other workers.

Preconcentration before direct analysis has also been investigated. Pellizzari[103][104] has reported the collection of some higher-molecular-weight aldehydes on Tenax-GC. After thermal desorption and reconcentration in a cryogenic trap, analysis is performed by gas chromatography/mass spectrometry. The method provided poor quantification. Gold et al.[48] successfully captured acrolein on molecular sieves. The sieves were desorbed with water, which was then injected onto a column packed with hydrophobic Tenax-GC. The method has not been used by other workers.

Derivatization is an alternative technique that has been extensively investigated. Levaggi and Feldstein[78] introduced a method in which samples were collected with impingers containing 1% sodium bisulfite solution. Aldehydes react with the bisulfite to form adducts. Formaldehyde and acrolein are analyzed by chromotropic acid and 4-hexylresorcinol methods, respectively. To analyze the remaining aldehyde, the bisulfite solution is injected onto a packed column in a gas chromatograph. Samples must be kept cold to prevent deterioration. The Intersociety Committee[10] has adapted the technique as a tentative method for the C_1-C_5 aldehydes, but there are no reported uses in the literature. A problem not explicitly discussed is the rapid degradation of column performance due to the in situ production of sulfur dioxide and sodium hydroxide as the adduct thermally decomposes.

Much work has been aimed at using 2,4-dinitrophenylhydrazone (DNPH) derivatives of aldehydes, well known for many years for their use in the qualitative identification of aldehydes. DNPH reacts with aldehydes in aqueous solution to form precipitates. In most attempts, this precipitate is redissolved in an organic solvent, which is then injected into a gas chromatograph. The resulting chromatograms show double peaks for each derivative corresponding to the syn- and anti-isomers formed around the nitrogen-carbon double bond characteristic of the derivative. These peaks are not symmetrical, because of steric influences during formation of the derivative. The peaks observed for the derivatives of propionaldehyde, acrolein, and acetone overlap and are difficult to separate. The most consistent problem is the verification that quantitative derivatization of the available aldehydes has occurred.

DNPH was applied by Hoshika and Takata[55] to the analysis of automobile exhaust and cigarette smoke. Papa and Turner[101,102] also applied it to automobile exhaust. In a two-step process, preliminary separation of DNPH aldehyde derivatives by preparative gas chromatography was followed by analytic gas chromatography.[131] In analyzing food samples, a number of workers have used glutaric acid and flash-exchange gas chromatography to regenerate free aldehydes from DNPH derivatives.[50,51,65,67,80,109]

A variety of alternative derivatizing reagents have been investigated. Gas chromatography of aldehydic derivatives of phenylhydrazine[71] and 2,4,6-trichlorophenyl hydrazine[64] has been studied. These reagents are close analogues of DNPH. The aldehydic derivatives of dimethylhydrazine,[63] hydroxylamine,[141] and tetramethyl ammonium acetyl hydrazide (Girard-T Reagent)[45,46,97] have been analyzed with gas chromatography. Like DNPH, these reagents all involve reaction of a free amine with the formyl group to form a nitrogen-carbon double-bonded derivative.

Direct analysis of aldehydes with high-performance liquid chromatography (HPLC) has not been thoroughly investigated, primarily because of the lack of a detector with sufficient sensitivity. To circumvent this problem, aldehydes can be made to react with DNPH to form a derivative with a strong ultraviolet-absorption spectrum. This approach has been investigated by Carey and Persinger,[29] Mansfield et al.,[87] Selim,[126] and others.

Ion chromatography is a new technique (DIONEX, Inc., Sunnyvale, Calif.) that has application to formaldehyde analysis. It combines liquid chromatography with an ion-exchange column to separate charged species. A conductivity detector provides excellent sensitivity. Formaldehyde is captured on specially impregnated charcoal and then desorbed with aqueous peroxide. The resulting formate ion can then be analyzed by ion chromatography. Two major difficulties with the method are ensuring quantitative recovery of formaldehyde from the charcoal and preventing the peroxide reagent from oxidizing other materials to formate ion.

ELECTROCHEMICAL METHODS

In addition to the usual techniques of analyzing organic materials, aldehydes can be analyzed by electrochemical methods. Both polarographic methods and amperometric titrations have been used.

Lupton and Lynch[83] developed polarographic techniques for the analysis of aldehydes in a wide range of samples. McLean and Holland[89] adapted their technique to a portable system for rapid analysis of aldehydes in automotive exhaust sampled by bubbling into water. The polarograph was rendered portable by replacing the dropping mercury electrode with a quiescent mercury pool a few millimeters in diameter. Analysis used the method of standard additions. The procedure is not specific, however, even for aldehydes. The authors suggested differential-pulse polarography for separation of the aldehydes, but this has not been tested.

Ikeda[61] developed a short-circuit argentometric amperometric titration for determining formaldehyde with a rotating platinum electrode. Equimolar amounts of acetaldehyde produced substantial interference, and other aldehydes may as well. The method is suitable for measuring quantities of formaldehyde as low as 0.1 mg.

CURRENT APPLICATIONS OF ANALYTICAL METHODS

The standard techniques for analysis of aldehydes in use today were developed for application in specific sampling situations. These situations and the techniques used are discussed below.

AIR

Ambient Air

The wet-chemistry spectrophotometric methods of analysis have been used extensively for the analysis of aldehydes in ambient air. The method based on MBTH has been applied to studies of total aliphatic aldehydes in the ambient atmosphere and from emission sources.[5,7,108,123] As mentioned earlier, it has been recommended for the determination of total aliphatic aldehydes by the Intersociety Committee.[10]

Invariably in atmospheric or emission samples the principal aldehyde detected is formaldehyde. The most extensively used procedure is based on chromotropic acid.[6-8,117,123] The Intersociety Committee[10] and NIOSH[139] have recommended the use of chromotropic acid. Schiff's reagent (basic fuchsin) and pararosaniline also have been suggested for atmospheric determinations.[73,84,110]

The high toxicity of acrolein has prompted analyses of this aldehyde in atmospheric and emission samples. The only sufficiently sensitive colorimetric method available for analysis of acrolein is based on 4-hexylresorcinol,[2,31] a reagent that has been used[7,112,123] and is recommended by the Intersociety Committee.[10] A single investigation used a gas-chromatographic technique to determine acrolein in ambient air.[18]

Gasoline and Diesel Exhaust

The MBTH technique has been applied to automobile (gasoline)[124,125] and diesel[111] exhaust emission to determine the concentration of total aliphatic aldehydes. Two titrimetric procedures have also been used for auto-exhaust measurements.[39]

The chromotropic acid method has been used widely for measuring formaldehyde in automobile[4-6,8,75,117,125] and diesel[14,40,81,82] exhaust. The Schryver method, involving the reaction of formaldehyde with phenylhydrazine followed by oxidation with potassium ferricyanide to form a red derivative, also has been used in studies of formaldehyde emitted in automobile and diesel exhaust.[38,39]

The acrolein content of automobile exhaust has been determined with the 4-hexylresorcinol method.[3,13,31,75] Diesel-exhaust emission has also been studied with this technique.[14,81,82,111]

Because of the relatively high concentrations of acrolein encountered in automobile and diesel exhaust, this pollutant can be effectively measured by gas-chromatographic techniques. Acrolein has been determined directly and as a derivative.[18,40,42,59,60,124]

Nonoccupational Indoor Air

Interest in measuring aldehyde concentrations in nonoccupational indoor environments is a relatively recent phenomenon. Workers in Europe were among the first to determine aldehyde and formaldehyde concentrations in residences. In the United States, attention has been focused on formaldehyde emitted from urea-formaldehyde products used in the construction of homes, especially mobile homes. Table 5-3 (in Chapter 5) summarizes the studies performed to date.

Formaldehyde concentrations were determined for residences in Denmark,[11] the Netherlands, and the Federal Republic of Germany.[148] Maximal concentrations observed in European dwellings reached 2.3 ppm, but average concentrations were 0.4 ppm or less.[11,148] Interestingly, maximal formaldehyde concentrations in residences built without formaldehyde-releasing materials in the Netherlands reached only 0.08 ppm, and average concentrations were only 0.03 ppm.[148] Chromotropic acid was used most often in these European studies.

The MBTH technique was used to measure total aliphatic aldehydes in a pair of mobile homes and a sample residence in Pittsburgh.[93] Breysse[24] used the chromotropic acid method recommended by NIOSH to sample 608 mobile homes in the state of Washington in which residents had complained of irritation; the peak formaldehyde concentration observed in an occupied home was 1.77 ppm, and the mean was less than 0.5 ppm. Garry et al.[47] used the chromotropic acid method with a shortened sampling time to assess formaldehyde concentrations in Minnesota mobile homes. The state of Wisconsin also has used the chromotropic acid method to sample in mobile homes in which residents had registered complaints (M. Woodbury, personal communication).

Systematic studies of formaldehyde and total aliphatic aldehydes as pollutants in nonoccupational indoor environments have been performed by the Lawrence Berkeley Laboratory (LBL) (Lin et al.[79] and Miksch et al., unpublished manuscript). LBL has used the MBTH technique to determine total aliphatic aldehyde concentrations, and the chromotropic acid technique and a modified pararosaniline technique have been used to measure formaldehyde. Sampling sites have included conventional and energy-efficient homes (occupied and unoccupied) and public buildings, such as schools, office buildings, and hospitals.

Occupational Indoor Air

Only occupational air-quality formaldehyde standards are recommended or promulgated by several agencies and professional organizations in the United States. OSHA[140] has promulgated an 8-h time-weighted average (TWA) standard of 3 ppm. The American Conference of Governmental Industrial Hygienists[9] has promulgated a threshold limit value (TLV) standard of 2 ppm. NIOSH[138] has recommended an exposure standard of no greater than 1 ppm for any 30-min sampling period.

The Intersociety Committee,[10] of which the ACGIH is a member, and NIOSH[139] both recommend a method of analysis for formaldehyde based on the use of chromotropic acid. Despite this recommendation, workers investigating formaldehyde concentrations in occupational environments have used a variety of techniques summarized in Table 5-2 (in Chapter 5).

Shipkovitz[127] investigated formaldehyde in textile plants where fabric was treated with formaldehyde-containing resins. Samples were generated by drawing air through bubblers containing sodium bisulfite solution and were analyzed by iodometric titration. The method was reported to have a sensitivity of 0.5 ppm, but was not specific for formaldehyde.

Collection in sodium bisulfite had been used earlier by the California Department of Public Health[21] to analyze air at a textile garment factory. The method of analysis was not reported. In the same year, however, the California Department of Public Health investigated airborne formaldehyde in a clothing store by using midget impingers containing a solution of MBTH.[91] As with sodium bisulfite, this reagent is not specific for formaldehyde.

A modified chromotropic acid procedure was used by Schuck et al.[121] to determine formaldehyde concentrations during a study of the ocular effects induced by smog components. Subjects were exposed to formaldehyde in a smog chamber. Chromotropic acid was used to determine formaldehyde concentrations between 0.04 and 10.9 ppm at two laminating plants using phenol-resorcinol glues.[44] A survey of six funeral homes used a modified chromotropic acid procedure, in which air was bubbled into 0.1% chromotropic acid in concentrated sulfuric acid, to determine exposure to formaldehyde during the embalming process.[69] Reported concentrations of 0.09-5.26 ppm may have been in error on the high side, inasmuch as no prefilter was used on air-sampling lines to remove paraformaldehyde dust which was also present.

WATER

Most of and perhaps all the methods that have been used to identify or measure aldehydes in samples of air and biologic tissue should be applicable to water samples; however, little research has been performed to determine the relative accuracy, precision, or sensitivity of the methods for measuring aldehydes in water.

Because of the high toxicity of acrolein to aquatic life, Kissel and co-workers[70] evaluated several chemical analytical procedures by comparing the analytical data with data from bioassays on acrolein. Eight methods--three derivatization and five direct--were evaluated. The derivatization methods were the bromide-iodide-thiosulfate titration method of Pressman and Lucas,[106] the 2,4-dinitrophenylhydrazine colorimetric method of Bowmer and co-workers,[22] and the aminophenol fluorescence method of Alarcon.[1] Direct measurements were performed by ultraviolet spectrophotometry,[149] gas-liquid chromatography,[130] nuclear-magnetic-resonance spectroscopy, differential-pulse polarography,[56] and direct fluorescence. The effect of different buffering systems on the toxicity and chemical analysis of acrolein was also investigated.

Analytical data produced by the derivatization methods and by gas-liquid chromatography did not correlate well with the bioassay data. Often, no biologic responses were observed for test solutions in which these methods indicated the presence of toxic concentrations of acrolein. The direct ultraviolet spectrophotometric method was also judged unsuitable, because it produced extraneous peaks, which were often more intense than the acrolein peaks and tended to mask them.

Data produced by nuclear-magnetic-resonance spectroscopy, differential-pulse polarography, and direct fluorescence correlated very well with the bioassay data. Assuming that the bioassay provided a more realistic measure of the concentrations of acrolein in solution, the authors concluded that these direct methods were suitable for monitoring acrolein in water. How well these methods will work for other aldehydes is unknown. Unfortunately, this study appears to be the only published one concerning the suitability of possible methods for measuring an aldehyde in water.

Gas chromatography, in combination with mass spectroscopy, appears to be gaining favor over other techniques for identifying and measuring aldehydes--not only in water, but in plant material where the chemical composition can be highly complex, which necessitates the isolation of the aldehyde from other components. Problems have been encountered, however, when conventional gas-chromatographic procedures (such as use of packed columns) are used. Such columns are incapable of resolving all the aldehydes present. For this reason, some investigators have resorted to using thin-layer chromatography in conjunction with gas chromatrography to obtain greater resolution.

The advent of the glass capillary column has essentially eliminated the need for thin-layer/gas chromatography. Hollowell (personal communication) used a gas chromatograph (Carlo Elba) equipped with a 50-m^3 glass capillary column and a splitless injector system in conjunction with a mass spectrometer (Finnigan, Model 3200) to identify and measure aldehydes in drinking water of various sources. The aldehydes were removed from the water samples and concentrated in XAD columns (resin series not designated). The eluting solvent was not reported, but presumably was benzene or ethyl acetate.

Although the gas-chromatographic system provided excellent separation of the aldehydes, the electron-impact mass-spectroscopic technique was not suitable for determining the exact structure (and therefore identifying) a number of compounds with an apparent alkanal structure. With the system, Piet was able to identify 13 aldehydes; their concentrations ranged from 0.005 to 0.3 ppb.

PLANT MATERIAL

Two analytical methods have been used for measuring aldehydes, one involving gas chromatography and the other, DNPH formation. For fruit products, an open tubular gas-chromatographic column with programed temperature control has been used to separate volatile components. A stainless-steel tube coated with GE SF-96 (50-silicone) was used with grape juice. A copper tube packed with 10% Carbowax 20M on Haloport F was used in research with ginkgo leaves. A combination of infrared spectra, retention data, and mass spectrometry was used to identify and measure particular aldehydes.

In 1972, Major and Thomas[86] compared the amount of 2-hexenal in ginkgo leaves as measured by gas chromatrography and by the weight of 2,4-DNPH. He added the ether extracts of steam-distilled leaves to a solution of 2,4-DNPH, hydrochloric acid, and methyl alcohol. After 2 h, the solvents were evaporated to a small volume, and the crystallized 2,4-DNPH was checked for purity by melting point and by thin-layer chromatography (TLC) on silica gel with 6:1 hexane:ether as the developer. Recovery by DNPH was inferior to that by the gas-chromatographic method.

Winter and Sundt,[147] leaders in investigation of strawberry flavors, have objected to the use of gas chromatography in aldehyde analysis, because it operates at a relatively high temperature and product modifications may result under these conditions. They favor paper chromatography, because the volatile constituents are fixed rapidly by derivative formation and thus protected from further changes. They have identified the isolated derivatives by melting point and by infrared spectroscopy.

In 1976, Braddock and Kesterson[23] used a more sophisticated 2,4-DNPH method than reported by Major and Thomas. Cold-pressed citrus oils dissolved in hexane were applied to a bed of a 2,4-DNPH reaction column. A volume of about 500 ml was eluted and aliquots were chromatographed on consecutive columns of Celite-Seasorb and alumina. Column effluents were evaporated to dryness and taken up in chloroform, and the absorbence was determined for estimation of quantities of 2,4-DNPH by the extinction coefficients. Effluents from the alumina columns were separated by thin-layer chromatography into individual aldehyde 2,4-DNPHs. Each aldehyde was scraped from the TLC plates and measured by its extinction coefficient. Tentative identification of individual aldehyde 2,4-DNPHs was by comparison of R_f values with standard derivatives. Aldehyde 2,4-DNPHs scraped from TLC plates were identified positively by comparing mass spectra of known derivatives with the unknowns.

REFERENCES

1. Alarcon, R. A. Fluorometric determination of acrolein and related compounds with m-aminophenol. Anal. Chem. 40:1704-1708, 1968.
2. Altshuller, A. P. Detection of acrolein in automobile exhaust and the atmosphere. Int. J. Air Water Pollut. 6:169-170, 1962.
3. Altshuller, A. P., I. R. Cohen, M. E. Meyer, and A. F. Wartburg, Jr. Analysis of aliphatic aldehydes in source effluents and in the atmosphere. Anal. Chim. Acta 25:101-117, 1961.
4. Altshuller, A. P., S. L. Kopszynski, W. Lonneman, and D. Wilson. Photochemical reactivities of exhausts from 1966 model automobiles equipped to reduce hydrocarbon emissions. J. Air Pollut. Control Assoc. 17:734-737, 1967.
5. Altshuller, A. P., and L. J. Leng. Application of the 3-methyl-2-benzothiazolone hydrazone method for atmospheric analysis of aliphatic aldehydes. Anal. Chem. 35:1541-1542, 1963.
6. Altshuller, A. P., L. J. Leng, and A. F. Wartburg, Jr. Source and atmospheric analyses for formaldehyde by chromotropic acid procedures. Int. J. Air Water Pollut. 6:381-385, 1962.
7. Altshuller, A. P., and S. P. McPherson. Spectrophotometric analysis of aldehydes in the Los Angeles atmosphere. J. Air Pollut. Control Assoc. 13:109-111, 1963.

8. Altshuller, A. P., D. L. Miller, and S. F. Sleva. Determination of formaldehyde in gas mixtures by the chromotropic acid method. Anal. Chem. 33:621-625, 1961.

9. American Conference of Governmental Industrial Hygienists. TLV_s. Threshold Limit Values for Chemical Substances and Physical Agents in the Workroom Environment with Intended Changes for 1977. Cincinnati: American Conference of Governmental Industrial Hygienists, 1977.

10. American Public Health Association Intersociety Committee. Methods of Air Sampling and Analysis. Washington, D.C.: American Public Health Association, Inc., 1972. 480 pp.

11. Andersen, I., G. R. Lundqvist, and L. Mølhave. Formaldehyde in the atmosphere in Danish homes. Ugeskr. Laeg. 136:2133-2139, 1974. (in Danish; Eng. summary)

12. Bailey, B. W., and J. M. Rankin. New spectrophotometric method for determination of formaldehyde. Anal. Chem. 43:782-784, 1971.

13. Barber, E. D., and J. P. Lodge, Jr. Paper chromatographic identification of carbonyl compounds as their 2,4-dinitro-phenylhydrazones in automobile exhaust. Anal. Chem. 35:348-350, 1963.

14. Battigelli, M. C. Sulfur dioxide and acute effects of air pollution. J. Occup. Med. 10:500-511, 1968.

15. Becker, K. H., U. Schurath, and T. Tatarczyk. Fluorescence determination of low formaldehyde concentrations in air by dye laser excitation. Appl. Opt. 14:310-313, 1975.

16. Bell, R. P. The reversible hydration of carbonyl compounds. Advan. Phys. Org. Chem. 4:1-29, 1966.

17. Bell, R. P., and P. G. Evans. Kinetics of the dehydration of methylene glycol in aqueous solution. Proc. Roy. Soc., Series A 291:297-323, 1966.

18. Bellar, T. A., and J. E. Sigsby, Jr. Direct gas chromatographic analysis of low molecular weight substituted organic compounds in emissions. Environ. Sci. Technol. 4:150-156, 1970.

19. Belman, S. The fluorimetric determination of formaldehyde. Anal. Chim. Acta 29:120-126, 1963.

20. Berge, A., and B. Mellegaard. Formaldehyde emission from particleboard--A new method for determination. For. Prod. J. 29(1):21-25, 1979.

21. Blejer, H. P., and B. H. Miller. Occupational Health Report of Formaldehyde Concentrations and Effects on Workers at the Bayly Manufacturing Company, Visalia. Study Report No. S-1806. Los Angeles: State of California Health and Welfare Agency, Department of Public Health, Bureau of Occupational Health, 1966. 6 pp.

22. Bowmer, K. H., A. R. G. Lang, M. L. Higgins, A. R. Pillay, and Y. T. Tchan. Loss of acrolein from water by volatilization and degradation. Weed Res. 14:325-328, 1974.

23. Braddock, R. J., and J. W. Kesterson. Quantitative analysis of aldehydes, esters, alcohols and acids from citrus oils. J. Food Sci. 41:1007-1010, 1976.

24. Breysse, P. A. The environmental problems of urea-formaldehyde structures--formaldehyde exposure in mobile homes. Presented at the American Medical Association Congress on Occupational Health, October 26, 1979.
25. Bricker, C. E., and H. R. Johnson. Spectrophotometric method for determining formaldehyde. Ind. Eng. Chem., Anal. Ed. 17:400-402, 1945.
26. Campbell, E. E., G. O. Wood, and R. G. Anderson. Development of Air Sampling Techniques--LASL Project No. R-059. Quarterly Report. September 16 - December 31, 1972. Los Alamos Scientific Laboratory Report LA-5164-PR. Los Alamos, New Mexico: Atomic Energy Commission, 1973. 6 pp.
27. Cantor, T. R. Experience with the determination of atmospheric aldehydes, pp. 514-515. In Automation in Analytical Chemistry. Technicon Symposia 1966. Vol. I. White Plains: Mediad, 1967.
28. Cares, J. W. Determination of formaldehyde by the chromatropic acid method in the presence of oxides of nitrogen. Am. Ind. Hyg. Assoc. J. 29:405-410, 1968.
29. Carey, M. A., and H. E. Persinger. Liquid chromatographic determination of derivatives that contain the dinitrophenyl group. J. Chromat. Sci. 10:537-543, 1972.
30. Chrastil, J., and J. T. Wilson. A sensitive colorimetric method for formaldehyde. Anal. Biochem. 63:202-207, 1975.
31. Cohen, I. R., and A. P. Altshuller. A new spectrophotometric method for the determination of acrolein in combustion gases and in the atmosphere. Anal. Chem. 33:726-733, 1961.
32. Colket, M. B., D. W. Naegeli, F. L. Dryer, and I. Glassman. Flame ionization detection of carbon oxides and hydrocarbon oxygenates. Environ. Sci. Technol. 8:43-46, 1974.
33. Cotton, F. A., and G. Wilkinson. Advanced Inorganic Chemistry: A Comprehensive Text. 3rd ed. New York: John Wiley & Sons, Inc., 1972. 1145 pp.
34. Dagani, D., and M. C. Archer. Colorimetric determination of actaldehyde in the presence of formaldehyde. Anal. Biochem. 87:455-459, 1978.
35. Dickinson, R. G., and N. W. Jacobsen. A new sensitive and specific test for the detection of aldehydes: Formation of 6-mercapto-3-substituted-s-triazolo[4,3-b]-s-tetrazines. Chem. Commun. 24:1719-1720, 1970.
36. Driscoll, J. N. Evaluation of a new photoionization detector for organic compounds. J. Chromat. 134:49-55, 1977.
37. Eegriwe, E. Reactions and reagents for the detection of organic compounds. Z. anal. Chem. 110:22025, 1937.
38. Elliott, M. A., G. J. Nebel, and F. G. Rounds. The composition of exhaust gases from diesel, gasoline and propane powered motor coaches. J. Air Pollut. Control Assoc. 5:103-108, 1955.
39. Ellis, C. F. Chemical Analyses of Automobile Exhaust Gases for Oxygenates. U.S. Bureau of Mines Report of Investigations 5822. Pittsburgh: U.S. Department of the Interior, Bureau of Mines, 1961. 35 pp.

40. Ellis, C. F., R. F. Kendall, and B. H. Eccleston. Identification of some oxygenates in automobile exhausts by combined gas liquid chromatography and infrared techniques. Anal. Chem. 37:511-516, 1965.
41. Flath, R. A., D. R. Black, D. G. Guadagni, W. H. McFadden, and T. H. Schultz. Identification and organoleptic evaluation of compounds in Delicious apple essence. J. Agric. Food Chem. 15:29-35, 1967.
42. Fracchia, M. F., F. J. Schuette, and P. K. Mueller. A method for sampling and determination of organic carbonyl compounds in automobile exhaust. Environ. Sci. Technol. 1:915-922, 1967.
43. Freed, D. J., and A. M. Mujsce. In situ generation of standards for gas chromatographic analysis. Anal. Chem. 49:139-141, 1977.
44. Freeman, H. G., and W. C. Grendon. Formaldehyde detection and control in the wood industry. For. Prod. J. 21(9):54-57, 1971.
45. Gadbois, D. F., J. M. Mendelsohn, and L. J. Ronsivalli. Modification of Girard-T reagent method for concentrating carbonyl compounds for gas chromatographic analysis. Anal. Chem. 37: 1776-1778, 1965.
46. Gadbois, D. F., P. G. Scheurer, and F. J. King. Analysis of saturated aldehydes by gas-liquid chromatography using methylolphthalimide for regeneration of their Girard-T derivatives. Anal. Chem. 40:1362-1365, 1968.
47. Garry, V. F., L. Oatman, R. Pleus, and D. Gray. Formaldehyde in the home. Some environmental disease perspectives. Minn. Med. 63:107-111, 1980.
48. Gold, A., C. E. Dubé, and R. B. Perni. Solid sorbent for sampling acrolein in air. Anal. Chem. 50:1839-1841, 1978.
49. Gronsberg, E. S. Determination of acrolein and vinyl butyl ether in the atmosphere. Gig. Truda Prof. Zabol. 12(7):54-56, 1968. (in Russian)
50. Halvarson, H. Quantitative gas chromatographic analysis of micro amounts of volatile carbonyl compounds via their DNPH derivatives. J. Chromat. 57:406-409, 1971.
51. Halvarson, H. The qualitative and quantitative evaluation of the low-molecular-weight monocarbonyls in meat products. J. Chromat. 66:35-42, 1972.
52. Harrenstein, M. S. Measurements of Aldehyde Concentrations in the Exhaust of an Internal Combustion Engine Fueled by Alcohol/Gasoline Blends. Graduate Thesis H296M. Coral Gables, Fla.: University of Miami, 1978. 99 pp.
53. Hauser, T. R., and R. L. Cummins. Increasing sensitivity of 3-methyl-2-benzothiazolone hydrazone test for analysis of aliphatic aldehydes in air. Anal. Chem. 36: 679-681, 1964.
54. Hemenway, D. R., M. C. Costanza, and S. M. MacAskill. Review of the 4-hexylresorcinol procedure for acrolein analysis. Am. Ind. Hyg. Assoc. J. 41:305-308, 1980.
55. Hoshika, Y., and Y. Takata. Gas chromatographic separation of carbonyl compounds as their 2,4-dinitrophenylhydrazones using glass capillary columns. J. Chromat. 120:379-389, 1976.

56. Howe, L. H. Differential pulse polarographic determination of acrolein in water samples. Anal. Chem. 48:2167-2169, 1976.
57. Hrubesh, L. W. Microwave rotational spectroscopy. Technique for specific pollutant monitoring. Radio Sci. 8:167-175, 1973.
58. Hrubesh, L. W., A. S. Maddux, D. C. Johnson, R. L. Morrison, J. N. Nielson, and M. Malachosky. Portable microwave multi-gas analyzer development. Lawrence Livermore Laboratory Report No. UCID-17867. Livermore, Cal.: University of California, Lawrence Livermore Laboratory, 1978. 110 pp.
59. Hughes, K. J., and R. W. Hurn. A preliminary survey of hydrocarbon-derived oxygenated material in automobile exhaust gases. J. Air Poll. Contr. Assoc. 10:367-373, 1960.
60. Hughes, K. J., R. W. Hurn, and F. G. Edwards. Separation and identification of oxygenated hydrocarbons in combustion products from automotive engines. Gas Chromat., Proc. 2nd Int. Symp., East Lansing, Michigan, 1959. Pp. 171-182, 1961.
61. Ikeda, S. Rapid determination of formaldehyde by short-circuit argentometric amperometric titration using a rotating platinum microelectrode. Anal. Chem. 46:1587-1588, 1974.
62. Jacobsen, N. W., and R. G. Dickinson. Spectrometric assay of aldehydes as 6-mercapto-3-substituted-$\underline{s}$-triazolo(4,3-b)-$\underline{s}$-tetrazines. Anal. Chem. 46:298-299, 1974.
63. Johnson, C. B., A. M. Pearson, and L. R. Dugan, Jr. Gas chromatographic analysis of the dimethylhydrazones of long chain aldehydes. Lipids 5:958-963, 1970.
64. Johnson, D. C., and E. G. Hammond. A sensitive method for the determination of carbonyl compounds. J. Am. Oil Chem. Soc. 48:653-656, 1971.
65. Jones, L. A., and R. J. Monroe. Flash exchange method for quantitative gas chromatographic analysis of aliphatic carbonyls from their 2,4-dinitrophenylhydrazones. Anal. Chem. 37:935-938, 1965.
66. Kadaba, P. K., P. K. Bhagat, and G. N. Goldberger. Application of microwave spectroscopy for simultaneous detection of toxic constituents in tobacco smoke. Bull. Environ. Contam. Toxicol. 19:104-112, 1978.
67. Kallio, H., R. R. Linko, and J. Kaitaranta. Gas-liquid chromatographic analysis of 2,4-dinitrophenylhydrazones of carbonyl compounds. J. Chromat. 65:355-360, 1972.
68. Katz, M., Ed. Methods of Air Sampling and Analysis. 2nd ed. Washington, D. C.: American Public Health Association, 1977. 984 pp.
69. Kerfoot, E. J., and T. F. Mooney, Jr. Formaldehyde and paraformaldehyde study in funeral homes. Am. Ind. Hyg. Assoc. J. 36:533-537, 1975.
70. Kissel, C. L., J. L. Brady, A. M. Guerra, J. K. Pau, B. A. Rockie, and F. F. Caserio, Jr. Analysis of acrolein in aged aqueous media. Comparison of various analytical methods with bioassays. J. Agric. Food Chem. 26:1338-1343, 1978.

71. Korolczuk, J., M. Daniewski, and Z. Mielniczuk. Gas chromatographic determination of carbonyl compounds as their phenylhydrazones. J. Chromat. 88:177-182, 1974.
72. Krug, E. L. R., and W. E. Hirt. Interference of nitrate in the determination of formaldehyde by the chromotropic acid method. Anal. Chem. 49:1865-1867, 1977.
73. Lahmann, E., and K. Jander. Formaldehyd-Bestimmungen in Strassenluft. Gesundherts-Ingenieur. 89:18-21, 1968.
74. Larson, G. P., P. P. Mader, F. H. Ziets, and E. E. McMahon. Quantitative determination of aldehyde, pp. 31-37. In Test Procedures and Methods in Air Pollution Control. Los Angeles: Air Pollution Control District, County of Los Angeles, 1950. 60 pp.
75. Leach, P. W., L. J. Leng, T. A. Bellar, J. E. Sigsby, Jr., and A. P. Altshuller. Effects of HC/NO_x ratios on irradiated auto exhaust. Part II. J. Air Pollut. Control Assoc. 14:176-183, 1964.
76. Ledbury, W., and E. W. Blair. The partial formaldehyde vapour pressures of aqueous solutions of formaldehyde. Part II. J. Chem. Soc. 127:2832-2839, 1925.
77. Levaggi, D. A., and M. Feldstein. The collection and analysis of low molecular weight carbonyl compounds from source effluents. J. Air Pollut. Control Assoc. 19:43-45, 1969.
78. Levaggi, D. A., and M. Feldstein. The determination of formaldehyde, acrolein, and low molecular weight aldehydes in industrial emissions on a single collection sample. J. Air Pollut. Control Assoc. 20:312-313, 1970.
79. Lin, C.-I., R. N. Anaclerio, D. W. Anthon, L. Z. Fanning, and C. D. Hollowell. Indoor/outdoor measurements of formaldehyde and total aldehydes. Presented at the 178th National American Chemical Society Meeting, Washington, D.C., September 9-14, 1979. 11 pp.
80. Linko, R. R., H. Kallio, and K. Rainio. Gas-liquid chromatographic analysis of 2,4,-dinitrophenylhydrazones of monocarbonyl compounds in carrots using glass capillary columns. J. Chromat. 155:191-194, 1978.
81. Linnell, R. H., and W. E. Scott. Diesel exhaust analysis. Preliminary results. Arch. Environ. Health 5:616-625, 1962.
82. Linnell, R. H., and W. E. Scott. Diesel exhaust composition and odor studies. J. Air Pollut. Control Assoc. 12:510-515, 545, 1962.
83. Lupton, J. M., and C. C. Lynch. Polarographic examination of carbonyl compounds. J. Am. Chem. Soc. 66:697-700, 1944.
84. Lyles, G. R., F. B. Dowling, and V. J. Blanchard. Quantitative determination of formaldehyde in the parts per hundred million concentration level. J. Air Pollut. Control Assoc. 15:106-108, 1965.
85. Major, R. T., P. Marchini, and A. J. Boulton. Observation on the production of α-hexenal by leaves of certain plants. J. Biol. Chem. 238:1813, 1963.

86. Major, R. T., and M. Thomas. Formation of 2-hexenal from linolenic acid by macerated Ginkgo leaves. Phytochem. 11:611-617, 1972.
87. Mansfield, C. T., B. T. Hodge, R. B. Hege, Jr., and W. C. Hamlin. Analysis of formaldehyde in tobacco smoke by high performance liquid chromatography. J. Chromat. Sci. 15:301-302, 1977.
88. Mari, R., M. Fève, and M. Dzierzinsky. Colorimetric reaction between phenylhydrazine, formaldehyde, and oxygen in alkaline solution. Determination of formaldehyde. Bull. Soc. Chim. France 1961: 1395-1399, 1961.
89. McLean, J. D., and J. F. Holland. Development of a portable polarograph for determination of aldehydes in automotive exhaust and production plant samples. Environ. Sci. Technol. 9:127-131, 1975.
90. Meyer, B. Urea-Formaldehyde Resins, p. 128. Reading, Mass.: Addison-Wesley Publishing Company, 1979.
91. Miller, B. H., and H. P. Blejer. Report of an Occupational Health Study of Formaldehyde Concentrations at Maximes, 400 E. Colorado Street, Pasadena, California. Study No. S-1838. Los Angeles: State of California Health and Welfare Agency, Department of Public Health, Bureau of Occupational Health, 1966. 5 pp.
92. Morgan, G. B., C. Golden, and E. C. Tabor. new and improved procedures for gas sampling and analysis in the National Air Sampling Network. J. Air Pollut. Control Assoc. 17:300-304, 1967.
93. Moschandreas, D. J., J. W. C. Stark, J. E. McFadden, and S. S. Morse. Indoor Air Pollution in the Residential Environment. Vol. 1. Data Collection, Analysis and Interpretation. U.S. Environmental Protection Agency Report No. EPA 600/7-78-229a. Research Triangle Park, N.C.: U.S. Environmental Protection Agency, Office of Research and Development, Environmental Monitoring and Support Laboratory, 1978. 201 pp.
94. Nash, T. Colorimetric estimation of formaldehyde by means of the Hantzsch reaction. Biochem. J. 55:416-421, 1953.
95. Nauman, R. V., P. W. West, F. Tron, and G. C. Gaeke, Jr. A spectrophotometric study of the Schiff reaction as applied to the quantitative determination of sulfur dioxide. Anal. Chem. 32:1307-1311, 1960.
96. Olansky, A. S., and S. N. Deming. Optimization and interpretation of absorbance response in the determination of formaldehyde with chromotropic acid. Anal. Chim. Acta 83: 241-249, 1976.
97. Osman, S. F., and J. L. Barson. Solvent fractionation of Girard T derivatives of carbonyl compounds using dimethyl sulfoxide. Anal. Chem. 39:530-531, 1967.
98. Pagnotto, L. D. Gas and vapor sample collectors, pp. R1-R18. In Air Sampling Instruments for Evaluation of Atmospheric Contaminants. 5th ed. Cincinnati: American Conference of Governmental Industrial Hygienists, 1978.

99. Palmes, E. D., A. F. Gunnison, J. DiMattio, and C. Tomczyk. Personal sampler for nitrogen dioxide. Am. Ind. Hyg. Assoc. J. 37:570-577, 1976.
100. Papa, L. J. Colorimetric determination of carbonyl compounds in automotive exhaust as 2,4-dinitrophenylhydrazones. Environ. Sci. Technol. 3:397-398, 1969.
101. Papa, L. J., and L. P. Turner. Chromatographic determination of carbonyl compounds as their 2,4-dinitrophenylhydrazones. I. Gas chromatography. J. Chromat. Sci. 10: 744-747, 1972.
102. Papa, L. J., and L. P. Turner. Chromatographic determination of carbonyl compounds as their 2,4-dinitrophenylhydrazones. II. High pressure liquid chromatography. J. Chromat. Sci. 10: 747-750, 1972.
103. Pellizzari, E. D. Development of Analytical Techniques for Measuring Ambient Atmospheric Carcinogenic Vapors. U.S. Environmental Protection Agency Report No. EPA-600/2-75-076. Research Triangle Park, N.C.: U.S. Environmental Protection Agency, Office of Research and Development, Environmental Sciences Research Laboratory, 1975. 201 pp.
104. Pellizzari, E. D. Development of Method for Carcinogenic Vapor Analysis in Ambient Atmospheres. U.S. Environmental Protection Agency Report No. EPA-650/2-74-121. Research Triangle Park, N.C.: U.S. Environmental Protection Agency, Office of Research and Development, National Environmental Research Center, Chemistry and Physics Laboratory, 1974. 148 pp.
105. Pinigina, I. A. Use of 2,4-dinitrophenylhydrazine for determining carbonyl compounds in the air. Gig. Sanit. 37(4):78-81, 1972. (in Russian)
106. Pressman, D., and H. J. Lucas. Hydration of unsaturated compounds. XI. Acrolein and acrylic acid. J. Am. Chem. Soc. 64:1953-1957, 1942.
107. Pyne. A. W., and E. L. Wick. Volatile components of tomato. J. Food Sci. 30:192-200, 1965.
108. Radian Corporation. Houston Area Oxidants Study. Report No. HCP-6. Austin, Texas: Radian Corporation, 1977.
109. Ralls, J. W. Higher recoveries of carbonyl compounds in flash exchange gas chromatography of 2,4-dinitrophenylhydrazones. Anal. Chem. 36:946, 1964.
110. Rayner, A. C., and C. M. Jephcott. Microdetermination of formaldehyde in air. Anal. Chem. 33:627-630, 1961.
111. Reckner, L. R., W.E. Scott, and W. F. Biller. The composition and odor of diesel exhaust. Proc. Am. Petrol. Inst. 45: 133-147, 1965.
112. Renzetti, N. A., and R. J. Bryan. Atmospheric sampling for aldehydes and eye irritation in Los Angeles smog--1960. J. Air Pollut. Control Assoc. 11:421-424, 427, 1961.
113. Roffael, E. Formaldehyde Release from Particleboard--Methods of Determination. Presented at the Consumer Product Safety Commission Technical Workshop on Formaldehyde, Washington, D.C., April 1980.

114. Sawicki, E., and R. A. Carnes. Spectrophotofluorimetric determination of aldehydes with dimedone and other reagents. Mikrochim. Acta 1968:148-159, 1968.

115. Sawicki, E., and T. R. Hauser. Spot test detection and colorimetric determination of aliphatic aldehydes with 2-hydrazinobenzothiazole. Application to air pollution. Anal. Chem. 32:1434-1436, 1960.

116. Sawicki, E., T. R. Hauser, and S. McPherson. Spectrophotometric determination of formaldehyde and formaldehyde-releasing compounds with chromotropic acid, 6-amino-1-naphthol-3-sulfonic acid (J acid), and 6-anilino-1-naphthol-3-sulfonic acid (phenyl J acid). Anal. Chem. 34:1460-1464, 1962.

117. Sawicki, E., T. R. Hauser, T. W. Stanley, and W. Elbert. The 3-methyl-2-benzothiazolone hydrazone test. Sensitive new methods for the detection, rapid estimation, and determination of aliphatic aldehydes. Anal. Chem. 33:93-96, 1961.

118. Sawicki, E., T. W. Stanley, and J. Pfaff. Spectrophotofluorimetric determination of formaldehyde and acrolein with J acid. Comparison with other methods. Anal. Chim. Acta 28:156-163, 1963.

119. Scaringelli, F. P., A. E. O'Keefe, E. Rosenberg, and J. P. Bell. Preparation of known concentrations of gases and vapors with permeation devices calibrated gravimetrically. Anal. Chem. 42:871-876, 1970.

120. Schiff, H. Eine neue Reithe organisher Diamine. Ann. Chem. 140:92, 1866. (in German)

121. Schuck, E. A., E. R. Stephens, and J. T. Middleton. Eye irritation response at low concentrations of irritants. Arch. Environ. Health 13:570-575, 1966.

122. Schuette, F. J. Plastic bags for collection of gas samples. Atmos. Environ. 1:515-519, 1967.

123. Scott Research Laboratories, Inc. Atmospheric Reaction Studies in the Los Angeles Basin. Phase I. Vol. II. Washington, D.C.: U.S. Public Health Service, National Air Pollution Control Administration, 1969. 542 pp.

124. Seizinger, D. E., and B. Dimitriades. Oxygenates in Automotive Exhaust Gas. Estimation of Levels of Carbonyls and Noncarbonyls in Exhaust from Gasoline Fuels. Bureau of Mines Report of Investigations 7675. Bartlesville, Okla.: U.S. Bureau of Mines, 1972. 75 pp.

125. Seizinger, D. E., and B. Dimitriades. Oxygenates in Automotive Exhausts. Effect of an Oxidation Catalyst. Bureau of Mines Report of Investigations 7837. Bartlesville, Okla.: U.S. Bureau of Mines, 1973. 26 pp.

126. Selim, S. Separation and quantitative determination of traces of carbonyl compounds as their 2,4-dinitrophenylhydrazones by high pressure liquid chromatography. J. Chromat. 136:271-277, 1977.

127. Shipkovitz, H. D. Formaldehyde Vapor Emissions in the Permanent-Press Fabric Industry. Report No. TR-52. Cincinnati: U.S. Department of Health, Education, and Welfare, Public Health

Service, Consumer Protection and Environmental Health Service, Environmental Control Administration, September 1968. 18 pp.
128. Slawinska, D., and J. Slawinski. Chemiluminescent flow method for determination of formaldehyde. Anal. Chem. 47:2101-2109, 1975.
129. Sleva, S. F. Determination of formaldehyde-chromotropic acid method, pp. H-1--H-5. In Interbranch Chemical Advisory Group. Selected Methods for the Measurement of Air Pollutants. DHEW (PHS) Publication No. 999-AP:11. Washington, D.C.: U.S. Government Printing Office, 1965.
130. Smith, C. W., Ed. Acrolein. Heidelberg, Germany: Huethig, 1975. (in German; English abstract in Chem. Abstr. 87:102823j, 1977)
131. Soukup, R. J., R. J. Scarpellino, and E. Danielczik. Gas chromatographic separation of 2,4-dinitrophenylhydrazone derivatives of carbonyl compounds. Anal. Chem. 36:2255-2256, 1964.
132. Stevens, K. L., J. L. Bomben, and W. H. McFadden. Volatiles from grapes. Vitis vinifera (Linn) cultivar Grenache. J. Agric. Food Chem. 15:378-380, 1967.
133. Szelejewska, I. Spectrophotometric determination of formaldehyde in presence of acrolein. Chem. Anal. (Warsaw) 20:325-330, 1975. (in Polish; English summary)
134. Tanebaum, M., and C. E. Bricker. Microdetermination of free formaldehyde. Anal. Chem. 23:354-357, 1951.
135. Teranishi, R., J. W. Corse. W. H. McFadden, D. R. Black, and A. I. Morgan, Jr. Volatiles from strawberries. I. Mass spectral identification of the more volatile compounds. J. Food Sci. 28:478-483, 1963.
136. Tsang, W., and J. A. Walker. Instrument for the generation of reactive gases. Anal. Chem. 49:13-17, 1977.
137. Tuazon, E. C., R. A. Graham, A.M. Winer, R. R. Easton, J. N. Pitts, Jr., and P. L. Hanst. A kilometer pathlength Fourier-transform infrared system for the study of trace pollutants in ambient and synthetic atmospheres. Atmos. Environ. 12:865-875, 1978.
138. U.S. Department of Health, Education, and Welfare, Public Health Service, National Institute for Occupational Safety and Health. Criteria for a Recommended Standard...Occupational Exposure to Formaldehyde. DHEW (NIOSH) Publication No. 77-126. Washington, D.C.: U.S. Government Printing Office, 1976. 165 pp.
139. U.S. Department of Health, Education, and Welfare, Public Health Service, National Institute for Occupational Safety and Health. Manual of Analytical Methods, pp. 125-1--125-9. 2nd ed. Vol. 1. Washington, D.C.: U.S. Government Printing Office, 1973.
140. U.S. Department of Labor, Occupational Safety and Health Administration. Safety and health regulations for construction. Recodification of air contaminant standards. Fed. Reg. 40:23072-23073, 28 May 1975.
141. Vogh, J. W. Isolation and analysis of carbonyl compounds as oximes. Anal. Chem. 43:1618-1623, 1971.

142. Walker, J. F. Formaldehyde. 3rd ed. Huntington, N.Y.: Robert E. Krieger Publishing Co., 1975. 728 pp.
143. West, P. W., and G. C. Gaeke. Fixation of sulfur dioxide as disulfitomercurate (II) and subsequent colorimetric estimation. Anal. Chem. 28:1816-1819, 1956.
144. West, P. W., and T. P. Ramachandran. Spectrophotometric determination of nitrate using chromotropic acid. Anal. Chim. Acta 35:317-324, 1966.
145. West, P. W., and B. Z. Sen. Spectrophotometric determination of traces of formaldehyde. Z. anal. Chem. 153:177-183, 1956.
146. Wilson, K. W. Fixation of atmospheric carbonyl compounds by sodium bisulfite. Anal. Chem. 30:1127-1129, 1958.
147. Winter, M., and E. Sundt. Flavors. V. Analysis of the raspberry flavoring material. 1. Volatile carbonyl constituents. Helv. Chim. Acta 45:2195-2211, 1962. (in French)
148. World Health Organization Working Group. Health Aspects Related to Indoor Air Quality. Report on a WHO Working Group. Copenhagen, Denmark: World Health Organization, 1979. 34 pp.
149. Yamate, N., and T. Matsumura. Determination of acrolein in ambient air by square wave polarography. Eisei Shikenjo Hokoku (Japan) 93:130-132, 1975. (in Japanese; English abstract in Chem. Abstr. 85:129642v, 1975)
150. Yunghans, R. S., and W. A. Munroe. Continuous monitoring of ambient atmospheres with the Technicon AutoAnalyzer, pp. 279-284. In Automation in Analytical Chemistry. Technicon Symposia 1965. Vol. 1. White Plains, New York: Mediad Inc., 1966.

CHAPTER 7

HEALTH EFFECTS OF FORMALDEHYDE

There is an increasing body of evidence that the exposure of the human population to formaldehyde vapors may be the source of the many complaints related to irritation of the eyes and respiratory tract, headache, tiredness, and thirst; these symptoms have been reported by occupants of homes, schools, and industrial buildings--mainly by residents of homes in which formaldehyde has been detected at high concentrations. Owing to the common use of formaldehyde in building materials and in foam insulation, there is a potential for exposure of employees engaged in the manufacture of these products and for exposure of the general public using the products. Furthermore, there are many workers in a great variety of occupations who, through the use of formaldehyde and its associated products, may be exposed to formaldehyde at high concentrations in the course of a day's work (see Table 7-1). Energy-conservation measures that have become widely used in recent years, including reduced ventilation rates, have led to increased indoor formaldehyde concentrations.[21,26] We have considered in some detail (in Chapter 5) these and the many other sources of formaldehyde pollution in our environment today. In view of the widespread use of formaldehyde and the large number of people who are exposed to it, we must be concerned about the potential health effects associated with these exposures.

Because of the unique importance of formaldehyde among the many aldehydes in use today, we devote this chapter to its consideration. The health effects of the several other important aldehydes are discussed in Chapter 8. Eye irritation and respiratory tract irritation are common results of human exposure to formaldehyde at relatively low concentrations. Documented cases of hypersensitivity with bronchial asthma due specifically to formaldehyde are few; more commonly, asthma is aggravated by the irritating properties of formaldehyde. Aqueous solutions damage the eye and irritate the skin on direct contact. Repeated exposure to dilute solutions may lead to allergic contact dermatitis. Systemic formaldehyde poisoning by ingestion is uncommon, because its irritancy makes ingestion unlikely. We discuss here the preliminary findings of a Chemical Industry Institute of Toxicology (CIIT) study with regard to formaldehyde induction of nasal cancer in rats and mice. The human carcinogenic, mutagenic, and teratogenic potential of formaldehyde is

TABLE 7-1

Potential Occupational Exposures to Formaldehyde[a]

Anatomists	Glass etchers
Agricultural workers	Glue and adhesive makers
Bakers	Hexamethylenetetramine makers
Beauticians	Hide preservers
Biologists	Histology technicians
Bookbinders	Home construction workers
Botanists	
	Ink makers
Crease-resistant-textile finishers	Lacquerers and lacquer makers
Concrete users	Laundry workers
Dentists	Medical personnel
Deodorant makers	Mirror workers
Dialysis technicians	
Disinfectant makers	Oil-well workers
Disinfectors	
Dress-goods store personnel	Paper makers
Dressmakers	Pentaerythritol makers
Drugmakers	Photographic-film makers
Dry cleaners	
Dyemakers	Resin makers
	Rubber makers
Electric-insulation makers	
Embalmers	Soil sterilizers and greenhouse workers
Embalmin-fluid makers	
Ethylene glycol makers	Surgeons
Fertilizer makers	Tannery workers
Fireproofers	Taxidermists
Formaldehyde-resin makers	Textile mordanters and printers
Formaldehyde workers	Textile waterproofers
Fumigators	
Fungicide workers	Varnish workers
Furniture dippers and sprayers	
Fur processors	Wood preservers

[a] Modified from NIOSH.[198]

not known, but it has exhibited mutagenic activity in a wide variety of organisms.

ASSESSMENT OF ADVERSE HEALTH EFFECTS

Adverse health effects due to formaldehyde may occur after exposure by inhalation, ingestion, or skin contact. It is difficult to ascribe specific health effects to specific concentrations of formaldehyde to which people are exposed, because they vary in their subjective responses and complaints. Moreover, persons with disease may be more responsive to low concentrations than hyposensitive persons who do not respond to the same concentrations. Thus, the threshold for response will not be constant among all segments of the population. Also, studies done in homes, both mobile and conventional, where the subjective complaints of consumers reportedly can be ascribed to formaldehyde (especially when only formaldehyde is measured) may not be completely valid, because other pollutants acting independently may cause the same symptoms or synergistically may enhance the perception of symptoms. (See Chapter 5 for factors that affect the outgassing of formaldehyde.) Interpretation of the health effects of formaldehyde must consider not only the concentration, but also the duration of exposure of subjects. For example, in some studies, exposures lasted only a few minutes;[81,90,132,175,207] in others, they lasted several hours[72,137,173,183] or days.[215,217] A short-term inhalation study cannot accurately predict the effects of formaldehyde on persons who reside in homes where there is a continuous low-dose exposure. Tolerance may develop after several hours of exposure[15,102,173] and modify the response to formaldehyde. In some persons not previously sensitized, repeated exposure to formaldehyde may result in the development of hypersensitivity.

Analytical procedures for formaldehyde vary in both sensitivity and specificity (see Chapter 6).[4,5,29,31,38,56,82,113,116,136,158,158,159,166,177,178,188,216]

BIOCHEMISTRY AND METABOLISM OF FORMALDEHYDE

Formaldehyde is a normal metabolite and a vital ingredient in the synthesis of essential biochemical substances in man and thus in small quantities is not toxic.[39,109] Formaldehyde controls a rate-limiting step in the processing of methyl groups derived from the metabolic dealkylation of O-, N-, and S-methyl compounds during their detoxification and excretion.[209] With ample dietary supplies of tetrahydrofolic acid, vitamin B_{12}, and such sulfhydryl compounds as cysteine and methionine, small amounts of formaldehyde are readily metabolized.

Formaldehyde also is involved in lipid metabolism in the decomposition of peroxides by catalase.[203]

The biochemical transformations of endogenous and exogenous formaldehyde are similar and involve coenzymes and hydrogen transport systems that are normally present in all animals and bacteria.[39,99,109] Interspecies variations in the metabolism of formaldehyde may account for differences in reaction rates in these systems.[55,75,147,205,209] Formaldehyde oxidation, for example, is greater in human liver than in rat liver; this may explain the unique susceptibility of man to methanol poisoning.[186]

The main reaction of formaldehyde appears to be an initial oxidation to formic acid in the liver and erythrocytes.[39,55,99,103,109,124] Once formic acid is formed, it can undergo three reactions: oxidation to carbon dioxide and water, elimination in the urine as a sodium salt, or entrance into the metabolic one-carbon pool. Formaldehyde may also enter the one-carbon pool directly.

In man, the formation of formate from formaldehyde appears to involve an initial reaction with glutathione to form a hemiacetal.[75,184] The enzyme formaldehyde dehydrogenase (FDH) then oxidizes the hemiacetal to formic acid, with NAD as a hydrogen acceptor.[55,184] In humans, FDH is a multifunctional complex of enzymes that converts methanol to formic acid without releasing formaldehyde as an intermediate,[75,184,202,206] inasmuch as it is difficult to isolate FDH alone.

The molecular weight of human FDH is 81,400, and that of rats is 111,000.[75,202] Human liver FDH activity is 50% greater than that of rat liver, in terms of enzyme units per gram of liver.[75] The actual product of the human FDH reaction is not free formic acid, but S-formylglutathione, which hydrolyzes slowly in human liver preparations to formate.[202]

Tran et al.[191] investigated the uptake of [^{14}C]formaldehyde and its conversion to carbon dioxide by erythrocytes from chronic alcoholics and nonalcoholics. The ingestion of ethanol initially decreased the rate of carbon dioxide production from formaldehyde in both groups, but a greater decrease was noted in the alcoholics' erythrocytes. A few hours later, the erythrocytes from alcoholics had a carbon dioxide production rate well above their baseline values; the rate returned to normal several days later. These findings could be explained on the basis that ethanol interfered with tetrahydrofolic acid activity during metabolism. The potential interference with tetrahydrofolic acid activity brings up the theoretical possibility that formaldehyde affects folate uptake by cells. Tetrahydrofolic acid is important, in that an induced folate deficiency may result in a number of medical conditions, including hematologic abnormalities and neurologic and growth effects in infants.[12,17,121,128,170] A folate-dependent one-carbon pathway was found to be primarily responsible for formate oxidation in monkeys poisoned with methanol.[130] Formate elimination from the blood of folate-deficient monkeys was about half that of controls.

It has been reported that formaldehyde causes the eye effects and formic acid some of the acidosis seen in methanol poisoning.[33] Although in vitro studies indicate that formaldehyde has significant effects on retinal oxidative phosphorylations,[41] it is rapidly metabolized to formic acid in humans, dogs, cats, rabbits, guinea pigs, rats, and monkeys.[131,160] Formaldehyde is eliminated from the blood with a half-life of 1-2 min. In a study of formate-poisoned monkeys, there was no detectable increase in formaldehyde concentration in samples of blood, urine, cerebrospinal fluid, vitreous humor, freeze-clamped liver (at the temperature of liquid nitrogen), kidney, optic nerve, or brain,[124,131] at a time when formate concentrations were high. In a recent report of methanol poisoning in humans, formate accumulation was marked; that indicates that formic acid plays a major role in the acidosis in human poisonings.[129]

Some adverse effects of formaldehyde may be related to its high reactivity with amines and formation of methylol adducts with nucleic acids, histones, proteins, and amino acids. The methylol adducts can react further to form methylene linkages among these reactants.[11,48] It appears that before formaldehyde reacts with amino groups in RNA, the hydrogen bonds forming the coiled RNA are broken.[64,146] Formaldehyde reacts with DNA less frequently than with RNA, because the hydrogen bonds holding DNA in its double helix are more stable.[64,172] Reaction of formaldehyde with DNA has been observed, by spectrophotometry and electron microscopy, to result in irreversible denaturation. In reactions with transfer RNA, formaldehyde interferes with amino acid acceptance.[11,172] The equilibrium reaction of formaldehyde with DNA involves thermally activated opening and closing of hydrogen bonds between matching base pairs in the helix.[172] If permanent cross links are formed between DNA reactive sites and formaldehyde, these links could interfere with the replication of DNA and may result in mutations. When human fetal lung fibroblasts were incubated with tracer amounts of ^{14}C-labeled formaldehyde and acetaldehyde,[155] a pulse of 10 min with formaldehyde followed by a 6-min and 24-h chase showed migration of carbon-14 into the nucleus. Fractionation of the nucleus revealed that the RNA fraction had the highest absolute and specific activity, whereas the DNA and protein fractions had considerably lower activity. All the counts from formaldehyde were found in the adenine and guanine components of RNA. The DNA count was distributed among adenine, guanine, and thymine.

EFFECTS IN ANIMALS

ACUTE TOXICOLOGY STUDIES

When administered orally, formaldehyde (formalin) is slightly toxic in rats, with LD_{50} values reported in the range of 500-800 mg/kg.[179,193] When administered by inhalation, it is moderately toxic in rats, with 3-min and 4-h LC_{50}s of 815 and 479 ppm, respectively.[138,141] Pulmonary edema was the predominant pathologic change. Similar results were obtained in cats and mice.

Formaldehyde causes mild to moderate irritation when applied to rabbit skin at 0.1-20% (Haskell Laboratory, Du Pont Company, unpublished data). Formaldehyde was also administered to nine guinea pigs intradermally or topically over a 2-wk period. After a 2-wk rest period, they were challenged with formaldehyde; five of the animals had become sensitized. Dermal sensitization by airborne formaldehyde has not been reported.

Formaldehyde is a severe eye irritant. Experimental application of 0.005 ml of 15% formalin to rabbit eyes caused a severe reaction--corneal and conjunctival edema and iritis--graded 8 on a complex injury-grading scale of 1-10.[32] Exposure of rabbits to formaldehyde vapors at 40-70 ppm caused slight tearing and eye discharge, but not corneal injury.[78]

EXTENDED TOXICOLOGY STUDIES

Continuous 90-d inhalation studies have been conducted with several species of laboratory animals. In one study, rats, guinea pigs, rabbits, monkeys, and dogs were exposed to formaldehyde at 3.7 ppm.[40] One of the exposed rats died, but there were no overt signs of toxicity. Various degrees of interstitial inflammation were seen in the lungs of all the exposed animals, and there was focal chronic inflammation in the hearts and kidneys of the rats and guinea pigs. It was uncertain whether these changes were compound-related. In another study, groups of 25 rats were continuously exposed at 1.6, 4.55, or 8.07 ppm for 45-90 d.[50] The only adverse effect at 1.6 ppm was discoloration of the hair. The 4.55-ppm group was exposed for 45 d and had a decrease in rate of weight gain. The 8.07-ppm group was exposed for 60 d and had respiratory and eye irritation, a decrease in food consumption, and a decrease in liver weight.

In a noncontinuous inhalation study, groups of 20 mice and 20 rats were exposed to formaldehyde at 4, 12.7, or 38.6 ppm, 6 h/d, 5 d/wk, for 13 wk (Chemical Industry Institute of Technology, unpublished data). No adverse effects were observed in the 4-ppm group. At 12.7 ppm, a decrease in body weight and evidence of nasal erosion in two exposed rats were observed. Ulceration and necrosis of the nasal mucosa seen at 38.6 ppm resulted in termination of exposure after 2 wk. Groups of 60 mice were exposed at 41 or 82 ppm, 1 h/d, three times a week, for 35 wk.[91] The 41-ppm group was then exposed at 123 ppm for 29 wk. All the groups tolerated the exposure reasonably well, and the average weight of the mice rose normally. Pathologic examination of the tracheal epithelium revealed basal cell hyperplasia, squamous cell metaplasia, and atypical metaplasia. Extension of metaplasia into the major bronchi was infrequent, except in the animals that were exposed at 123 ppm. In these animals, the metaplastic changes in the epithelium appeared to extend farther into the major bronchi with increasing exposure. Exposure of a similar group of mice at 163 ppm was terminated after 11 d, because of severe pathologic changes and deaths.

The Formaldehyde Institute is sponsoring studies at Biodynamics, Inc., on effects of virtually continuous inhalation of formaldehyde in monkeys, hamsters, and rats. These are daily 22-h exposures at 3, 1, and 0.2 ppm that are repeated for 26 wk. Results of gross and microscopic evaluation of animals exposed at 0.2 and 1.0 ppm (now completed) showed no treatment-related effects. Final results (C. F. Reinhardt, personal communication) on animals exposed at 3 ppm have shown no adverse effects in hamsters; in rats and monkeys, there is histologic evidence of squamous metaplasia of the nasal mucosa in exposed animals. The hamsters showed no histologic changes at any of the exposure concentrations.

RESPIRATORY SYSTEM EFFECTS

Formaldehyde is readily soluble in the mucous membranes of animals. Respiratory tract uptake is almost 100% in dogs.[52] When inhaled by guinea pigs for 1 h at 0.3-50 ppm, formaldehyde increased airway resistance and decreased compliance. These effects were reversible at concentrations less than 40 ppm and were not seen 1 h after exposure. Guinea pigs exposed for 1 h at 3.5 ppm had a 40% increase in airflow resistance and a 12% decrease in compliance. The increase in resistance was dose-related over the range of 0.25-50 ppm; tracheal cannulation doubled the increase in resistance. The combination of formaldehyde and sodium chloride aerosol (0.04 μm in diameter) at 10 mg/m^3 further increased airway resistance.[6]

In another study, 25 rats each were exposed continuously for 3 mo at 0.0098, 0.028, 0.82, and 2.4 ppm. At 2.4 ppm, there was a significant decrease in cholinesterase activity; at 2.4 and 0.82 ppm, there were proliferation of lymphocytes and histiocytes in the lungs and some peribronchial and perivascular hyperemia. There were no significant findings at the two lower concentrations.

The effects of formaldehyde exposure on respiratory rate were studied in mice. Exposure for 10 min at 3.1 ppm, 3 h/d, for 3 d before exposure at a higher challenge concentration (0.55-13.4 ppm) produced the same response as in a previously unexposed group. Similar exposure at concentrations higher than 3.1 ppm caused an increased response. However, accommodation occurred during each exposure period, with the respiration rate approaching normal.[98] Other research has shown formaldehyde to decrease ciliary transport within 10 min at concentrations of 20-100 ppm.[42,45]

CARDIOVASCULAR SYSTEM EFFECTS

Large doses of formaldehyde have a vasopressor effect (increased blood pressure) in anesthetized mice. Smaller doses lead to a depressor response. Qualitatively, the responses are similar to that seen with acetaldehyde.[53] Dogs do not have such responses to formaldehyde. Other results from the same study suggest that an initial decrease in blood pressure is caused by alterations in the sympathetic nervous system. A later, more marked decrease may be the result of a direct effect on vascular smooth muscle.[107]

MUTAGENIC POTENTIAL

Numerous studies have been conducted to determine the mutagenicity of formaldehyde, and Auerbach et al.[11] have reviewed the subject extensively. Formaldehyde has exhibited mutagenic activity in a wide variety of organisms, but the mechanism of formaldehyde mutagenesis has not been resolved. Formaldehyde may cause mutations by reacting directly with DNA; by forming mutagenic products on reaction with amino groups on simple amines, amino acids, nucleic acids, or proteins; or by forming peroxides that can react directly with DNA or indirectly by free-radical formation.

Mutagenic activity has been detected in E. coli[14] and Pseudomonas fluorescens,[58] but not in the Ames strains of Salmonella typhimurium.[106] Sasaki and Endo reported that the mutagenicity of formaldehyde was very weak and appeared only within a limited range of concentration in which the Ames test was modified slightly by preincubating for 15 min at 37°C before charging the plates.[165] Weak mutagenic activity was observed when the fungi Neurospora crassa and Aspergillus nidulans were treated. The increase in mutagenic activity observed in these studies after treatment in the presence of catalase inhibitors suggested that peroxides were involved in the induction of mutations. Formaldehyde induced mitotic recombination in Saccharomyces cerevisiae.[34] Recently, formaldehyde was shown to induce mutations and cause DNA damage and repair in Saccharomyces.[35,117,118] The studies concerning formaldehyde mutagenesis in Drosophila have been reviewed by several authors.[11,156,181] Mutations were induced in male larvae fed formaldehyde-containing food and in adults given injections of aqueous solutions of formaldehyde. The exposure of adults or larvae to formaldehyde vapors has not produced mutations. In one of five species of grasshoppers, formaldehyde caused chromosomal damage.[120] Germinating barley seeds soaked in formaldehyde solutions did not give evidence of mutations on maturation.[54]

The mutagenic potential of formaldehyde in mammalian systems has not been thoroughly studied. An increase in mutation frequency was observed when formaldehyde was tested in the L5178Y mouse lymphoma assay.[37,76] Formaldehyde increased the mutation frequency in each of the four experiments conducted. However, a clear dose-response relationship was evident in only one of four experiments. No mutagenic activity was reported when formaldehyde was tested in the Chinese hamster ovary cell/HGPRT assay.[93] The data and a description of the treatment conditions have not yet been published. No effect was observed in limited dominant-lethal studies in which Swiss mice were given intraperitoneal injections of formaldehyde,[59] but many other mutagens were inactive in this series of tests.

Formaldehyde has mutagenic activity in a variety of microorganisms and in some insects. Work is necessary to ascertain its mutagenic potential in in vitro cultures of germinal or somatic mammalian cells. Such information would be used in evaluating the hazard to humans exposed to formaldehyde.

EMBRYOTOXIC AND TERATOGENIC POTENTIAL

Formaldehyde has not been shown to be teratogenic in animals. Pregnant dogs were fed diets containing formaldehyde (formalin in 40% solution) at 125 or 375 ppm on days 4-56 of gestation. None of the 212 pups examined showed anomalies. Some of these pups were returned to the breeding colony, and their offspring showed no abnormalities.[95]

Rats were continuously exposed at 0.01 or 0.8 ppm for 20 d. Halfway through the exposure period, the animals were mated. No gross abnormalities were observed in the offspring, but there was an increase in gestation time. The number of fetuses decreased with increased formaldehyde concentration. However, the actual numbers of offspring in the 0.8-ppm, 0.01-ppm, and control groups were 208, 235, and 135, respectively. No explanation was given for the large increase in offspring from the exposed rats, compared with controls.[73] In another study, rats were exposed at 4.1 ppm for 4 h/d on days 1-19 of pregnancy. No effect on the course of pregnancy or malformations in the fetuses were seen.[171] No alteration of reproductive function was seen in male rats given formaldehyde at 0.1 ppm in their drinking water and 0.4 ppm in the air for 6 mo.[80]

In a gavage study, pregnant outbred albino mice were fed formaldehyde on days 6-15 of gestation.[123] The mice were sacrificed on day 18; the general health and reproductive status of the dams were evaluated, and the fetuses were examined for external, visceral, and skeletal malformations. The formaldehyde, which contained 12-15% methanol as a preservative, was lethal to 22 of 34 dams treated with 185 mg/kg per day and one of 35 dams treated with 148 mg/kg per day. These doses did not produce statistically significant teratogenic effects in the fetuses of the surviving dams (two-sided $p < 0.05$, compared with controls).

When dogs were fed hexamethylenetetramine (which decomposes to formaldehyde and ammonia in acid media) at 600 and 1,250 ppm on days 4-56 of gestation, no evidence of teratogenesis was observed. And long-term feeding studies in rats given 1,600 ppm showed no effect on reproductive capacity.[95]

CARCINOGENIC POTENTIAL

A 90-d pilot study of formaldehyde was conducted by the CIIT (unpublished data). Rats and mice were exposed to atmospheres containing formaldehyde at 4, 12.7, or 40 ppm. The exposures were conducted approximately 6 h/d, 5d/wk, for 13 wk (12 wk for the highest concentration). Other animals served as controls and were exposed only to clean, filtered air. Exposure at 40 ppm resulted in ulceration or necrosis of nasal turbinate mucosa in significant numbers of animals of each species. Rats of both sexes had a high incidence of tracheal mucosal ulceration and necrosis; only a few male mice exhibited this lesion. Pulmonary congestion was prominent in both male and female rats and male mice at the high dosage. Female mice in the control and high-dosage groups had a similar incidence of pulmonary congestion. Secondary lesions encountered in rats exposed at 40 ppm were apparently related to bacterial septicemia after severe damage to respiratory tract mucosa. It was concluded that exposure at 40 ppm was lethal, but that exposure at 12.7 ppm was not lethal and should be tolerable for an extended period. The pilot study was followed by a study of Fischer 344 rats and B6C3F1 mice described in the following abstract:[185]

> Groups of 120 male and 120 female rats were exposed by inhalation to 0, 2, 6, or 15 ppm formaldehyde vapor 6 hr/day, 5 days/week, for 18 months of a 24-month study. The present communication describes interim findings based on data available after 18 months of exposure. Squamous cell carcinomas occurred in the nasal cavities of 36 rats exposed to 15 ppm formaldehyde. The tumors ranged from small early carcinomas of the nasal turbinate to large invasive osteolytic neoplasms which extended into the subcutis of the premaxilla. Similar tumors were not detected in rats exposed for 18 months to 2 or 6 ppm or in mice exposed to 2, 6, or 15 ppm formaldehyde. Rhinitis, epithelial dysplasia, and squamous metaplasia occurred in rats from all exposure levels of formaldehyde; however, the severity and extent of the lesions were dose related. In contrast, papillary hyperplasia and squamous atypia occurred only in animals exposed to 15 ppm formaldehyde.

This is the first experimental study to implicate formaldehyde as a potential carcinogen, but the significance of these preliminary findings can be evaluated only after completion of the study and analysis of the pathologic findings. (The CIIT reported at the Formaldehyde Symposium on November 20-21, 1980, in Raleigh, N.C., that nasal cancer had been observed in rats exposed at 6 ppm for 24 mo and in mice exposed at 15 ppm for 24 mo.)

Mice (strain C3H) exposed to formaldehyde at 83 ppm, for 1 h/d, 3 d/wk, for 35 wk or at 41.5 ppm for 1 h/d, 3 d/wk, for 35 wk and at 125 ppm for an additional 29 wk had basal cell hyperplasia and squamous cell metaplasia in the tracheobronchial epithelium, but no tumors.[91] Hamsters exposed at 10 ppm for 5 h/d, 5 d/wk, for their lifetime (average, 18 mo) had increased cell proliferation and hyperplasia in the lungs (P. Nettesheim, unpublished data); weekly 5-h exposures at 50 ppm for lifetime (18 mo) produced squamous metaplasia, but no tumors. In neither of these studies was nasal tissue specifically examined.

Injection-site sarcomas developed in two of 10 rats given weekly injections of 0.4% aqueous formaldehyde for 15 mo.[208] Fibrosarcomas were observed in the liver and omentum in two other rats. These results are not useful, because of lack of controls and inappropriateness of the route of administration.

A. R. Sellakumar et al. (personal communication) exposed Sprague-Dawley rats to hydrogen chloride at a mean concentration of 10.6 ppm and formaldehyde at 14.7 ppm for 6 h/d, 5 d/wk, for their lifetime. Before dilution to the stated concentrations in the exposure chamber, the initial reaction mixture had average hydrogen chloride and formaldehyde concentrations of about 6,500 and 1,000 ppm, respectively; alkylating-agent activity of 1.8 ppm was also detected, possibly as a result of the interaction of hydrogen chloride and formaldehyde in the gas phase. Alkylating-agent activity in the animal exposure chamber, as measured by chromatography, was 0.028 ppm. Of the 99 exposed animals, 25 developed squamous cell carcinomas of the nasal epithelium.[169] No squamous cell tumors were observed in controls. One of the alkylating agents identified in the chamber was bis(chloromethyl) ether (BCME), at a concentration of less than 1 ppb. BCME is a potent carcinogen; esthesioneuroepitheliomas of the nose, squamous cell carcinomas of the lung and nasal turbinates, and adenocarcinomas of the lung and nasal cavity have been produced in rats after 10-100 exposures to BCME at 0.1 ppm for 6 h/d, 5 d/wk.[108]

Published reports indicate that BCME should not be formed in substantial amounts during chronic animal studies if concentrations of both hydrogen chloride and formaldehyde are less than 100 ppm at ambient temperature and humidity.[97,189] However, Frankel et al.[67] studied the reactions of formaldehyde and hydrogen chloride in the formation of BCME in glass vessels. They found that BCME is formed at less than 0.5 ppb when formaldehyde and hydrogen chloride are each present at 20 ppm, at less than 0.4 to 8.3 ppb (average, 2.7 ppb) when they are present at 100 ppm, and at 5-59 ppb when they are present at 300 ppm. It was estimated that it would take longer than 18 h to reach a steady state and concluded that further study was needed to define the reaction kinetics. (See Chapter 5 for discussion of the potential for the formation of BCME in the atmosphere.)

The carcinogenic potential of hexamethylenetetramine (HMT), which can decompose in an acid medium to release formaldehyde and ammonia, has been examined.[47] Mice and rats were given fresh solutions of HMT in drinking water every 24 h at 0.5-5% for 30-60 wk and at 1-5% for 2-104 wk, respectively. Mice were observed for up to 130 wk, and rats for up to 3 yr. At 5% HMT, there was 50% mortality in the rats after 2 wk. No significant effects on growth or survival were observed in any of the other groups of rats or in the mice. Histologic examination indicated that no effects were attributable to HMT. No carcinogenic activity was observed.

EFFECTS IN HUMANS

The principal effect of low concentrations of formaldehyde observed in humans is irritation of the eyes and mucous membranes. Table 7-2 summarizes data on human responses to airborne formaldehyde at various concentrations. It shows a wide range in formaldehyde concentrations reported to cause specific health effects. The severity of symptoms appears to be dose-related at extremes of concentration. In general, at low concentrations, below 0.05 ppm, no effects were reported. Objective changes in laboratory tests (i.e., optical chronaxy, EEG, etc.) without manifest symptoms were reported at concentrations beginning at 0.05 ppm, but more often at 1.5 ppm and

TABLE 7-2

Reported Health Effects of Formaldehyde at Various Concentrations

Health Effects Reported	Approximate Formaldehyde Concentration, ppm	References
None reported	0-0.5	65, 132, 198
Neurophysiologic effects	0.05-1.50	65, 132, 198
Odor threshold	0.05-1.0	15, 20, 65, 68, 112, 175, 207, 215, 217
Eye irritation	0.01-2.0[a]	61, 78, 133, 137, 163, 168, 175, 198, 207, 217
Upper airway irritation	0.10-25	3, 9, 15, 20, 60, 102, 107, 134, 137, 173, 192, 198, 215, 217, 218
Lower airway and pulmonary effects	5-30	68, 71, 85, 86, 107, 151, 152, 167, 173, 198, 204, 215, 218
Pulmonary edema, inflammation, pneumonia	50-100	16, 152, 218
Death	100+	16, 152

[a] The low concentration (0.01 ppm) was observed in the presence of other pollutants that may have been acting synergistically.

higher. The odor of formaldehyde is generally perceived by about 1 ppm, but some people can detect 0.05 ppm. Variable nonspecific complaints--such as increased thirst, dizziness, headache, tiredness, and difficulty in sleeping--are difficult to evaluate; however, they were generally reported when concentrations exceeded 1 ppm. Symptoms of eye irritation were reported at concentrations as low as 0.05 ppm. At concentrations at or above 1 ppm, nose, throat, and bronchial irritation was noted. Such irritation was readily reported when the concentration reached 5 ppm. When concentrations exceeded 50 ppm, severe pulmonary reactions occurred, including pneumonia, bronchial inflammation, and pulmonary edema, sometimes resulting in death.

Table 7-2 clearly shows the variability and overlap of responses among subjects. Some persons develop tolerance to olfactory, ocular, or upper respiratory tract irritation. Such factors as smoking habits, socioeconomic status, preexisting disease, various host factors, and interactions with other pollutants and aerosols are expected to modify these responses.

EYE

Eye irritation is a common complaint of persons exposed to formaldehyde vapor.[133,168,175,207,217] Formaldehyde is detectable at 0.01 ppm, and at 0.05-0.5 ppm it produces a more definable sensation of eye irritation.[61,163,198] Occupational exposures at 0.9-1.6 ppm to formaldehyde released from paper pulp treated previously with urea-formaldehyde or melamine-formaldehyde resulted in complaints of itching eyes, dry and sore throats, disturbed sleep, and unusual thirst on awaking in the morning.[137] Eye, nose, and throat irritation was reported by three of 16 subjects exposed for 5 h/d for 4 d at 0.3 mg/m^3 (0.2 ppm) and 15 of 16 subjects exposed at 1.0 mg/m^3 (0.7 ppm) in a chamber.[9] Sim and Pattle[175] exposed 12 men in an exposure chamber at 13.8 ppm for 30 min. There was considerable nasal and eye irritation when the men first entered the chamber. However, the eye irritation was reportedly not severe, and the symptoms wore off after about 10 min in the chamber. Other studies reported that eye irritation may occur at concentrations below 1 ppm.[133,168,175,207,217] Marked irritation with watering of the eyes occurs at a concentration of 20 ppm in air.[198] Eye damage from formaldehyde vapor at low concentration is thought not to occur, because of the protective closure of the eye that results from discomfort.[78] Increased blink rates were noted at concentrations of 0.3-0.5 ppm in persons studied in so-called pure air irradiated in smog chambers.[168] Blink rate, although used as an objective measure of eye irritation, appears variable for any given subject. The irritant effects of formaldehyde seem to be accentuated when it is mixed with other gases. In 14 smog-chamber tests, there was an average eye-irritation index of 4.9 ± 1.0 units (on a scale of 0-24; 0-16, none to severe irritation, and over 16, lacrimation in more than 50% of the subjects). It was concluded that the human subjects tested could readily detect and react to formaldehyde at as low as 0.01 ppm.

A difference in the concentration-response curves for formaldehyde was seen in the presence of photooxidation products of ethylene and propylene. A linear relationship was noted between eye irritation and formaldehyde concentration over a range of 0.3-1 ppm. It seemed that formaldehyde and peroxyacetylnitrate accounted for 80% and 20%, respectively, of the eye irritation associated with photochemical air pollution. In the usual smog-chamber experiments, dilute mixtures of nitric oxide, nitrogen dioxide, and hydrocarbons in air are irradiated. The Committee is not certain about the extent to which nitric acid, formic acid, and similar compounds shown to be present since the earlier studies were done contributed to the eye irritation observed in those experiments.

Accidental splash exposures of human eyes to aqueous solutions of formaldehyde have resulted in a wide variety of injuries, depending on concentration and treatment. These range from discomfort and minor transient injury to delayed but permanent corneal opacity and loss of vision. Immediate flushing with water spared the eyes of one worker who received a splash injury from 40% formaldehyde solution.[100] A similarly exposed coworker whose eyes were not flushed with water lost vision in both eyes. Results of other accidental exposures to aqueous formaldehyde in humans and experimental ocular studies in animals were described by Grant.[78] Potts has shown that intravenous administration of formaldehyde (at 0.9 g/kg) has a pronounced action on retinal function, as indicated by changes in alpha and beta waves of the electroretinogram that were correlated with ophthalmoscopic retinal edema.[153] The changes would be missed if histology alone were used to detect them. In a NIOSH study, a complete visual test battery and ophthalmologic examination of workers exposed at 1.5 ppm revealed no effects of formaldehyde on the eye.[210]

In summary, human eyes are very sensitive to formaldehyde, detecting atmospheric concentrations of 0.01 ppm in some cases (when mixed with other pollutants) and producing a sensation of irritation at 0.05-0.5 ppm. Lacrimation is produced at 20 ppm, but damage is prevented by closure of the eyes in response to discomfort. Aqueous solutions of formaldehyde accidentally splashed into the eyes must be immediately flushed with water to prevent serious injury, such as lid and conjunctival edema, corneal opacity, and loss of vision. Table 7-3 summarizes some of the studies concerning eye irritation.

OLFACTORY SYSTEM

The odor threshold of formaldehyde is usually around 1 ppm, but may be as low as 0.05 ppm.[15,20,65,68,112,173,175,207,217] Olfactory fatigue with increased olfactory thresholds of rosemary, thymol, camphor, and tar was reported among plywood and particleboard workers and is presumed to be associated with formaldehyde exposure.[215]

TABLE 7-3

Eye Irritation Effects of Formaldehyde

Formaldehyde Concentration, ppm	Exposure	Effects on Eyes	Reference
Chamber—single:			
0.03–3.2	20–35 min; gradually increasing concentration	Increase in blink rate; irritation	210
13.8	30 min	Irritation (and nose irritation)	175
20	Less than 1 min	Discomfort and lacrimation	13
Chamber—repeated:			
0.25	5 h/d for 4 d	19% "slight discomfort"	9
0.42	5 h/d for 4 d	31% "slight discomfort" and conjunctival irritation	9
0.83–1.6	5 h/d for 4 d	94% "slight discomfort" and conjunctival irritation	9
Occupational:			
4–5	--	Irritation, lacrimation, and discomfort in 30 min	63
0.9–2.7	--	Tearing	15
0.3–2.7	--	Prickling and tearing	173
0.9–1.6	--	Intense irritation and itching	152
0.13–0.45	--	Stinging and burning	210
Indoor residential:			
0.067–4.82	--	Tearing	212
0.02–4.15	--	Irritation	212
0.03–2.5	--	Irritation	24

RESPIRATORY TRACT

The human nose adjusts the temperature and water-vapor content of air and removes a large proportion of foreign gases and dusts,[154] and the nasal mucociliary system clears foreign material deposited on it. Nasal congestion from injury may lead to partial mouth-breathing; when nasal functions are impaired or the nose is otherwise bypassed for mouth-breathing, the burden of conditioning and cleaning the air falls on the lung. If the nasal defense system is disturbed or if mouth-breathing occurs, greater concentrations of formaldehyde will reach the lungs, and other noxious materials that are ordinarily cleared from the airways may be retained. In this regard, the differences in breathing of rats and mice should be noted. Rats and mice are obligatory nose breathers; therefore, nasal defense mechanisms may be more important in these animals. Thus, with respect to target organs for formaldehyde, it may be inappropriate to extrapolate results of rat and mouse formaldehyde-inhalation experiments directly to humans.

Upper Airway Irritation

Symptoms of upper airway irritation include the feeling of a dry throat, tingling sensation of the nose, and sore throat, usually associated with tearing and pain in the eyes. Irritation occurs over a wide range of concentrations, usually beginning at approximately 0.1 ppm, but reported more frequently at 1-11 ppm[15,20,60,102,107,137,215,217] (see Table 7-2). Tolerance to eye and upper airway irritation may occur after 1-2 h of exposure.[15,102,173] However, even if tolerance develops, the irritation symptoms can return after a 1- to 2-h interruption of exposure.[3,15,102,134,173,192] As in the case of eye irritation, some persons seem to tolerate higher concentrations, 16-30 ppm--perhaps subjects who developed tolerance.

When 16 healthy young subjects were exposed to formaldehyde at 0.25, 0.42, 0.83, or 1.6 ppm for 5 h/d for 4 d, nasal-mucus flow rate was decreased at all concentrations except 0.83 ppm.[9] Subjective responses to formaldehyde included slight conjunctival irritation and dryness of the throat and the upper third of the nose.

Helwig reported that schoolchildren and teachers developed eye and respiratory tract irritation, gastrointestinal disturbances, increased thirst, and apathy after moving into a prefabricated school building.[84] The "new-building odor" was particularly strong after weekends and holidays. Measurements of airborne formaldehyde made with Drager tubes revealed concentrations of 5 ppm or more on one occasion. Mild dysrhythmias were present in 20 children who underwent EEG studies. No details were given regarding the medical complaints or the number of children who developed adverse reactions while attending classes. The author felt that plastic polymers used in chipboard might also produce similar effects. Children who moved to another building after graduation no longer had any symptoms.

Eye and upper respiratory tract irritation were noted in some employees of funeral homes that used formaldehyde and paraformaldehyde in the embalming process; airborne concentrations in the embalming rooms were 0.25-1.39 ppm.[102] A garment factory had airborne concentrations of 0.9-2.7 ppm;[15] eye and upper respiratory tract irritation were more common in areas where large quantities of partially completed permanent-press materials accumulated.

The incidence of chronic rhinitis and pharyngitis was higher among formaldehyde-exposed workers in a wood-processing facility than in a control group.[198,218] A majority of workers complained of throat irritation, diminished smell, and dryness of the nose and pharynx. Examination of the nose and throat revealed hypertrophic or subtrophic nasal mucosa and subtrophic or atrophic pharyngitis in almost half the exposed workers. The incidence of pathology was highest in workers with the most exposure to formaldehyde. Formaldehyde concentrations reportedly ranged from 0.5 to 8.9 ppm, although occasional brief excursions above this limit were also observed. This study of wood-processing employees did not include measurements of other airborne contamintants, such as wood dust. In another study, reduced mucociliary function of the nasal mucosa and increased olfactory threshold to rosemary, thymol, camphor, and tar were observed in formaldehyde-exposed workers, compared with controls, regardless of evidence of nasal pathology.[215]

Nasal cancers in humans have been reported in some highly select occupations, such as wood-working and work with nickel.[1,199] Because of the shape of and the high linear velocity of air in the anterior part of the nose, a large portion of dust that enters the nose is deposited in this portion. But the main nasal passage has a large surface area and is narrow, and air in this portion has low linear velocity; gases are therefore absorbed here. There may be a direct or indirect local effect of chemical agents or an interference with repair mechanisms at the sites of deposit or absorption. Further research is necessary concerning the morphology of the nasal turbinates and the histopathology of the nasal mucosa in rats, mice, and humans before definitive comparisons can be made with respect to exposure to specific chemicals, such as formaldehyde.

In summary, irritation of the nose and throat caused by formaldehyde may occur at concentrations of 0.1 ppm, but occurs more frequently at 1-11 ppm. Examinations of the nose and throat reveal chronic changes that are more severe in persons exposed to the higher concentrations. Exposure to formaldehyde can cause alterations in the nasal defense mechanisms that include a decrease in mucociliary clearance and a loss of olfactory sensitivity. There are no studies that show that formaldehyde is carcinogenic in humans. The potential for animal carcinogenicity is discussed elsewhere in this report.

Lower Airway and Pulmonary Effects

Lower airway irritation that is characterized clinically by cough, chest tightness, and wheezing is reported often in people exposed to formaldehyde at 5-30 ppm.[68,71,107,152,198,215,218] Chest x rays of persons apparently exposed to formaldehyde at high concentrations are usually normal, except for occasional reports of accentuated bronchovascular markings, but pulmonary-function test results may be abnormal.[218] Acute respiratory distress was reported in a physician after several hours of formaldehyde exposure.[152] Physical examination of the physician's chest revealed diffuse rales and occasional rhonchi. A chest x ray was interpreted as showing early pulmonary edema. It is not known whether this case constitutes an example of a hypersensitivity reaction to formaldehyde or acute chemical pneumonitis. No specific information was given on the exposure to formaldehyde.

Pulmonary-function studies of rubber workers exposed to a hexamethylene-tetramine-resorcinol resin showed decreased pulmonary function.[71] However, no association could be demonstrated between concentrations of airborne resorcinol, formaldehyde, hydrogen cyanide, or ammonia and change in pulmonary function. In a study of employees who manufacture filters with fibers that are impregnated with phenol-formaldehyde, a reduction in the ratio of FEV_1 to FVC, expressed as percent, and maximal expiratory flow at 50% of vital capacity were noted on Monday morning, compared with values of the previous Friday, for employees who had worked more than 5 yr.[167] Detailed measurements of formaldehyde were not made, but two surveys reported concentrations of 0.4-0.8 ppm and 9.14 ppm. The work environment included other pulmonary irritants, such as phenol and acrylic fiber breakdown products. Chronic cough and sputum production occurred more often in those currently employed in production for over 5 yr than in those never involved in production, although little change in pulmonary-function test results was noted during the course of a workweek or workday.

The prevalence of respiratory illness and complaints among employees in eight textile plants was more than 15% for four plants and 5-15% for the other four.[173] These results were obtained from medical records and were not confirmed through medical examination of the employees. Airborne formaldehyde concentrations were 0-2.7 ppm, with an average of 0.68 ppm. Workers reported that formaldehyde concentrations varied considerably with changes in temperature and humidity. It is not known whether the airborne formaldehyde concentrations were representative of seasonal fluctuations.

Pulmonary edema, pneumonitis, and death could result from very high formaldehyde concentrations, 50-100 ppm.[16,152,218] It is not known what concentrations are lethal to humans, but concentrations exceeding 100 ppm would probably be extremely hazardous to most and might be fatal in sensitive persons.

Asthma

Allergic contact dermatitis caused by formaldehyde sensitivity is well-recognized, but there have been relatively few documented cases of occupational asthma attributable to formaldehyde and proved by bronchial inhalation challenge tests.[85,86,110,143,148,151,164,167,194,204] In the cases reported by Hendrick and Lane, nurses in a renal hemodialysis unit developed asthma as a result of continued exposure to formaldehyde that was used to sterilize the artificial-kidney machines.[85,86] In all, eight of 28 persons studied had experienced asthmatic attacks or bronchitis. In five of the eight, attacks had been recurrent for at least 3 yr, and only one had ever experienced such symptoms before joining the unit. Bronchial provocation tests were positive in only two persons. In those two, wheezing began approximately 2-3 h after exposure, and the results of measured pulmonary function tests fell by as much as 50%. Reactions persisted for from 10 h to 10 d, depending on the exposure; concentrations in the air were not reported. The asthmatic reaction could be inhibited by beclomethasone aerosol.

Mechanism of Airway Responses to Formaldehyde

Formaldehyde has been shown to cause bronchial asthma in humans.[85,86,110,143,148,151,164,167,194,204] Although asthmatic attacks are in some cases due specifically to formaldehyde sensitization or allergy, formaldehyde seems to act more commonly as a direct airway irritant in persons who have bronchial asthmatic attacks from other causes. Persons with bronchial asthma respond to numerous agents, such as exogenous irritants and allergens, respiratory infections, cold air, smoke, dust, and stress.[22,74] The asthmatic person seems to represent an extreme on the scale of respiratory sensitivity to inhaled irritants. The data suggest a dose-response relationship, with increasing numbers of asthmatics having attacks as air pollution worsens. Thus, the airways of asthmatics respond to many nonspecific inhaled irritants, including formaldehyde.

The exact mechanism of the asthma syndrome related to formaldehyde exposure is not known. It has been suggested that an immunologic basis is sometimes operative. However, no studies have demonstrated the presence of specific circulating immunoglobulins (IgE or IgG) in affected persons.

Nonimmunologic mechanisms may explain the effects of formaldehyde on the airways. Although formaldehyde at low concentrations may cause asthmatic symptoms in some sensitized subjects, in irritant concentrations it produces bronchoconstriction in even normal persons. The effect of lower concentrations on airways may be similar to those of chemicals, such as toluene diisocyanate (TDI), that at low concentrations not ordinarily considered irritating do produce an adverse airway response unrelated to allergy, possibly on a pharmacologic basis.[27,28,150,176,213] An abnormality of the beta-adrenergic receptor system has been proposed as an explanation for asthma due to TDI.[27,28] Other possible pharmacologic mechanisms may be similar to those associated with cotton dust, cotton extracts have been reported to cause histamine release from basophils.[19]

Inhalation of formaldehyde vapors may itself act directly on smooth muscle or nerve endings, causing airway hyperreactivity, an important component of bronchial asthma.[44,111,149,182] Methacholine and histamine challenge tests have demonstrated this hyperreactivity with other environmental pollutants.[21,22,26,28,74]

Recently, alterations in the bronchial mucosal epithelial barrier have been proposed as a theory to explain the effects of environmental agents on airways.[18,88] Normally, the bronchial mucosa provides a barrier, preventing entry of high-molecular-weight protein into the submucosal layer. Environmental agents can increase both the permeability of the bronchial epithelium and the response to histamine at subthreshold concentrations. The disruption of the bronchial epithelial barrier, perhaps the tight junction between cells, permits the environmental and pharmacologic agents better access to the underlying tissue and the capability of reaching afferent nerve fibers that are directly beneath the tight junctions of the epithelial cells. This greater accessibility to the nerve fibers leads to the apparent increased reactivity of airways. In addition, formaldehyde may be able to act directly on bronchial smooth muscle beneath the epithelial barrier.[101] Nonspecific mast-cell degranulation from formaldehyde, resulting in release of vasoactive substances and causing smooth-muscle contraction, is another possible nonimmunologic mechanism.

Summary

A number of lower airway and pulmonary effects may occur from formaldehyde exposure. In most normal persons exposed to formaldehyde, concentrations greater than 5 ppm will cause cough and possibly a feeling of chest tightness. It is possible that normal persons will experience these symptoms at 2-3 ppm, but data are not available on this. In some susceptible persons, concentrations below 5 ppm can cause these symptoms, including wheezing. In persons with bronchial asthma, the irritation caused by formaldehyde may precipitate an acute asthmatic attack, possibly at concentrations below 5 ppm. Rarely does a person with asthma become sensitized (allergic) specifically to formaldehyde and thereby respond to concentrations lower than 0.25 ppm. This reaction is not due to formaldehyde's irritant properties, but is related to some poorly understood immunologic (or possibly nonimmunologic) mechanism. In concentrations greater than 50 ppm, severe lower respiratory tract effects can occur, with involvement not only of the airways, but also of alveolar tissue. Acute injury of this type includes pneumonia and noncardiac pulmonary edema.

SKIN

Skin contact with formaldehyde has been reported to cause a variety of cutaneous problems in humans, including irritation, allergic contact dermatitis, and urticaria.[144,162,180] Allergic contact dermatitis from formaldehyde is relatively common, and formaldehyde is one of the more frequent causes of this condition both in the United States[66] and in other areas.[69] The North American Contact Dermatitis Group reported that formaldehyde is the tenth leading cause of skin reactions among dermatitis patients patch-tested for allergic contact dermatitis. Approximately 4% of 1,200 patients had positive skin reactions when tested with 2% formalin (0.8% formaldehyde) under an occlusive patch.[142] Minor epidemics of allergic contact dermatitis have been described in diverse situations, for example, among nurses who handled thermometers that had been immersed in a 10% solution of formaldehyde[161] and among those who were exposed to formaldehyde in hemodialysis units.[180]

In many cases, either the initiation or the elicitation of the allergy has been caused by contact with formaldehyde or formalin, but it may also result from formaldehyde-releasing agents used in cosmetics, medications, and germicides, from incompletely cured resins, and from the decomposition of formaldehyde-containing resins used in textiles.[115] People with cutaneous allergy to formaldehyde have particular problems because there are so many sources of formaldehyde exposure in ordinary daily life. For example, the FDA lists 846 cosmetic formulations containing formaldehyde.[201] The skin reaction rate from cosmetic formulations containing formaldehyde has not been excessive, because it is used mainly as a preservative in shampoos, whose contact time with skin is short. Formaldehyde-releasing cosmetic preservatives, such as Quaternium-15, have shown a greater reaction frequency than formaldehyde itself (unpublished data from Cosmetics Technology Division, Bureau of Foods, FDA).

Humans can come into contact with low concentrations of formaldehyde from many sources, and repeated contact with them may be sufficient to provoke responses in people with allergic contact sensitization. It is important to mention that previously "normal" people can become sensitized. These sources include components of plastics, glues, antifungal disinfectants, preservatives, paper, fabrics, leather, coal and wood smoke, fixatives for histology, and photographic materials.[70] Available data do not permit the determination of a degree of exposure to formaldehyde-containing products that would be safe once sensitization has occurred.

Occupational dermatitis from urea-formaldehyde dusts and powders (containing free formaldehyde) in the workplace was reported by Harris.[81] Exposed skin—e.g., on the face, lips, and neck and in interdigital areas—was affected, as well as such permeable skin sites as the scrotum and eyelids and intertriginous areas, such as the armpit and flexure areas of the arms.

The response of formaldehyde-sensitive persons is related to the extent of exposure (see Table 7-4).[126] However, most sensitized persons can tolerate topical axillary products containing formaldehyde at up to about 30 ppm.[96] With increasing concentration, one sees a higher frequency of responders,[127] probably because skin penetration by formaldehyde varies from one person to another and even from one site to another on the same person. Thus, different amounts of formaldehyde may reach different target sites. The dose needed to elicit a response depends on these factors and others, such as occlusion, temperature, contact time, and vehicle.

TABLE 7-4

Elicitation (Occluded) of Skin Reactions in
Five Formaldehyde-Sensitized Subjects[a]

Formaldehyde Challenge Concentration, %	No. Responding (n = 5)
1	4
0.5	2
0.2	1
0.1	1
0.01	1

[a]Data from Marzulli and Maibach.[126]

Allergic contact dermatitis is a manifestation of cell-mediated immunity. The standard diagnostic test for this condition is the epidermal patch test. In the case of formaldehyde, interpretation may be complicated by the irritant potential of the substance. Patch testing is now generally conducted with a 2% concentration of formalin. Before the early 1970s, a 5% solution in water was commonly used; many of the reported results of earlier patch testing may have been spurious.[142] Patch testing for skin sensitization to formaldehyde resin is performed with a 5-10% concentration of the resin in petrolatum.[2]

So-called predictive tests for skin sensitization are used--first on animals, then on man--to identify the allergenic potential of new substances or formulations entering the marketplace. Guinea pigs are the favored animal species. The Draize intradermal technique[49] and one of the published adjuvant techniques[125] are animal methods often used before human investigation in evaluating skin hypersensitivity. The Draize technique is likely to underestimate the human response, whereas adjuvant (Freund's complete adjuvant) techniques are likely to overestimate it. In human predictive testing, two techniques are useful: the "modified Draize"[127] and the "maximization"[104] methods. Results obtained for formaldehyde with some of these techniques are compared in Table 7-5.

Although formaldehyde has been reported to cause contact urticaria, it is not yet clear whether this is immunologically mediated.[144] Formaldehyde is a potent sensitizer and irritant, repeated exposure to it may also result in dermatitis.

In summary, formaldehyde is a skin irritant and skin sensitizer. Formaldehyde plastics sensitize skin by contact with formaldehyde resin or by releasing formaldehyde from incompletely cured plastic dusts or particles. Aqueous formaldehyde solutions (e.g., cosmetic formulations) elicit a skin response (under occlusive cover) in some sensitized people at concentrations as low as 0.01%, but underarm products containing up to 0.003% formaldehyde are tolerated by most sensitized persons. Formaldehyde-releasing preservatives, such as Quaternium-15, may sensitize to formaldehyde or to the parent material. Occupational exposure to free formaldehyde in urea-formaldehyde dusts and powders may also result in dermatitis.

CENTRAL NERVOUS SYSTEM

Central nervous system responses to formaldehyde have been tested in a variety of ways, including by determination of optical chronaxy,[132] electroencephalographically,[65] and by the sensitivity of the dark-adapted eyes to light.[132] Responses are reported in some persons at 0.05 ppm and are maximal at about 1.5 ppm. Formaldehyde at less than 0.05 ppm probably has little or no objective adverse effect.[198] Fel'dman and Bonashevskaya reported that formaldehyde at 0.032 ppm produced no electroencephalographic changes and did not reach the odor threshold in five extremely sensitive

TABLE 7-5

Predictive Skin-Sensitization Test Results with
Aqueous Formaldehyde

Species	Method	Positive-Response Frequency, %
Guinea pig	Draize intradermal	5[119]
	Adjuvant (maximization)	80[119]
Human	Maximization	72[119]
	Modified Draize	4.5-7.8[127]

subjects.[65] Melekhina demonstrated sensitivity of the dark-adapted eye to light at about 0.08 ppm.[132]

ALIMENTARY TRACT

Ingestion of formaldehyde has been reported to cause headache, upper gastrointestinal pain,[23,51,57,105,122,198] allergic reactions,[198] corrosive effects on gastrointestinal and respiratory tracts,[57,105,114] and systemic damage.[57,105,114] Accidental or suicidal poisoning with formaldehyde usually involves the ingestion of aqueous solutions; death occurs after the swallowing of as little as 30 ml of formalin.[16,105] Gastrointestinal tract damage is most marked in the stomach and lower esophagus, with the tongue, oral cavity, and pharynx generally not severely affected.[198] The small intestine may occasionally be involved; perforated appendix is a rare complication. When the chemical infiltrates around the epiglottis, injury to the larynx and trachea may occur.[16,105,198] After ingestion, there may be loss of consciousness, vascular collapse, pneumonia, hemorrhagic nephritis, and spontaneous abortion.[16,105] One autopsy report of a fatal ingestion described hardening of organs adjacent to the stomach (lung, liver, spleen, and pancreas), hyperemia and edema of the lungs, bilateral diffuse bronchopneumonia, fatty degeneration of the liver with subcapsular hemorrhage, renal tubular necrosis, and involvement of the brain.[16,105,157]

Other avenues of acute poisoning include intravesical instillation of formalin for control of intractable bladder hemorrhage[36] and accidental irrigation of the colon with aqueous formaldehyde.[94] Paresthesia, soft-tissue necrosis, and sequestration of bone have occurred when formaldehyde preparations have been used for devitalizing dental pulps.[79,83,135] An outbreak of hemolytic anemia among patients at a hemodialysis unit was traced to formaldehyde leaking from water filters impregnated with a melamine-formaldehyde resin.[145]

EFFECTS ON REPRODUCTIVE SYSTEM

Menstrual abnormalities and complications of pregnancy were reported to occur more frequently in Russian women employed in the textile industry and in contact with urea-formaldehyde resins.[174] The unique role, if any, of formaldehyde in this study is not clear, because of the lack of information, e.g., on other potentially toxic compounds in the workplace that might adversely affect the reproductive system, on the composition and comparability of the populations that were the source of the reported data, and on various demographic, socioeconomic, and physiologic factors. Nevertheless, it is pertinent to summarize what was reported. Formaldehyde concentrations were 1.5-4.5 $\mu g/m^3$ for high-exposure trimmers, 0.3-0.7 $\mu g/m^3$ for sorters, and 0.05-0.1 $\mu g/m^3$ for others. About 70% of the women were under 40 yr old. Menstrual disorders were

encountered more often in women with greater exposures (trimmers) and in direct relationship to duration of employment. Oligodysmenorrhea was the major menstrual disorder: 24.3 ± 2.2% of the trimmers, 20.2 ± 2.2% of the sorters, and 9.2 ± 1.1% of the controls.
Complications of pregnancy were more prevalent in the more exposed group. Anemia, as a complication, was noted twice as often in the exposed group. Other complications--such as intrauterine asphyxia, premature rupture of the membranes, late toxemia, threatened abortion, and premature deliveries--were analyzed and said to be more frequent in the exposed groups, but no substantial analysis was reported. There was also a greater percentage of newborns with low birthweight in the exposed groups. Of the infants born to women who had contact with formaldehyde, 26.9 ± 4.9% weighed 2,500-2,990 g at birth, compared with 11.3 ± 1.3% of the infants born to women in the control grup ($p < 0.05$).

BLOOD

Hemolysis has been observed among patients undergoing chronic hemodialysis. It resulted in contamination of several lots of dialysis water with an excess formaldehyde concentration of 10 mmol/L.[145] Water filters treated with melamine-formaldehyde resin were the source of the contaminated formaldehyde. A concentration as low as 0.1 mM caused decreased ATP content when incubated with blood cells. There is also evidence that formaldehyde sterilization of dialyzers may cause antibody-mediated hemolysis that contributes to renally induced anemia.[62]

CONSUMER COMPLAINTS IN RESIDENTIAL ENVIRONMENTS

Over the last several years, increasing numbers of complaints have caused concern about the health hazards of residing in homes where formaldehyde is released into the living space. The Consumer Product Safety Commission (CPSC) received more than 1,000 complaints from users of mobile homes and conventional homes insulated with UF foam by March 1980.[43,134,195,197] The Department of Housing and Urban Development reported an increase in complaints about formaldehyde during this same period. On August 1, 1979, the CPSC issued a consumer advisory on UF insulation, citing possible health problems associated with this type of insulation.[134,195]

A number of studies have been undertaken to determine the magnitude and extent of formaldehyde exposure of persons in the residential environment.[3,24,30,77,87,186,192,196,211] In 1975, Anderson et al.[10] reported formaldehyde concentrations ranging from 0.08 to 2.24 mg/m^3, with an average of 0.62 mg/m^3, in 25 rooms in 23 conventional Danish homes with chipboard in their interior construction. In 1977, Breysse[24] reported four cases (investigated in 1961) in which people in conventional buildings had complained of eye and upper respiratory irritation in association with exposure to formaldehyde from particleboard and chipboard. In a compilation of periodic investigations (1968-1977) of complaints, he noted 74 mobile homes, six of which were unoccupied, in which 92 persons experienced adverse reactions "allegedly" resulting from exposure to formaldehyde. The range of concentrations reported was 0-2.5 ppm; fewer than 10% were above 1.0 ppm. The prevalence of symptoms in the 92 people was reported as follows: eye irritation, 80 persons; nose irritation, 12; respiratory tract irritation, 58; headache, 51; nausea, 12; and drowsiness, 26. Severity of symptoms was not correlated with formaldehyde concentration. However, it should be pointed out that people questioned noted relief of symptoms when they left their homes for the weekend and return of symptoms when they went home.

In November 1977, the Connecticut Department of Health and Consumer Protection began receiving complaints from state residents who had UF foam insulation installed in their homes.[77] By September 1978, 84 complaints had been received. The Department tested the 84 homes and found formaldehyde in the air in 75. The sensitivity of the testing system was reported to be less than 0.05 ppm. Health symptoms were reported by 224 residents of 74 homes, in which detectable concentrations of formaldehyde ranged between 0.5 and 10 ppm, with a mean of 1.8 ppm. The symptoms of the residents included eye, nose, and throat irritation; GI tract symptoms; headache; skin problems; and some miscellaneous complaints, such as fatigue, aches, and swollen glands. In 37%, however, symptoms occurred when formaldehyde was not detectable by the methods used. When formaldehyde was detectable (0.5-10 ppm), 49% of the occupants had eye irritation, 37% nose and throat irritation, 46% headache, and 22% GI tract symptoms; in homes with no detectable formaldehyde, 26% had eye symptoms, 41% nose and throat irritation, 26% headache, and 42% GI tract symptoms.

Since January 1978, the Wisconsin Division of Health has collected air samples and environmental data on 100 mobile homes, conventional homes, and offices that have particleboard in their construction and foam insulation.[46] Air samples were collected in midget impingers and analyzed with the chromotropic acid procedure. Health information was obtained from the occupants of these structures. Formaldehyde ranged from undetectable to 4.18 ppm. The median concentration was 0.35 ppm (0.47 ppm for mobile homes and 0.10 ppm for conventional homes). The symptoms observed included eye and upper respiratory tract irritation, headache, fatigue, nausea, vomiting, diarrhea, and respiratory problems. As formaldehyde concentrations increased, the percentage of persons experiencing eye irritation increased frm 60% to 92%. Among infants and young children, vomiting, diarrhea, and respiratory problems were identified as particularly important conditions. The relationship between smoking and formaldehyde concentration in the dwelling was examined; smoking did not significantly increase formaldehyde concentration in the home at the time of concentration measurement.

Consumer reports have also been summarized by the CPSC.[196] In-depth investigations of 15 persons were conducted by the CPSC field staff and by private contractors. Most of the reported symptoms were related to eye and throat irritation. Five persons were admitted to the hospital for medical problems attributed to formaldehyde. The CPSC also collected more than 100 reports from newspaper clippings, consumer complaints, and state reports that were not investigated in detail.

NONSPECIFIC SUBJECTIVE SYMPTOMS IN EXPOSED POPULATIONS AND EFFECTS ON INFANTS AND CHILDREN

Various subjective and nonspecific complaints have consistently been reported, including disturbed sleep, thirst, headache, and nausea.[0, 24, 68, 81, 84, 92, 137, 173, 190, 198, 212, 215]

Recently, there has been concern about the effects of formaldehyde on infants and children.[214] The Wisconsin Division of Health conducted a survey between January 1, 1978, and November 1, 1979, that consisted of analysis of information collected with a questionnaire completed by 249 persons, representing 96 homes and 260 occupants. Two frequent findings were "nosebleed" and "rash" in infants and young children. Nine of 23 infants (less than a year old) required hospitalization; four were hospitalized for vomiting, diarrhea, or both and five for respiratory problems. Three of the latter five also had vomiting, diarrhea, or both. The mean formaldehyde concentration in the homes of the hospitalized infants was 0.68 ± 0.66 ppm. In each case, symptoms reportedly disappeared when the infant was removed from the home and returned when the infant went home.

OCCUPATIONAL STANDARDS FOR FORMALDEHYDE

Occupational exposure limits issued by various countries are listed in Table 7-6. The present OSHA standard for formaldehyde is 3 ppm, as a time-weighted average concentration over an 8-h workshift. In 1974, the ACGIH recommended a ceiling limit of 2 ppm, mainly because irritation might occur above this concentration. NIOSH has recommended a workplace ceiling limit of 1 ppm.[198]

RESIDENTIAL STANDARDS FOR FORMALDEHYDE

Occupational standards for formaldehyde have been determined in the United States and other countries, but the recommendations are for maximal time-weighted 8-h average concentrations for the workplace and for ceiling or peak concentrations. In the United States, there is no standard for formaldehyde for 24-h continuous nonoccupational exposure, as in the home. The American Industrial Hygiene Association has recommended an outdoor ambient-air standard of 0.10 ppm.[7] A panel of the National Research Council has stated that airborne formaldehyde in spacecraft for manned space flights should not exceed 0.10 ppm for an exposure of 90 d to 6 mo.[140] The American Society of Heating, Refrigerating and Air-Conditioning Engineers has recommended 0.20 ppm as a 24-h residential exposure limit.[8] West

TABLE 7-6

Occupational Standards for Formaldehyde in Effect, 1976[a]

Country	Standard mg/m^3	ppm	Type
United States:			
Federal Standard	---	3	TWA
	---	5	Ceiling
	---	10	30-min ceiling
ACGIH TLV	2.5	2	Ceiling
ANSI Z-37	---	3	TWA
	---	5	Ceiling
	---	10	30-min ceiling
Florida	---	5	Ceiling
Hawaii	---	10	Ceiling
Massachusetts	---	3	Ceiling
Mississippi	---	5	Ceiling
Pennsylvania	---	5	TWA
	---	5	5-min ceiling
South Carolina	---	5	Ceiling
Bulgaria	5	--	Ceiling
Czechoslovakia	2	--	Ceiling
	5	--	Peak
Federal Republic of Germany	6	5	Ceiling
Finland	6	5	Ceiling
German Democratic Republic	5	--	Ceiling
Great Britain	12	10	Ceiling
Hungary	1	--	Ceiling
Italy	5	--	Ceiling
Japan	6	5	Ceiling
Poland	5	--	Ceiling
Rumania	3	--	Ceiling
UAR	---	20	Ceiling
USSR	0.5	0.4	Ceiling
Yugoslavia	6	5	Ceiling

[a] Modified from NIOSH.[198]

Germany, Denmark, and The Netherlands have residential standards of 0.10, 0.12, and 0.10 ppm, respectively (C.D. Hollowell, personal communication; Hollowell et al.[89]). Sweden has recommended that a standard be set in the range of 0.10-0.70 ppm.[192]

SIGNIFICANCE OF ADVERSE HEALTH EFFECTS IN REGARD TO POPULATION AT RISK

The total number of people who are exposed to formaldehyde and who manifest adverse health effects is difficult to determine. There is evidence that such responses may occur in a substantial proportion of the exposed population in the United States. The variability in response among exposed persons makes it particularly difficult to assess the problem.

People are exposed to formaldehyde from occupational sources, consumer products, outdoor ambient air, and indoor air.

In the occupational setting, about 1.4 million persons are directly or indirectly exposed to formaldehyde. It is not possible to determine exactly the exposure in each industry. However, owing to the irritant nature of formaldehyde, most workplaces probably have concentrations of less than 3 ppm--more often around 1 ppm or less for an 8-h workday.

Some 11 million persons live in homes that contain either UF foam insulation or particleboard made with UF resins. When measurements have been performed, a wide range of formaldehyde concentrations from 0.01 ppm to 10.6 ppm, have been reported. Most homes have shown less than 0.5 ppm with a range of 0.1-0.2 ppm being more prevalent. Because people spend up to 70% of their time indoors, the exposure to formaldehyde released from UF foam or particleboard could be substantial.

Formaldehyde concentrations measured in ambient air are lower than in the occupational or indoor residential situation. Outdoor concentrations vary, but are rarely more than 0.1 ppm and usually less than 0.05 ppm. However, the probability of high outdoor exposure to formaldehyde for the 220 million people in the United States does not appear to be substantial, except for unusual circumstances of traffic, fuel use, or automobile density. Consumer exposures are mainly by direct contact, and contact dermatitis is an important consideration, as has been discussed.

Little is known about the magnitude of the population that is more susceptible to the effects of inhaling formaldehyde vapor. Asthmatics may constitute a segment of the general population that is more susceptible; inhalation even at low concentrations may precipitate acute symptoms. Airway hyperactivity may explain the susceptibility of asthmatics to formaldehyde at low concentrations. Using data gathered from over 1,500 methacholine challenge tests, one can estimate the prevalence of airway hyperreactivity in the population at large.[190] About 9 million people in the United States have bronchial asthma. Essentially all will react positively to methacholine challenge tests and thus be considered to have hyperreactive airways.[190] The degree of airway reactivity is variable and depends on a number of factors.[22] It has been estimated that 30% of atopic nonasthmatic people--perhaps 10 million--have positive methacholine tests.[190] Townley et al. reported that 5% of nonatopic persons--another 8.5 million--have positive methacholine tests.[190] Therefore, on the basis of calculations reported for positive methacholine challenge tests, it can be estimated that about 25 million persons in the United States, or 10-12% of the population, may be considered to have some degree of airway hyperreactivity. This population could potentially be more susceptible to formaldehyde.

Information on other assumed susceptible populations is limited. The U.S. Department of Health, Education, and Welfare, in a 1977 report on prevention, control, and elimination of respiratory disease, estimated that 10 million persons in the United States had chronic obstructive lung disease (excluding asthma).[200] A small percentage of them will have positive methacholine challenge tests. Britt et al.[25] suggested that the presence of methacholine sensitivity and evidence of airway hyperreactivity are risk factors for the development of chronic obstructive pulmonary disease (COPD). Perhaps patients with COPD who manifest airway hyperreactivity constitute a susceptible population, inasmuch as they react more acutely to airborne irritants, including formaldehyde.

On the basis of sensitivity to methacholine, some atopic persons, some nonatopic subjects, and some COPD patients may constitute a potential formaldehyde-susceptible population. This population could also have greater eye and upper respiratory tract sensitivity. However, many apparently normal people react to the irritant properties of formaldehyde, and this makes it more difficult to determine the susceptible population.

In another attempt to estimate the susceptible population (particularly in relation to eye, nose, and throat sensitivity), information on a small number of healthy young adults exposed to formaldehyde at various concentrations for short periods was considered.[139] At 1.5-3.0 ppm, more than 30% of the subjects tested reported mild to moderate eye, nose, and throat (ENT) irritation symptoms, and 10-20% had strong reactions. When test subjects were exposed at 0.5-1.5 ppm, slight or mild ENT irritation was noted in more than 30%, but 10-20% still had more marked reactions. Approximately 20% of the subjects had slight ENT irritation in response to formaldehyde at 0.25-0.5 ppm. Finally, at the lowest concentration tested, less than 0.25 ppm, some exposed subjects ("less than 20 percent") still reported minimal to slight ENT discomfort. These data might be interpreted as suggesting that there are subjects, perhaps 10-20% of those tested, who are more responsive and may react acutely to formaldehyde at very low concentrations.

Data on the proportion of the population susceptible to the irritant effects of formaldehyde seem to be consistent. The estimated prevalence of airway hyperreactivity (based on methacholine challenge testing) in the general population is 10-12% and about 10-20% of the subjects in the study just described showed excessive ENT sensitivity. We may get further information from mobile-home surveys from which environmental and clinical data are available. No measurements of other airborne contaminants were made, so the importance of other substances in the household environment is not known. Irritation symptoms were reported by 30-50% of subjects when formaldehyde concentrations were greater than 0.5 ppm. When the concentration was less than 0.5 ppm, irritation symptoms were reported in fewer than 30% of subjects. Finally, in a more controlled study in which irritation symptoms were investigated, mild irritation responses (doubling of blinking rate) occurred in 11% subjects tested at 0.5 ppm.

In summary, fewer than 20% but perhaps more than 10% of the general population may be susceptible to formaldehyde and may react acutely at very low concentrations, particularly if they are above 1.5 ppm. People report mild ENT discomfort and other symptoms at less than 0.5 ppm, with some noting symptoms at as low as 0.25 ppm. Low-concentration formaldehyde exposures may produce ENT symptoms and possibly lower-airway complaints. In some susceptible persons, an "allergic" reaction to formaldehyde may occur at very low concentrations, causing bronchoconstriction and asthmatic symptoms. This particular type of reaction to formaldehyde appears to be uncommon; its exact prevalence cannot now be estimated.

REFERENCES

1. Acheson, E. D., R. H. Cowdell, E. Hadfield, and R. G. Macbeth. Nasal cancer in woodworkers in the furniture industry. Br. Med. J. 2:587-596, 1968.
2. Adams, R. Occupational Contact Dermatitis. Philadelphia: J.B. Lippincott Co., 1969. 262 pp.
3. Ad Hoc Task Force--Epidemiology Study on Formaldehyde. Epidemiological Studies in the Context of Assessment of the Health Impact of Indoor Air Pollution. Summary and Recommendations. Bethesda, Md.: Consumer Product Safety Commission, May 10, 1979.
4. Altshuller, A. P., T. A. Bellar, and S. P. McPherson. Hydrocarbons and Aldehydes in the Los Angeles Atmosphere. Presented at Air Pollution Control Association Annual Meeting, May 2, 1962, Chicago, Illinois. Cincinnati: U.S. Department of Health, Education, and Welfare, Division of Air Pollution, Public Health Service, 1962.
5. Altshuller, A. P., L. G. Leng, and A. F. Wartburg. Source and atmospheric analyses for formaldehyde by chromotropic acid procedure. Int. J. Air Water Pollut. 63:381-385, 1962.
6. Amdur, M. O. The response of guinea pigs to inhalation of formaldehyde and formic acid alone and with a sodium chloride aerosol. Int. J. Air Pollut. 3:201-220, 1960.
7. American Industrial Hygiene Association. Community air quality guides. Am. Ind. Hyg. Assoc. J. 29:505-512, 1968.
8. American Society for Heating, Refrigerating and Air-Conditioning Engineers. Standards for Natural and Mechanical Ventilation. ASHRAE Standard 62-73. New York: ASHRAE, Inc., 1979.

9. Andersen, I. Formaldehyde in the indoor environment--health implications and the setting of standards, pp. 65-77, and discussion, pp. 77-87. In P. O. Fanger and O. Valbjørn, Eds. Indoor Climate. Effects on Human Comfort, Performance, and Health in Residential, Commercial, and Light-Industry Buildings. Proceedings of the First International Indoor Climate Symposium, Copenhagen, August 30-September 1, 1978. Copenhagen: Danish Building Research Institute, 1979.
10. Andersen, I., G. R. Lundqvist, and L. Molhave. Indoor air pollution due to chipboard used as a construction material. Atmos. Environ. 9:1121-1127, 1975.
11. Auerbach, C., M. Moutschen-Dahmen, and J. Moutschen. Genetic and cytogenetical effects of formaldehyde and related compounds. Mutat. Res. 39:317-362, 1977.
12. Babior, B. M. Folate and aplasia of bone marrow. N. Eng. J. Med. 298:506-507, 1978.
13. Barnes, E. C., and H. W. Speicher. The determination of formaldehyde in air. J. Ind. Hyg. Toxicol. 24:10-17, 1942.
14. Bilimoria, M. H. The detection of mutagenic activity of chemicals and tobacco smoke in a bacterial system. Mutat. Res. 31:328, 1975.
15. Blejer, H. P., and B. H. Miller. Occupational Health Report of Formaldehyde Concentrations and Effects on Workers at the Bayly Manufacturing Company, Visalia. Study Report No. S-1806. Los Angeles: State of California Health and Welfare Agency, Department of Public Health, Bureau of Occupational Health, 1966. 6 pp.
16. Böhmer, K. Formalin poisoning. Dtsch. Z. Gesamte Gerichtl. med. 23:7-18, 1934. (in German)
17. Botez, M. I., J.-M. Peyronnard, J. Bachevalier, and L. Charron. Polyneuropathy and folate deficiency. Arch. Neurol. 35:581-584, 1978.
18. Boucher, R. C., P. D. Pare, and J. C. Hogg. Relationship between airway hyperreactivity and hyperpermeability in Ascaris-sensitive monkeys. J. Allergy Clin. Immunol. 64:197-201, 1979.
19. Bouhuys, A., and K. P. van de Woestijne. Respiratory mechanics and dust exposure to byssinosis. J. Clin. Inv. 49:106-118, 1970.
20. Bourne, H. G., Jr., and S. Seferian. Formaldehyde in wrinkle-proof apparel produces...tears for milady. Ind. Med. Surg. 28:232-233, 1959.
21. Boushey, H. A., D. W. Empey, and L. A. Laitinen. Meat wrapper's asthma. Effects of fumes of polyvinyl chloride on airways function. Physiologist 18:148, 1975.
22. Boushey, H. A., M. J. Holtzman, J. R. Sheller, and J. A. Nadel. Bronchial hyperreactivity. Am. Rev. Respir. Dis. 121:389-413, 1980.
23. Bower, A. J. Case of poisoning by formaldehyd. J. Am. Med. Assoc. 52: 1106, 1909.

24. Breysse, P. A. Formaldehyde exposure following urea-formaldehyde insulation. Environ. Health Safety News 26, 1978. 13 pp.
25. Britt, E. J., B. Cohen, H. Menkes, E. Bleecker, S. Permutt, R. Rosenthal, and P. Norman. Airways reactivity and functional deterioration in relatives of COPD patients. Chest 77(Suppl.):260-261, 1980.
26. Butcher, B. T., R. M. Karr, C. E. O'Neil, M. R. Wilson, V. Dharmarajan, J. E. Salvaggio, and H. Weill. Inhalation challenge and pharmacologic studies of toluene diisocyanate (TDI)-sensitive workers. J. Allergy Clin. Immunol. 64:146-152, 1979.
27. Butcher, B. T., J. E. Salvaggio. C. E. O'Neil, H. Weill, and O. Garg. Toluene diisocyanate pulmonary disease: Immunopharmacologic and mecholyl challenge studies. J. Allergy Clin. Immunol. 59:223-227.
28. Butcher, B.T., J. E. Salvaggio, H. Weill, and M. M. Ziskind. Toluene diisocyanate (TDI) pulmonary disease: Immunologic and inhalation challenge studies. J. Allergy Clin. Immunol. 58:89-100, 1976.
29. Cantor, T. R. Experience with the determination of atmospheric aldehydes, pp. 514-515. In Automation in Analytical Chemistry. Technicon Symposia 1966. Vol. I. New York: Mediad, Inc., 1967.
30. Carbone, R. D. Formaldehyde Exposure in Mobile Homes. Master's Thesis. Seattle: University of Washington, 1978.
31. Cares, J. W. Determination of formaldehyde by the chromotropic acid method in the presence of oxides of nitrogen. Am. Ind. Hyg. Assoc. J. 29:405-410, 1968.
32. Carpenter, C. P., and H. F. Smyth, Jr. Chemical burns of the rabbit cornea. Am. J. Ophthal. 29: 1363-1372, 1946.
33. Casarett, L. J., and J. Doull. Toxicology: The Basic Science of Poisons, pp. 299, 512, and 513. New York: MacMillan Publishing Co., Inc., 1975.
34. Chanet, R., C. Izard, and E. Moustacchi. Genetic effects of formaldehyde in yeast. I. Influence of the growth stages on killing and recombination. Mutat. Res. 33:179-186, 1975.
35. Chanet, R., and R. C. von Borstel. Genetic effects of formaldehyde in yeast. III. Nuclear and cytoplasmic mutagenic effects. Mutat. Res. 62:239-253, 1979.
36. Chugh, K. S., P. C. Singhal, and S. S. Banerjee. Acute tubular necrosis following intravesical instillation of formalin. Urol. Int. 32:454-459, 1977.
37. Clive, D., and J. F. S. Spector. Laboratory procedure for assessing specific locus mutations at the TK locus in cultured L5178Y mouse lymphoma cells. Mutat. Res. 31:17-29, 1975.
38. Cohen, I. R., and A. P. Altshuller. 3-Methyl-2-benzothiazolone hydrazone method for aldehydes in air. Collection efficiencies and molar absorptivities. Anal. Chem. 38:1418, 1966.
39. Conners, T. A. Effects of drugs on structure, biosynthesis and catabolism of nucleic acids, proteins, carbohydrates and lipids, pp. 443-447. In Z. M. Bacq, R. Capek, R. Paoletti, and J.

Renson, Eds. Fundamentals of Biochemical Pharmacology. Oxford: Pergamon Press, 1971.
40. Coon, R. A., R. A. Jones, L. J. Jenkins, Jr., and J. Siegel. Animal inhalation studies on ammonia, ethylene glycol, formaldehyde, dimethylamine and ethanol. Toxicol. Appl. Pharmacol. 16:646-655, 1970.
41. Cooper, J. R., and M. M. Kini. Biochemical aspects of methanol poisoning. Biochem. Pharmacol. 11:405-416, 1962.
42. Cralley, L. V. The effect of irritant gases upon the rate of ciliary activity. J. Ind. Hyg. Toxicol. 24:193-198, 1942.
43. Crittenden, A. Built-in fumes plague homes. New York Times. Section 3. Business and Finance. May 7, 1978.
44. Curry, J. J. Comparative action of acetyl-beta-methyl choline and histamine on the respiratory tract in normals, patients with hay fever, and subjects with bronchial asthma. J. Clin. Invest. 26:430-438, 1947.
45. Dalhamn, T., and A. Rosengren. Effect of different aldehydes on tracheal mucosa. Arch. Otolaryngol. 93:496-500, 1971.
46. Dally, K. A., L. P. Hanrahan, and M. A. Woodbury. Formaldehyde exposure in nonoccupational environments. In press, 1980.
47. Della Porta, G., M. I. Colneghi, and G. Parmiani. Non-carcinogenicity of hexamethylenetetramine in mice and rats. Food Cosmet. Toxicol. 6:707-715, 1968.
48. Doenecke, D. Digestion of chromosomal proteins in formaldehyde treated chromatin. Hoppe-Seyler's Z. Physiol. Chem. 359:1343-1352, 1978.
49. Draize, J. Dermal toxicity, pp. 46-59. In Appraisal of the Safety of Chemicals in Foods, Drugs and Cosmetics. Topeka, Kansas: The Association of Food and Drug Officials of the United States, 1959.
50. Dubreuil, A., G. Bouley, J. Godin, and C. Boudène. Continuous inhalation of low-level doses of formaldehyde: Experimental study on the rat. Eur. J. Toxicol. 9:245-250, 1976. (in French; English summary)
51. Earp, S. E. The physiological and toxic actions of formaldehyde. With a report of three cases of poisoning by formalin. N.Y. Med. J. 104:391-392, 1916.
52. Egle, J. L., Jr. Retention of inhaled formaldehyde, propionaldehyde, and acrolein in the dog. Arch. Environ. Health 25:119-124, 1972.
53. Egle, J. R., Jr., and P. M. Hudgins. Dose-dependent sympathomimetic and cardioinhibitory effects of acrolein and formaldehyde in the anesthetized rat. Toxicol. Appl. Pharmacol. 28:358-366, 1974.
54. Ehrenberg, L., A. Gustafsson, and U. Lundqvist. Chemically induced mutation and sterility in barley. Acta Chem. Scand. 10: 492-494, 1956.
55. Einbrodt, H. J. Formaldehyde and formic acid level in blood and urine of people following previous exposure to formaldehyde. Zentralbl. Arbeitsmed. Arbeitsshutz Prophyl. 26:154-158, 1976.

56. Elfers, L. A., and S. Hochheiser. Estimation of Atmospheric Aliphatic-Aldehyde Concentration by Use of Visual Color Comparator. Raleigh, N.C.: U.S. Department of Health, Education, and Welfare, Public Health Service, Consumer Protection and Environmental Health Service, National Air Pollution Control Administration, 1969.
57. Ely, F. Formaldehyde poisoning. J. Am. Med. Assoc. 54:1140-1141, 1910.
58. Englesberg, E. The mutagenic action of formaldehyde on bacteria. J. Bacteriol. 63:1-11, 1952.
59. Epstein, S. S., E. Arnold, J. Andrea, W. Bass, and Y. Bishop. Decection of chemical mutagens by the dominant lethal assay in the mouse. Toxicol. Appl. Pharmacol. 23:288-325, 1972.
60. Ettinger, I., and M. Jeremias. A study of the health hazards involved in working with flameproofed fabric. N.Y. State Dept. Labor Div. Ind. Hyg. Mon. Rev. 34:25-27, 1955.
61. Fairhall, L. T. Industrial Toxicology. 2nd ed. Baltimore: Williams & Wilkins, 1957. 483 pp.
62. Fassbinder, W., U. Frei, and K.-M. Koch. Haemolysis due to formaldehyde-induced anti-N-like antibodies in haemodialysis patients. Klin. Wochenschr. 57:673-679, 1979.
63. Fassett, D. W. Aldehydes and acetals, pp. 1959-1989. In F. A. Patty, Ed. Industrial Hygiene and Toxicology. 2nd rev. ed. D. W. Fassett and D. D. Irish, Eds. Vol. II. Toxicology. New York: Interscience Publishers, 1963.
64. Feldman, M. Y. Reactions of nucleic acids and nucleoproteins with formaldehyde. Prog. Nucleic Acid Res. Mol. Biol. 13:1-49, 1973.
65. Fel'dman, Yu. G., and T. I. Bonashevskaya. On the effects of low concentrations of formaldehyde. Hyg. Sanit. 36(5):174-180, 1971.
66. Fisher, A. A. Contact Dermatitis. 2nd ed. Philadelphia: Lea and Febiger. 1973. 448 pp.
67. Frankel, L. S., K. S. McCallum, and L. Collier. Formation of bis(chloromethyl)ether from formaldehyde and hydrogen chloride. Environ. Sci. Technol. 8:356-359, 1974.
68. Freeman, H. G., and W. C. Grendon. Formaldehyde detection and control in the wood industry. For. Prod. J. 21(9):54-57, 1971.
69. Fregert, S. Manual of Contact Dermatitis. Copenhagen: Munksgaard, 1974. 107 pp.
70. Fregert, S., and H. J. Bandmann. Patch Testing. New York: Springer Verlag, 1975. 78 pp.
71. Gamble, J. F., A. J. McMichael, T. Williams, and M. Battigelli. Respiratory function and symptoms: An environmental-epidemiological study of rubber workers exposed to a phenol-formaldehyde type resin. Am. Ind. Hyg. Assoc. J. 37:499-513, 1976.
72. Glass, W. I. An outbreak of formaldehyde dermatitis. N. Z. Med. J. 60:423-427, 1961.

73. Gofmekler, V. A. Effect on embryonic development of benzene and formaldehyde in inhalation experiments. Hyg. Sanit. 33:327-332, 1968.
74. Golden, J. A., J. A. Nadel, and H. A. Boushey. Bronchial hyperirritability in healthy subjects after exposure to ozone. Am. Rev. Resp. Dis. 118:287-294, 1978.
75. Goodman, J. I., and T. R. Tephly. A comparison of rat and human liver formaldehyde dehydrogenase. Biochim. Biophys. Acta 252:489-505, 1971.
76. Gosser, L. B., and B. E. Butterworth. Mutagenicity evaluation of formaldehyde in the L5178Y mouse lymphoma assay. Wilmington: E. I. duPont de Nemours & Co., Haskell Laboratory for Toxicology and Industrial Medicine, 1977. 6 pp.
77. Governor's Task Force on Insulation. Report on U-F Foam Insulation. Hartford, Conn.: Connecticut Department of Consumer Protection, 1979.
78. Grant, W. M. Toxicology of the Eye, pp. 502-506. 2nd ed. Springfield, Ill.: Charles C Thomas, 1974.
79. Grossman, L. I. Paresthesia from N2 or N2 substitute. Report of a case. Oral Surg. Oral Med. Oral Pathol. 45:114-115, 1978.
80. Guseva, V. A. Gonadotropic effect of formaldehyde on male rats during its simultaneous introduction with air and water. Gig. Sanit. 37:102-103, 1972.
81. Harris, D. K. Health problems in the manufacture and use of plastics. Brit. J. Ind. Med. 10:255-268, 1953.
82. Hauser, T. R. Determination of aliphatic aldehydes. 3-Methyl-2-benzothiazolone hydrazone, hydrochloride (MBTH) method, pp. F-1 to F-4. In Selected Methods for the Measurement of Air Pollutants. DHEW Publication No. 99-AP-11. Cincinnati: U.S. Department of Health, Education, and Welfare, Public Health Service, Consumer Protection and Environmental Health Service, Interbranch Chemical Advisory Committee, 1969.
83. Heling, B., Z. Ram, and I. Heling. The root treatment of teeth with Toxavit. Report of a case. Oral Surg. Oral Med. Oral Pathol. 43:306-309, 1977.
84. Helwig, H. How safe is formaldehyde? Dtsch. Med. Woch. 102:1612-1613, 1977. (in German)
85. Hendrick, D. J., and D. J. Lane. Formalin asthma in hospital staff. Brit. Med. J. 1:607-608, 1975.
86. Hendrick, D. J., and D. J. Lane. Occupational formalin asthma. Brit. J. Ind. Med. 34:11-18, 1977.
87. Hilgemeier, M. W. Presentation on New Hampshire experiences with urea-formaldehyde foam, given at Ad Hoc Task Force Seminar on An Assessment of the Odor Problems from U-F Foam Insulations, Washington, D.C., December 1, 1978.
88. Hogg, J. C., P. D. Paré, and R. C. Boucher. Bronchial mucosal permeability. Fed. Proc. 38:197-201, 1979.
89. Hollowell, C. D., J. V. Berk, and G. W. Traynor. Impact of reduced infiltration and ventilation on indoor air quality in residential buildings. ASHRAE Trans. 85:816-826, 1979.

90. Horsfall, F. L., Jr. Formaldehyde hypersensitiveness--an experimental study. J. Immunol. 27:569-581, 1934.
91. Horton, A. W., R. Tye, and K. L. Stemmer. Experimental carcinogenesis of the lung. Inhalation of gaseous formaldehyde or an aerosol of coal tar by C3H mice. J. Nat. Cancer Inst. 30:31-43, 1963.
92. Høvding, G. Occupational dermatitis from pyrolysis products of polythene. Acta Derm. Venereol. 49:147-149, 1969.
93. Hsie, A. W., J. P. O'Neill, J. R. San Sebastian, D. B. Couch, J. C. Fuscoe, W. N. C. Sun, P. A. Brimer, R. Machanoff, J. C. Riddle, N. L. Forbes, and M. H. Hsie. Mutagenicity of carcinogens: Study of 101 agents in a quantitative mammalian cell mutation system, CHO/HGPRT. Fed. Proc. 37:1384, 1978.
94. Humpstone, O. P., and W. Lintz. A case of formalin poisoning. J. Am. Med. Assoc. 52:380-381, 1909.
95. Hurni, H., and H. Ohder. Reproduction study with formaldehyde and hexamethylenetetramine in beagle dogs. Food Cosmet. Toxicol. 11:459-462, 1973.
96. Jordan, W. P., Jr., W. T. Sherman, and S. E. King. Threshold responses in formaldehyde-sensitive subjects. J. Am. Acad. Dermatol. 1:44-48, 1979.
97. Kallos, G. J., and R. A. Solomon. Investigations of the formation of bis(chloromethyl)ether in simulated hydrogen chloride-formaldehyde atmospheric environments. Am. Ind. Hyg. Assoc. J. 34:469-473, 1973.
98. Kane, L. E., and Y. Alarie. Sensory irritation to formaldehyde and acrolein during single and repeated exposures in mice. Am. Ind. Hyg. Assoc. J. 38:509-522, 1977.
99. Karlson, P. Introduction to Modern Biochemistry. 2nd ed. Chapters 6, 8, 11, and 12. New York: Academic Press, 1965.
100. Kelecom, J. Les brûlures oculaires par le formol. Arch. Ophthal. (Paris) 22:259-262, 1962.
101. Kendall, A. I. The relaxation of histamine contractions in smooth muscle by certain aldehydes. Studies in bacterial metabolism. LXXXIII. J. Infectious Dis. 40:689-698, 1927.
102. Kerfoot, E. J., and T. F. Mooney, Jr. Formaldehyde and paraformaldehyde study in funeral homes. Am. Ind. Hyg. Assoc. J. 36:533-537, 1975.
103. Kitchens, J. F., R. E. Casner, G. S. Edwards, W. E. Harward III, and B. J. Macri. Investigation of Selected Potential Environmental Contaminants: Formaldehyde. U.S. Environmental Protection Agency Report No. EPA-560/2-76-009. Washington, D.C.: U.S. Environmental Protection Agency, Office of Toxic Substances, 1976. 204 pp.
104. Kligman, A. M., and W. Epstein. Updating the maximization test for identifying contact allergens. Contact Dermatitis 1:231-239, 1975.
105. Kline, B. S. Formaldehyd poisoning. With report of a fatal case. Arch. Intern. Med. 36:220-228, 1925.
106. Koops, A. In Vitro Microbial Mutagenicity Studies of Formaldehyde (37% A.I.). Wilmington, Del.: E. I. du Pont de

Nemours and Company, Haskell Laboratory for Toxicology and Industrial Medicine, 11 March 1976 (unpublished). 4 pp.

107. Kratochvil, I. The effect of formaldehyde on the health of workers employed in the production of crease resistant ready made dresses. Pr. Lek. 23:374-375, 1971. (in Czech; English abstract)

108. Kuschner, M., S. Laskin, R. T. Drew, V. Cappiello, and N. Nelson. Inhalation carcinogenicity of alpha halo ethers. III. Lifetime and limited period inhalation studies with bis(chloromethyl)ether at 0.1 ppm. Arch. Environ. Health 30:73-77, 1975.

109. La Du, B. N., H. G. Mandel, and H. L. Way, Eds. Fundamentals of Drug Metabolism and Drug Disposition, pp. 169-171, 206-208, 292-294. Baltimore: Williams & Wilkins, 1971.

110. Laffont, H., and J.-B. Noceto. A case of asthma due to sensitivity to formaldehyde. Algérie Méd. 65:777-781, 1961. (in French)

111. Lam, S., R. Wong, and M. Yeung. Nonspecific bronchial reactivity in occupational asthma. J. Allerg. Clin. Immunol. 63:28-34, 1979.

112. Leonardos, G., D. Kendall, and N. Barnard. Odor threshold determinations of 53 odorant chemicals. J. Air Pollut. Control Assoc. 19:91-95, 1969.

113. Levaggi, D. A., and M. Feldstein. The determination of formaldehyde, acrolein, and low molecular weight aldehydes in industrial emissions on a single collection sample. J. Air Pollut. Control Assoc. 20:312-313, 1970.

114. Levison, L. A. A case of fatal formaldehyde poisoning. J. Am. Med. Assoc. 42:1492, 1904.

115. Logan, W. S., and H. O. Perry. Contact dermatitis to resin-containing casts. Clin. Orthop. Relat. Res. 90:150-152, 1973.

116. Lyles, G. R., F. B. Dowling, and V. J. Blanchard. Quantitative determination of formaldehyde in the parts per hundred million concentration level. J. Air Pollut. Control Assoc. 15:106-108, 1965.

117. Magaña-Schwencke, N., B. Ekert, and E. Moustacchi. Biochemical analysis of damage induced in yeast by formaldehyde. I. Induction of single-strand breaks in DNA and their repair. Mutat. Res. 50:181-193, 1978.

118. Magaña-Schwencke, N., and E. Moustacchi. Biochemical analysis of damage induced in yeast by formaldehyde. III. Repair of induced cross-links between DNA and proteins in the wild-type and in excision-deficient strains. Mutat. Res. 70:29-35, 1980.

119. Magnusson, B., and A.M. Kligman. Usefulness of guinea pig tests for detection of contact sensitizers, pp. 551-560. In F. N. Marzulli and H. I. Maibach, Eds. Advances in Modern Toxicology. Vol. 4. Dermatotoxicology and Pharmacology. Washington, D.C.: Hemisphere Publishing Corporation, 1977.

120. Manna, G. K., and B. B. Parida. Formalin induced sex chromosome breakage in the spermatocyte cells of the grasshopper, *Tristria pulvinata*. J. Cytol. Genet. 2:86-91, 1967.

121. Mant, M. J., T. Connolly, P. A. Gordon, and E. G. King. Severe thrombocytopenia probably due to acute folic acid deficiency. Crit. Care Med. 7:297-300, 1979.

122. March, G. H. Formalin poisoning; recovery. Br. Med. J. 2:687, 1927.

123. Marks, T. A., W. C. Worthy, and R. E. Staples. Influence of Formaldehyde and Sonacide® (Potentiated Acid Glutaraldehyde) on Embryo and Fetal Development in Mice. Research Triangle Park, N.C.: Research Triangle Institute, 1980. (in press)

124. Martin-Amat, G., K. E. McMartin, S. S. Hayreh, M. S. Hayreh, and T. R. Tephly. Methanol poisoning. Ocular toxicity produced by formate. Toxicol. Appl. Pharmacol. 45:201-208, 1978.

125. Marzulli, F., T. R. Carson, and H. I. Maibach. Delayed contact hypersensitivity studies in man and animals, pp. 107-122. In Proceedings. Joint Conference on Cosmetic Sciences. Washington, D.C., April 21-23, 1968. Washington, D.C.: The Toilet Goods Association, Inc., 1968.

126. Marzulli, F. N., and H. I. Maibach. Antimicrobials: Experimental contact sensitization in man. J. Soc. Cosmet. Chem. 24:399-421, 1973.

127. Marzulli, F. N., and H. I. Maibach. The use of graded concentrations in studying skin sensitizers: Experimental contact sensitization in man. Food Cosmet. Toxicol. 12:219-227, 1974.

128. Matoth, Y., I. Zehavi, E. Topper, and T. Klein. Folate nutrition and growth in infancy. Arch. Dis. Childhood 54:699-702, 1979.

129. McMartin, K. E., J. J. Ambre, and T. R. Tephly. Methanol poisoning in human subjects. Role for formic acid accumulation in the metabolic acidosis. Am. J. Med. 68:414-418, 1980.

130. McMartin, K. E., G. Martin-Amat, A. B. Makar, and T. R. Tephly. Methanol poisoning. V. Role of formate metabolism in the monkey. J. Pharmacol. Exp. Therap. 201:564-572, 1977.

131. McMartin, K. E., G. Martin-Amat, P. E. Noker, and T. R. Tephly. Lack of a role for formaldehyde in methanol poisoning in the monkey. Biochem. Pharmacol. 28:645-649, 1979.

132. Melekhina, V. P. Hygienic evaluation of formaldehyde as an atmospheric air pollutant, pp. 9-18. In B. S. Levine, Ed. U.S.S.R. Literature on Air Pollution and Related Occupational Diseases. Vol. 9. A Survey. Washington, D.C.: U.S. Public Health Service, 1963-1964. (available from National Technical Information Service, Springfield, Va., as TT64-11574)

133. Miller, B. H., and H. P. Blejer. Report of an Occupational Health Study of Formaldehyde Concentrations at Maximes, 400 E. Colorado Street, Pasadena, California. Study No. S-1838. Los Angeles: State of California Health and Welfare Agency, Department of Public Health, Bureau of Occupational Health, 1966. 5 pp.

134. Mills, J. CPSC warns about health hazard of foam home material. Washington Post, Real Estate, August 11, 1979.
135. Montgomery, S. Paresthesia following endodontic treatment. J. Endodon. 2:345-347, 1976.
136. Morgan, G. B., C. Golden, and E. C. Tabor. New and improved procedures for gas sampling and analysis in the National Air Sampling Network. J. Air Pollut. Contr. Assoc. 17:300-304, 1967.
137. Morrill, E. E., Jr. Formaldehyde exposure from paper process solved by air sampling and current studies. Air Cond. Heat. Vent. 58(7):94-95, 1961.
138. Nagorny, P. A., Z. A. Sudakova, and S. M. Schablenko. On the general toxic and allergic action of formaldehyde. Gig. Tr. Prof. Zabol. 1:27-30, 1979.
139. National Research Council, Committee on Toxicology. Formaldehyde. An Assessment of Its Health Effects. Washington, D.C.: National Academy of Sciences, 1980. 38 pp.
140. National Research Council, Panel on Air Quality in Manned Spacecraft. Atmospheric Contaminants in Spacecraft. Washington, D.C.: National Academy of Sciences, 1972. 11 pp.
141. Neely, W. B. The metabolic fate of formaldehyde-^{14}C intraperitoneally administered to the rat. Biochem. Pharmacol. 13:1137-1142, 1964.
142. North American Contact Dermatitis Group. Epidemiology of contact dermatitis in North America: 1972. Arch. Dermatol. 108:537-540, 1973.
143. Nova, H., and R. G. Touraine. Asthma from formaldehyde. Arch. Mal. Prof. 18:293-294, 1957. (in French)
144. Odom, R. B., and H. I. Maibach. Chapter 15. Contact urticaria: A different contact dermatitis, pp. 441-453. In F. N. Marzulli and H. I. Maibach, Eds. Advances in Modern Toxicology. Vol. 4. Dermatotoxicology and Pharmacology. Washington, D.C.: Hemisphere Publishing Corporation, 1977.
145. Orringer, E. P., and W. D. Mattern. Formaldehyde-induced hemolysis during chronic hemodialysis. New Eng. J. Med. 294:1416-1420, 1976.
146. Paleček, E. Premelting changes in DNA conformation. Prog. Nucleic Acid Res. Mol. Biol. 18:151-213, 1976.
147. Palese, M., and T. R. Tephly. Metabolism of formate in the rat. J. Toxicol. Environ. Health 1:13-24, 1975.
148. Paliard, F., L. Roche, C. Exbrayat, and E. Sprunck. Chronic asthma due to formaldehyde. Arch. Mal. Prof. 10:528-530, 1949. (in French)
149. Parker, C. D., R. E. Bilbo, and C. E. Reed. Methacholine aerosol as test for bronchial asthma. Arch. Intern. Med. 115:452-458, 1965.
150. Pepys, J., C. A. C. Pickering, A. B. X. Breslin, and D. J. Terry. Asthma due to inhaled chemical agents--tolylene di-isocyanate. Clin. Allergy 2:225-236, 1972.
151. Popa, V., D. Teculescu, D. Stănescu, and N. Gavrilescu. Bronchial asthma and asthmatic bronchitis determined by simple chemicals. Dis. Chest 56:395-404, 1969.

152. Porter, J. A. H. Acute respiratory distress following formalin inhalation. Lancet 2:603-604, 1975.
153. Potts, A. M., J. Praglin, I. Farkas, L. Orbison, and D. Chickering. Studies on the visual toxicity of methanol. VIII. Additional observations on methanol poisoning in the primate test object. Am. J. Ophthalmol. 40(5, Pt. II):76-83, 1955.
154. Proctor, D. F. The upper airways. I. Nasal physiology and defense of the lungs. Am. Rev. Resp. Dis. 115:97-129, 1977.
155. Pruett, J. J., H. Scheuenstuhl, and D. Michaeli. The incorporation and localization of aldehydes (highly reactive cigarette smoke components) into cellular fractions of cultured human lung cells. Arch. Environ. Health 35:15-20, 1980.
156. Rapoport, I. A. Mutations under the influence of unsaturated aldehydes. Doklady Akad. Nauk S.S.S.R. 61:713-715, 1948. (in Russian)
157. Rathery, F., R. Piédelièvre, and J. Delarue. Death by absorption of formalin. Ann. Méd. Lég. Criminol. 20:201-206, 1940. (in French)
158. Rayner, A. C., and C. M. Jephcott. Microdetermination of formaldehyde in air. Anal. Chem. 33:627-630, 1961.
159. Renzetti, N. A., and R. J. Bryan. Atmospheric sampling for aldehydes and eye irritation in Los Angeles smog--1960. J. Air Pollut. Control Assoc. 11:421-424, 427, 1961.
160. Rietbrock, N. Kinetics and pathways of methanol metabolism. Naunyn-Schmeidebergs Arch. Pharmakol. Exper. Pathol. 263:88-105, 1969. (in German)
161. Rostenberg, A., Jr., B. Bairstow, and T. W. Luther. A study of eczematous sensitivity to formaldehyde. J. Invest. Dermatol. 19:459-462, 1952.
162. Roth, W. G. Tylosic palmar and plantar eczema caused by steaming clothes containing formalin. Berufsdermatosen 17:263-268, 1969.
163. Rumack, B. Position Paper. Urea-Formaldehyde Foam. Denver: Rocky Mountain Poison Center, 1978. 22 pp.
164. Sakula, A. Formalin asthma in hospital laboratory staff. Lancet 2:816, 1975.
165. Sasaki, Y., and R. Endo. Mutagenicity of aldehydes in Salmonella. Abstract No. 27. Mutat. Res. 54:251-252, 1978.
166. Sawicki, E., and R. A. Carnes. Spectrophotofluorimetric determination of aldehydes with dimedone and other reagents. Mikrochim. Acta 1968:148-159, 1968.
167. Schoenberg, J. B., and C. A. Mitchell. Airway disease caused by phenolic (phenol-formaldehyde) resin exposure. Arch. Environ. Health 30:574-577, 1975.
168. Schuck, E. A., E. R. Stephens, and J. T. Middleton. Eye irritation response at low concentrations of irritants. Arch. Environ. Health 13:570-575, 1966.
169. Sellakumar, A. R., R. E. Albert, G. M. Rusch, G. V. Katz, N. Nelson, and M. Kuschner. Inhalation carcinogenicity of formaldehyde and hydrogen chloride in rats. Abstr. No. 424. Am. Assoc. Can. Res. Proc. 21:106, 1980.

170. Shapira, Y., A. Ben Zvi, and M. Statter. Folic acid deficiency: A reversible cause of infantile hypotonia. J. Ped. 93:984-986, 1978.
171. Sheveleva, G. A. Investigation of the specific effect of formaldehyde on the embryogenesis and progeny of white rats. Toksikol. Novykh Prom. Khim. Veschestv 12:78-86, 1971.
172. Shikama, K., and K. I. Miura. Equilibrium studies on the formaldehyde reaction with native DNA. Eur. J. Biochem. 63:39-46, 1976.
173. Shipkovitz, H. D. Formaldehyde Vapor Emissions in the Permanent-Press Fabrics Industry. Report No. TR-52. Cincinnati: U.S. Department of Health, Education, and Welfare, Public Health Service, Consumer Protection and Environmental Health Service, Environmental Control Administration, 1968. 18 pp.
174. Shumilina, A. V. Menstrual and child-bearing functions of female workers having contact with formaldehyde under factory conditions. Gig. Trud. Prof. Zabol. 12:18-21, 1975.
175. Sim, V. M., and R. E. Pattle. Effect of possible smog irritants on human subjects. J. Am. Med. Assoc. 165:1908-1913, 1957.
176. Smith, A. B., S. M. Brooks, J. Blanchard, I. L. Bernstein, and J. Gallagher. Absence of airway hyperreactivity to methacholine in a worker sensitized to toluene diisocyanate (TDI). J. Occup. Med. 22: 327-331, 1980.
177. Smith, R. G., R. J. Bryan, M. Feldstein, B. Levadie, F. A. Miller, E. R. Stephens, and N. G. White. Tentative method of analysis for formaldehyde content of the atmosphere (MBTH, colorimetric method, applications to other aldehydes). Health Lab. Sci. 7(3):173-178, 1970.
178. Smith, R. G., R. J. Bryan, M. Feldstein, B. Levadie, F. A. Miller, E. R. Stephens, and N. G. White. Tentative method of analysis for low molecular weight aliphatic aldehydes in the atmosphere. Health Lab. Sci. 9(1):75-78, 1972.
179. Smyth, H. F., Jr., J. Seaton, and L. Fischer. The single dose toxicity of some glycols and derivatives. J. Ind. Hyg. Toxicol. 23:259-268, 1941.
180. Sneddon, I. B. Dermatitis in an intermittent haemodialysis unit. Brit. Med. J. 1:183-184, 1968.
181. Solyanik, R. G., Yu. V. Federov, and I. A. Rapoport. The mutagenic effect of some alkylating compounds on eastern equine encephalomyelitis virus. Sov. Genet. 8:412-413, 1972.
182. Spector, S. L., and R. S. Farr. A comparison of methacholine and histamine inhalations in asthmatics. J. Allergy Clin. Immunol. 56:308-316, 1975.
183. Stickney, R. N. Engineering, Safety, and Control for the Proper Handling of Formaldehyde. Presented at the Formaldehyde Seminar and Plant Tour held by the Borden Chemical Company, Fayetteville, N.C., March 11, 1958.
184. Strittmatter, P., and E. G. Ball. Formaldehyde dehydrogenase, a glutathione-dependent enzyme system. J. Biol. Chem. 213:445-461, 1955.

185. Swenberg, J. A., W. D. Kerns, R. I. Mitchell, E. J. Gralla, and K. L. Pavkov. Induction of squamous cell carcinomas of the rat nasal cavity by inhalation exposure to formaldehyde vapor. Cancer Res. 40:3398-3402, 1980.

186. Tabershaw, I. R., H. N. Doyle, L. Gaudette, S. H. Lamm, and O. Wong. A Review of the Formaldehyde Problems in Mobile Homes. Rockville, Md.: Tabershaw Occupational Medicine Associates, P.A., for National Particleboard Association, 1979. 19 pp.

187. Tani, T., S. Satoh, and Y. Horiguchi. The vasodilator action of formaldehyde in dogs. Toxicol. Appl. Pharmacol. 43:493-499, 1978.

188. Thomas, J. F., E. N. Sanborn, M. Mukai, and B. D. Tebbens. Identification of aldehydes in polluted atmospheres and combustion products. A.M.A. Arch. Ind. Health 20:420-428, 1959.

189. Tou, J. C., and G. J. Kallos. Study of aqueous HCl and formaldehyde mixtures for formation of bis (chloromethyl) ether. Am. Ind. Hyg. Assoc. J. 35:419-422, 1974.

190. Townley, R. G., A. K. Bewtra, N. M. Nair, F. D. Brodkey, G. D. Watt, and K. M. Burke. Methacholine inhalation challenge studies. J. Allergy Clin. Immunol. 64:569-574, 1979.

191. Tran, N., M. Laplante, and E. Lebel. Abnormal oxidation of ^{14}C-formaldehyde to $^{14}CO_2$ in erythrocytes of alcoholics and nonalcoholics after consumption of alcoholic beverages. J. Nucl. Med. 13:677-680, 1972.

192. Traynor, G. W., D. W. Anthon, and C. D. Hollowell. Indoor air quality: Gas stove emissions, p. 24. In Building ventilation and indoor air quality program. Chapter fom J. Kessel, Ed. Energy and Environment Division Annual Report 1978. Lawrence Berkeley Laboratory Report LBL-9284/EEB-Vent 79-6. Berkeley, Cal.: Lawrence Berkeley Laboratory, 1979.

193. Tsuchiya, K., Y. Hayashi, M. Onodera, and T. Hasegawa. Toxicity of formaldehyde in experimental animals. Concentrations of the chemical in the elution from dishes of formaldehyde resin in some vegetables. Keio J. Med. 24:19-37, 1975.

194. Turiar, C. Asthma through sensitivity to formaldehyde. Soc. Franc. d'Allergie, Séance du 18 Nov. 1952.

195. U.S. Consumer Product Safety Commission. CPSC issues consumer advisory on formaldehyde insulation. In News from CPSC. Washington, D.C.: U.S. Consumer Product Safety Commission, August 1, 1979.

196. U.S. Consumer Product Safety Commission, Directorate for Hazard Identification and Analysis--Epidemiology. Summaries of in-depth investigations, newspaper clippings, consumer complaints and state reports on urea-formaldehyde foam home insulation. Washington, D.C.: U.S. Consumer Product Safety Commission, July 1978.

197. U.S. Consumer Product Safety Commission. Urea-formaldehyde foam insulation; Proposed notice to purchasers. Fed. Reg. 45:33946, 1980.

198. U.S. Department of Health, Education, and Welfare, Public Health Service, Center for Disease Control, National Institute for

Occupational Safety and Health. Criteria for a Recommended Standard. Occupational Exposure to Formaldehyde. DHEW (NIOSH) Publication No. 77-126. Washington, D.C: U.S. Government Printing Office, 1976. 165 pp.

199. U. S. Department of Health, Education, and Welfare, Public Health Service, Center for Disease Control, National Institute for Occupational Safety and Health. Criteria for a Recommended Standard...Occupational Exposure to Inorganic Nickel. DHEW (NIOSH) Publication No. 77-164. Washington, D.C.: U.S. Government Printing Office, 1977.

200. U.S. Department of Health, Education, and Welfare, Public Health Service, National Institutes of Health. Chronic obstructive lung diseases, pp. 84-91. In Respiratory Diseases. Task Force Report on Prevention, Control, Education. DHEW Publication No. (NIH)77-1248. Washington, D.C.: U.S. Government Printing Office, 1977.

201. U.S. Food and Drug Administration, Bureau of Foods, Division of Cosmetics Technology. FDA cosmetics data file of voluntary formulation registration. October 1979.

202. Uotila, L., and M. Koivusalo. Formaldehyde dehydrogenase from human liver. J. Biol. Chem. 249:7653-7663, 1974.

203. Van Dijken, J. P., R. Otto, and W. Harder. Oxidation of methanol, formaldehyde and formate by catalase purified from methanol-grown Hansenula polymorpha. Arch. Microbiol. 106:221-226, 1975.

204. Vaughan, W. T. The Practice of Allergy, p. 677. St. Louis: The C. V. Mosby Company, 1939.

205. Volfová, O. Studies on methanol-oxidizing yeast. III. Enzyme. Folia Microbiol. 20:307-319, 1975.

206. von Wartburg, J. P., and P. M. Schurch. Atypical human liver alcohol dehydrogenase. Ann. N. Y. Acad. Sci. 151:936-946, 1968.

207. Walker, J. F. Formaldehyde, pp. 77-99. In A. Standen, Ed. Kirk-Othmer Encyclopedia of Chemical Technology. 2nd rev. ed. Vol. 10. New York: Wiley-Interscience Publishers, 1966.

208. Watanabe, F., T. Matsunaga, T. Soejima, and Y. Iwata. Study on the carcinogenicity of aldehyde. I. Experimentally produced rat sarcomas by repeated injections of aqueous solution of formaldehyde. GANN 45:451-452, 1954. (in Japanese)

209. Waydhas, C., K. Weigl, and H. Sies. The disposition of formaldehyde and formate arising from drug N-demethylations dependent on cytochrome $\underline{P}$-450 in hepatocytes and in perfused rat liver. Eur. J. Biochem. 89:143-150, 1978.

210. Wayne, L. G., R. J. Bryan, and K. Ziedman. Irritant Effects of Industrial Chemicals: Formaldehyde. DHEW (NIOSH) Publication No. 77-117. Washington, D.C.: U.S. Government Printing Office, 1976. [138] pp.

211. Wisconsin Division of Health, Bureau of Prevention. Formaldehyde Case File Summary. October 23, 1978. Madison, Wis.: Wisconsin Division of Health, 1978. 3 pp.

212. Wisconsin Division of Health, Bureau of Prevention. Statistics of Particle Board Related Formaldehyde Cases through December 15, 1978. Madison, Wis.: Wisconsin Division of Health, 1978. 4 pp.
213. Woodbury, J. W.. Asthmatic syndrome following exposure to tolylene diisocyanate. Ind. Med. Surg. 25:540-543, 1956.
214. Woodbury, M. A., and C. Zenz. Formaldehyde in the home environment. Prenatal and infant exposure. In press, 1980.
215. Yefremov, G. G. The state of the upper respiratory tract in formaldehyde production employees. Zh. Ushn. Nos. Gorl. Bolezn. 30: 11-15, 1970. (in Russian)
216. Yunghans, R. S., and W. A. Monroe. Continuous monitoring of ambient atmospheres with the Technicon AutoAnalyzer, pp. 279-284. In Automation in Analytical Chemistry. Technicon Symposia 1965. Vol. I. New York: Mediad, Inc., 1966.
217. Zaeva, G. N., I. P. Ulanova, and L. A. Dueva. Materials for revision of the maximal permissible concentrations of formaldehyde in the inside atmosphere of industrial premises. Gig. Tr. Prof. Zabol. 12:16-20, 1968. (in Russian)
218. Zannini, D., and L. Russo. Long-standing lesions in the respiratory tract following acute poisoning with irritating gases. Lav. Um. 9:241-254, 1957. (in Italian; English summary)

CHAPTER 8

HEALTH EFFECTS OF SOME OTHER ALDEHYDES

This chapter discusses the effects of aldehydes other than formaldehyde on biologic preparations, animals, and man. It represents an extensive review of the available published information to assess the health effects of aldehydes on humans and animals.

TOXICITY OF ALDEHYDES OTHER THAN FORMALDEHYDE

The total of occupational exposures to aldehydes in 1976 is shown in Table 8-1. Direct evidence of the human health effects of aldehydes is related predominantly to eye irritation, olfaction threshold, and irritation of the upper respiratory tract and skin. To a lesser degree, isolated human biochemical reactions have been monitored.

Toxicity information obtained from animal studies for common aliphatic and aromatic aldehydes is summarized in Tables 8-2, 8-3, and 8-4. The major pathophysiologic effects of aldehydes[31] are described briefly below.

IRRITATION OF THE SKIN, EYES, AND RESPIRATORY MUCOSA

Irritancy is a property of nearly all the aldehydes, but it occurs more commonly and is more important in the case of those with lower molecular weights and those with unsaturation in the aliphatic chain or with halogenated substituents. The general and parenteral toxicity of these compounds appears to be related primarily to irritation, although in some cases (such as fluoroacetaldehyde or fluorobutanal) metabolic conversion to the corresponding fluorinated acids produces an extraordinarily high degree of toxicity. The irritant properties of the dialdehydes have not been extensively studied, but in some instances concentrated solutions can severely irritate the skin and the eyes. The acetals and the aromatic aldehydes in general have a lower degree of irritant action, although there are some exceptions. Furfural has irritant properties, but is not nearly as active as acrolein.[31]

TABLE 8-1

Occupational Exposures to Aldehydes[a]

Aldehyde	No. Exposures
Acetaldehyde	1,744
Acrolein	7,301
Benzaldehyde	15,985
n-Butyraldehyde	1,259
Furfural	15,412
Glutaraldehyde	35,083
Glyoxal	3,848
Propionaldehyde	1,544

[a] Data from NIOSH.[111]

TABLE 8-2

Toxicity of Saturated Aliphatic Aldehydes

Compound	Rat Oral LD$_{50}$, mg/kg	Inhalation Species	Exposure Time	ppm	Irritant Effect Skin	Eye	Respiratory System	Chronic Inhalation Exposure of Rats Exposure	ppm	Effect	Carcino-genicity	Muta-genicity	Terato-genicity
Propionaldehyde	1,410[100]	Rat	30 min	LC$_{50}$ = 26,000[97]	None to mild[100]	Moderate[100]	Minor[96]	6 h/d × 6 d	1,300	Reduced weight gain[a]	(-)[80]	(-)[85]	?
								6 h/d × 20 d	90	None[35]			
Butyraldehyde	5,890[100]	Rat	30 min	LC$_{50}$ = 59,160[97]	Moderate[b]	Severe[100]	Moderate[96]	6 h/d × 12 d	1,000	None[35]	?	(-)[85]	?
Isobutyraldehyde	3,730[101]	Rat	4 h	ALC = 8,000[101]	Mild[101]	Moderate[101]	Moderate > 500 ppm[a]	6 h/d × 12 d	1,000	Slight nasal irritation[35]	?	(-)[85]	?
Hexanal	4,890[101]	Rat	4 h	ALC = 2,000[101]	Mild[101]	Mild[101]					?	?	?
Glutaraldehyde (pentanedial)	820[c]	Rat	4 h (no deaths)	Saturated air[102]	Moderate[c]	Severe[102]					?	(-)[85]	?
Chloral (trichloroacet-aldehyde)	725[55]	Rat	4 h	LC$_{50}$ = 73[a]	Severe[76]	Strong lacri-mator[a]	Severe[a]	4 h/d × 10 d	13	Death (6/10), severe lung injury[a]	?	(+)[65]	(+-)[110]
								2 h/d × 13 d	12.5	Death (7/10), weight loss, pulmonary edema[c]			
								4 h/d × 7 mo	10	Retarded weight gain[c]			
		Dog	4 h	LC$_{50}$ = 1,000[a]									
Aldol (3-hydroxy-butyraldehyde)	2,186[99]	Rat	30 min (no deaths)	Saturated[99]	Moderate[99]	Moderate[99]					?[11]	?	?

[a] E. I. duPont de Nemours & Company, Haskell Laboratory, unpublished data.

[b] Eastman-Kodak Company, unpublished data.

[c] Union Carbide Company, unpublished data, 1958.

TABLE 8-3

Toxicity of Unsaturated Aliphatic Aldehydes

Compound	Rat Oral LD_{50}, mg/kg	Inhalation Species	Exposure Time	ppm	Irritant Effect Skin	Eye	Respiratory System	Carcinogenicity	Mutagenicity	Teratogenicity
Crotonaldehyde	300[98]	Rat	30 min	LC_{50} = 1,400[97]		Severe[98]		?	(−)[85]	?
Methacrolein	140[99]	Rat	4 h	LC_{83} = 250[99]	Mild[99]	Severe[99]		?	?	?
Citral (3,7-dimethyl-2,6-octadienal)	4,960[46]							?	(−)[a]	?

---

[a] V. F. Simmon, SRI International, unpublished data, 1978.

TABLE 8-4

Toxicity of Aromatic Aldehydes

Compound	Rat Oral LD$_{50}$, mg/kg	Inhalation		Irritant Effect			Chronic Inhalation Exposure of Rats			Carcino-genicity	Muta-genicity	Terato-genicity
		Species	Exposure Time	ppm	Skin	Eye	Respiratory System	Exposure	ppm	Effect			
Benzaldehyde	1,300[46]	Rat	4 h (no deaths)	>1,500[a]	None[a]			4 h/d x 10 d	185	None[a]	?	(-)[85]	?
Terephthalaldehyde (1,4-benzenedi-carboxaldehyde)	7,500[a]				Mild[a]						?	?	?
Cinnamicaldehyde (phenylacrolein)	2,220[46]				Strong sensi-tizer by human patch tests[71]						(-)[106]	(-)[85]	?
Salicylaldehyde (2-hydroxybenz-aldehyde)	<2,000[b]				Moderate[b]	Moderate[b]					?	(-)[85]	?

[a] E. I. duPont de Nemours & Company, Haskell Laboratory, unpublished data.

[b] Dow Chemical Company, unpublished data.

SENSITIZATION

Direct sensitization to aldehyde vapors appears to be relatively rare, and sensitization to addition products, such as the bisulfites, almost never occurs. Because of the bifunctional nature of the dialdehydes, they should theoretically be capable of acting as skin sensitizers, but there have been few reports of this phenomenon.[31]

Sensitization to the unsaturated aldehydes may occur, but it is usually very difficult to separate the primary irritation from sensitization. Skin sensitization to the acetals and aromatic aldehydes appears to be infrequent. Pulmonary sensitization and asthma-like symptoms are rarely caused by the inhalation of aldehydes.[31]

ANESTHESIA

Chloral hydrate and paraldehyde have unquestionable anesthetic properties. The former may act through its metabolism to trichloroethanol, and the latter by depolymerization to acetaldehyde. When administered experimentally in large parenteral or oral doses, a number of the aliphatic aldehydes produce anesthesia-like symptoms. However, in industrial exposures, this action is minimized because the primary irritant action prevents substantial voluntary inhalation. In addition, the small quantities that can be tolerated by inhalation are usually metabolized so rapidly that no anesthetic symptoms occur. Some nausea, vomiting, headache, and weakness have been reported in chemists exposed to high concentrations of isovaleraldehyde, but these symptoms have not been interpreted as definite anesthetic reactions.[31]

ORGAN PATHOLOGY

The principal pathologic conditions produced in animals exposed to aldehyde vapors are damage to the respiratory tract and pulmonary edema. In general, multiple hemorrhages and alveolar exudate may be present, but are usually much less apparent than with gases like phosgene. The effects produced by ketene, acrolein, crotonaldehyde, and chloracetaldehyde are much more pronounced and similar to those of phosgene, and chlorine. High dosages of methylal and furfural have been reported to cause various changes in the liver, kidneys, and central nervous system, but there has been no confirmation of this type of action in human industrial exposures. The aldehydes are remarkably free of effects that lead to definite cumulative damage to tissues other than effects that may be associated with primary irritation or sensitization.[31]

METABOLISM AND MECHANISM OF IRRITATION

The simple aliphatic aldehydes are oxidized to their corresponding fatty acids, which normally undergo β-oxidation. Urinary metabolites are not generally detectable, because the fatty acids are further oxidized to carbon dioxide and water. Acetaldehyde is present in normal metabolism, and its importance as a metabolite of ethanol is well known. In general, the toxicity of aldehydes appears to decrease with increasing molecular weight. This relationship is shown by both the oral LD_{50} and the primary irritant action of the lower-molecular-weight substances, which makes them appear to be more potent.[31] As shown in Table 8-5, some of the higher aldehydes are less toxic than their corresponding alcohols, but the unsaturated aldehydes are more toxic than corresponding saturated ones. Although primary irritation and contact dermatitis are occasionally seen after occupational exposure to aldehydes, there is no evidence of serious cumulative effects.

Evidence from a human chamber-exposure study indicated that unsaturation greatly increases the primary irritant activity of an aldehyde (Table 8-6). Halogen substitution in aldehydes may also greatly increase the local tissue irritation.[31] Dixon[22] has postulated that the presence of an aldehyde group adjacent to a double bond has a polarizing effect on the latter, which makes the double bond capable of adding nucleophilic groups, such as sulfhydryl groups. If the sulfhydryl groups in enzymes found in nerve endings are attacked, it seems reasonable that this might be related to the physiologic response of lacrimation. There are, however, difficulties with this explanation, as pointed out by Dixon.[22] The lacrimatory action of such materials is usually very transient and ceases immediately on removal of the irritant. Dixon speculated that the nerve endings may respond to a change in the relative amount of the sulfhydryl compound present, but further evidence seems necessary on this point. It is interesting that, if exposure to a lacrimator is sufficiently prolonged, a point is reached at which lacrimation no longer occurs; this suggests complete saturation of some reactive site.[31]

The primary irritation is probably associated with the reactivity of aldehydes with proteins and amino acids. For example, methylol or hydroxymethyl derivatives may be formed from reaction of aliphatic aldehydes with amino groups.

Cyclic compounds may be formed through later reactions. In the case of the aromatic aldehydes, the products of reaction with amino groups appear to be Schiff bases ($C_6H_5CH=N-R$). Various types of cross-linking reactions can also occur with either aldehydes or dialdehydes, resulting in alteration in protein structures.[31]

The aromatic aldehydes are oxidized to their corresponding organic acids. The oxidation occurs relatively slowly in the liver, but it is usually complete, except where such substituents as hydroxy groups make the aldehydes capable of being excreted by alternative metabolic pathways, such as sulfate or glucuronic acid conjugations.[31]

TABLE 8-5

Oral Toxicities of Corresponding Alcohols, Aldehydes, and Acids[a]

R	RCH_2	Oral LD_{50} in Rats, g/kg		
		Alcohol, RCH_2OH	Aldehyde, RCHO	Acid, RCOOH
CH_3	Ethyl	9-10	1.93	3.53
C_2H_5	Propyl	1.87	1.41	--
$CH_2{:}CH$	Acrylyl	0.064	0.046	2.52
n-C_3H_7	Butyl	4.36	5.89	8.79
$CH_3CH{:}CH$	Crotonyl	--	0.3	1.0
iso-C_3H_7	iso-Butyl	2.46	3.73	--
n-C_5H_{11}	Hexyl	4.59	4.89	6.44
$(C_2H_5)_2CH$	2-Ethylbutyl	1.85	3.98	2.20
n-$C_4H_9(C_2H_5)CH$	2-Ethylhexyl	3.20	3.73	3.00

[a]Reprinted with permission from Williams.[117]

TABLE 8-6

Results of Human Exposure to Aldehydes in a Chamber[a]

Aldehyde	Chamber Concentration, ppm	Duration of Exposure, min	Symptoms
Acrolein	0.80	10	Extremely irritating and lacrimatory (20 s)[b]
	1.22	5	Extremely irritating and lacrimatory (5 s)[b]
Crotonaldehyde	4.1	15	Highly irritating and lacrimatory (30 s)[b]
Acetaldehyde	134	30	Slightly irritating
Propionaldehyde	134	30	Nonirritating
Butyraldehyde	230	10	Nonirritating
Isobutyraldehyde	207	30	Nonirritating

[a] Data from Sim and Pattle.[96]

[b] Average time after exposure at which lacrimation occurred.

Aldehydes occur naturally in foods and have been used extensively as flavoring agents. The rapid metabolism of the aliphatic and aromatic aldehydes by normal pathways undoubtedly accounts for the apparent safety of the large number of these substances that are ingested by humans and animals.

Several aldehydes--such as those listed in Table 8-7--have been shown to have antineoplastic activity.[87]

It has been demonstrated that several environmental irritants are ciliotoxic and mucus-coagulating agents. Aldehydes may thus facilitate the uptake of other airborne substances by the bronchial epithelium.

ACETALDEHYDE

Acetaldehyde is much less irritating to the human eye, nose, and throat than formaldehyde or acrolein. Animal studies have shown acetaldehyde to have low acute toxicity and no appreciable cumulative effects. It does not appear to have substantial mutagenic effects, but has been shown, in a single study, to have dose-dependent embryotoxic and teratogenic properties. Its carcinogenic potential has not been adequately studied. A major source of acetaldehyde in man is the metabolism of ethanol.

ACETALDEHYDE IN ANIMALS

Acute Toxicity

By the acute oral route of administration, acetaldehyde is slightly toxic, with reported LD_{50} values in rats of 1,930 mg/kg[74] and 5,300 mg/kg.[70] Its effect on the skin and eyes of laboratory animals has not been investigated. Human occupational exposure has shown that contact of the eye or skin with liquid acetaldehyde can produce painful, but not serious, burns.[61][64] The rapid evaporation of the liquid limits the duration of the contact.[64] Acetaldehyde appears to act as an anesthetic in acute inhalation exposures at high concentrations. When exposed at 20,000 ppm, rats became anesthetized after a brief period of pronounced excitement. Half the animals died after 30 min of exposure; pulmonary edema was the principal pathologic finding. The survivors recovered in about an hour.[97] In other laboratory animals (mice, rabbits, and guinea pigs), acetaldehyde is slightly to moderately toxic, with calculated 4-h LC_{50} values of approximately 1,100 ppm.[83] Subcutaneous-injection studies in rats and mice produced lethal effects at 500-600 mg/kg.[9]

TABLE 8-7

Antineoplastic Activity of Aldehydes[a]

Aldehyde	Active Against
Glyoxal	Leukemia L 1210 Sarcoma-180 Sarcoma-180
Methylglyoxal	Leukemia L 1210 Sarcoma-180 Walker-256
Kethoxal	Leukemia L 1210 Sarcoma-180 Walker-256 Jensen sarcoma Lymphosarcoma
1,4- and 1,5-dicarbonylaldehyde	Sarcoma-180
Polyaldehydes	Sarcoma-180
Pyridin-2-carboxaldehyde	Leukemia L 1210 Sarcoma-180 Levis-King carcinoma
Isoquinolin-1-carboxaldehyde	Leukemia L 1210 Sarcoma-180 Levis-King carcinoma
Benzaldehyde N-mustards	Leukemia L 1210 Walker-256 Dunnung leukemia
Salicylaldehyde and other aromatic aldehydes	Leukemia L 1210

[a] Data from Schauenstein et al.[87]

Extended Studies

In chronic and subchronic oral studies, acetaldehyde has not produced major toxic effects. In a 6-wk subchronic study, no adverse effects on behavior, weight, or condition of the blood were observed in groups of 10 rats given 550 or 1,100 mg/kg per day. Pathologic examination revealed a statistically significant increase in glycogen content of the liver in the 550-mg/kg group, but not in the 1,100-mg/kg group. Histopathologic examination of the liver revealed nonspecific inflammatory and dystrophic changes. A 6-mo oral study revealed no adverse effects on behavior or weight in groups of 10 rats given up to 50 mg/kg per day. Minimal changes in ECG pattern and blood morphology were reported. These effects were reversible; all animals returned to normal within a month.[70]

In one inhalation study, groups of 20 hamsters were exposed to acetaldehyde at 390, 1,340, or 4,560 ppm, 6 h/d, 5 d/wk for 18 wk. The highest concentration induced growth retardation, ocular and nasal irritation, increased numbers of erythrocytes in the blood, increased heart and kidney weights, and severe histopathologic changes in the respiratory tract. The latter consisted of inflammatory changes, hyperplasia, metaplasia, and necrosis of the respiratory epithelium. The upper respiratory tract was more severely injured than the lower. Changes at 1,340 ppm included increased kidney weights in males and slight hyperplastic and metaplastic changes of the tracheal epithelium. No adverse effects were reported at 390 ppm.[54]

Syrian golden hamsters exposed to acetaldehyde vapor at 1,500 ppm for 7 h/d, 5 d/wk, for 52 wk developed epithelial hyperplasia and metaplasia accompanied by nasal and tracheal inflammation. There was no evidence of carcinogenicity produced by acetaldehyde. Some animals were exposed to benzo[a]pyrene and diethylnitrosamine to study whether acetaldehyde was a cofactor in respiratory tract carcinogenesis. This part of the study produced insufficient evidence to determine any cofactor effects.[31a]

Respiratory-System Effects

The adverse effect of cigarette smoke on the lungs has been attributed in part to its acetaldehyde content (0.98-1.31 mg/cigarette).[38] This was studied in mice by exposing them to acetaldehyde at 1,390 ppm for 30 min twice a day, for 5 wk. A reduction in functional residual capacity of the lung similar to that seen in animals exposed to cigarette smoke was observed.[116] Other studies involving ciliotoxic and cytotoxic effects of tobacco smoke and its constituents have indicated that acetaldehyde is an important compound in this regard.[38,74]

Cardiovascular-System Effect

Acetaldehyde, in common with other aliphatic aldehydes, has a vasopressor effect--i.e., it increases blood pressure. In inhalation experiments, marked increases were seen in blood pressure at concentrations of 1,665 ppm and higher.[23] When acetaldehyde was administered by intraperitoneal injection, vasopressor effects were produced at 5-20 mg/kg. At higher doses, a decrease in blood pressure was a secondary response to a decrease in heart rate.[26]

Metabolism

Acetaldehyde is a toxic intermediate in the metabolism of ethanol. The main pathway involves an enzyme, alcohol dehydrogenase (ADH). Approximately 90% of the acetaldehyde formed from ethanol is oxidized by the liver to acetic acid, which is converted to carbon dioxide and water. The ADH-catalyzed oxidation of ethanol to acetaldehyde has been extensively reviewed.[30,58,59,63,108,118]

Carcinogenic Potential

The carcinogenic potential of acetaldehyde has not been defined by appropriate long-term animal studies. Spindle-cell sarcomas were produced in rats given repeated subcutaneous injections but metastasis to other tissues was not reported.[115] No marked pathologic changes were reported in a group of rats fed a diet of acetaldehyde-containing rice for more than 300 d. Additional details are not available.[68]

Mutagenic Potential

Acetaldehyde was not mutagenic in the standard Ames test with Salmonella typhimurium.[19] It has some mutagenic activity in the fruit fly Drosophila melanogaster, but this activity is much weaker than that produced by formaldehyde.[78] The chromosome-breaking potential of acetaldehyde has been indicated by the dose-dependent sister chromatid exchanges in Chinese hamster ovary cells[69] and human lymphocyte chromosomes.[81]

Embryotoxic and Teratogenic Potential

Acetaldehyde has shown embryotoxic and teratogenic effects in mice similar to those produced by ethanol. Pregnant mice were given intravenous injections of acetaldehyde at 40 or 80 mg/kg on days 7, 8, and 9 of gestation. Acetaldehyde increased the percentage of embryos resorbed and decreased their weight and their protein content. Teratogenic effects included anomalies in closure of the cranial and caudal regions of the neural tube.[73]

HUMAN INVESTIGATIONS

Because of the explosive hazards of acetaldehyde, it is usually handled in closed systems in industry and exposures are not apt to be continuous or large. Therefore, occupational-exposure data are lacking.

Acetaldehyde is readily detected well below 50 ppm. Some persons can notice it below 25 ppm. At 50 ppm, a majority of volunteers exposed for 15 min had some eye irritation; and at 200 ppm, all subjects had redness of the eyes and transient conjunctivitis.[31]

Eye irritation and, to a lesser extent, nose and throat irritation are the only signs noted during exposure to the usual concentrations encountered industrially.[31] The odor and eye-irritation thresholds are 0.07 ppm[4] and 50 ppm,[6] respectively. The TC_{LO} (lowest toxic concentration) for an observed health effect after inhalation is 134 ppm, and the threshold limit value is 200 ppm.[6] Acetaldehyde can cause narcosis, bronchitis, albuminuria, fatty degeneration of the liver, and pulmonary edema at high concentrations. Lethal doses cause progressive slowing of heart and respiratory rates followed by respiratory paralysis. There have been very few studies of persons after industrial exposure. In one case, chronic exposure of workers to acetaldehyde at 0.5-22 mg/m^3 (0.21-9.36 ppm) caused irritation of the mucous membranes.[21]

There have been many reports of the pharmacology of acetaldehyde in relation to the effects of alcohol. Human subjects were given intravenous infusions to increase blood concentrations to 0.2-0.7 mg% (about 10 times normal). At these concentrations, heart rate and respiratory ventilation were increased, and a "hangover" sensation is noted.[31]

Bittersohl[11] reported an increased prevalance of malignant tumors in aldehyde workers. Of the 220 people employed in a plant, 150 had been employed for more than 20 yr. Nine neoplasms were noted in males: two squamous-cell carcinomas of the oral cavity, one adenocarcinoma of the stomach, one adenocarcinoma of the cecum, and five squamous-cell carcinomas of the bronchial tree. The workers had mixed exposures that included acetaldehyde, butyraldehyde, crotonaldehyde, higher condensed aldehydes, butanol, hexatriol, hexatetrol, octadiol, and butadiene. Of the nine employees, eight smoked 5-10 cigarettes per day.

ACROLEIN

Acrolein is highly toxic by all routes of administration. Its vapors cause severe respiratory and ocular irritation. Contact with liquid acrolein can produce skin or eye necrosis. Serious injury is produced even by a 1% aqueous solution. Acrolein has not been shown to be carcinogenic or embryotoxic, but appears to be mutagenic in some nonmammalian systems.

ACROLEIN IN ANIMALS

Acute Toxicology

By the oral route of administration, acrolein is highly toxic, with reported LD_{50} values in rats, mice, and rabbits of 46 mg/kg,[100] 40 mg/kg,[13] and 7.1 mg/kg (Shell Chemical Company, unpublished data), respectively. Acrolein is easily absorbed through the skin of rabbits in lethal amounts (LD_{50}, 168-562 mg/kg) (Shell Chemical Company, unpublished data; Union Carbide Company, unpublished data). When applied undiluted, acrolein causes necrosis; even a 1% aqueous solution can produce a burn on the abdominal skin of rabbits. Likewise, instillation of a 1% solution into the rabbit eye caused severe injury.[100]

The marked toxicity of acrolein has also been shown by inhalation exposures. In a 30-min exposure of rats, the LC_{50} was 131 ppm;[97] in a 4-h exposure, it was 8 ppm.[100]

Correspondingly low values have been reported for other species of animals. Acrolein vapors are also very irritating to the eyes, nose, and throat of laboratory animals, as well as man. Exposure of cats and rats at approximately 12 ppm caused severe symptoms of eye and respiratory tract irritation.[44,66]

Injection studies have demonstrated the high acute toxicity of acrolein, with lethal doses of 30-50 mg/kg.[97]

Extended Studies

Acrolein was added to the drinking water of rats, and that water was administered as their only source of water for up to 90 d. The unpalatability of the water was manifest in reduced body weights and increased kidney weights. The death of animals given water containing acrolein at 600-1,800 ppm was due to lack of water intake. No adverse pathologic or hematologic changes were observed (G.W. Newell, unpublished data; Union Carbide Company, unpublished data).

Continuous inhalation of acrolein by rats for up to 90 d had little or no effect at concentrations of 0.06-0.22 ppm.[12,40,60] Dogs and monkeys appeared to be more sensitive to the vapors of acrolein. In one of these studies,[60] no abnormal behavior was seen at 0.22 ppm, but pathologic examination revealed acrolein-related changes in the tracheas of the monkeys and lungs of the dogs. Higher concentrations (1-1.8 ppm) produced visible signs of ocular and nasal irritation throughout the 90-d period. In another study,[12] exposure of rats at 0.55 ppm produced signs of irritation that subsided after 3-4 wk. The mean body weight of exposed animals was also significantly lower than that of the controls. At 1-2 ppm, acrolein induced minimal biochemical, pathologic, and functional injury of the lower respiratory tract. The changes--which included a decrease in urinary vanillyl-mandelic acid and inflammatory infiltrates with mild perivascular edema in the respiratory tract--reached a peak during the first month of exposure and then subsided.[39]

During a continuous-inhalation study, groups of hamsters, rats, and rabbits were exposed to acrolein at 0.4, 1.4, and 4.9 ppm for 6 h/d, 5 d/wk for 13 wk. Marked changes in the nasal epithelium were evident in all species at the highest concentration. Histopathologic examination revealed necrotizing rhinitis and squamous hyperplasia and metaplasia of the epithelium. Rats appeared to be the most susceptible species examined, with acrolein-related abnormalities associated with 0.4 ppm exposure. This concentration was nontoxic to both hamsters and rabbits.[34] Lyon et al.[60] exposed rats, guinea pigs, monkeys, and dogs to acrolein at 0.7 and 3.7 ppm for 8 h/d, 5 d/wk, for 6 wk or 24 h/d for 90 d at 0.21-1.8 ppm. No clinical signs of toxicity were observed up to 0.7 ppm. Dogs and monkeys were visibly affected by respiratory irritation at the greater exposures. Repeated exposures at 0.7 ppm produced chronic inflammatory changes, and 3.7 ppm caused squamous metaplasia of the lungs in monkeys. Continuous exposure at 0.22 ppm resulted in moderate emphysema, acute congestion of the lungs, and squamous metaplasia and basal cell hyperplasia of the trachea.

In a chronic study, 36 hamsters were exposed at 4.0 ppm, 7 h/d, 5 d/wk for 52 wk. The irritation produced by this concentration of acrolein was initially manifest by salivation, nasal discharge, restlessness, and the animals' keeping their eyes closed. The animals apparently became acclimated to the vapors and behaved normally after the second week of exposure, except for increased restlessness (compared with the controls). At the end of the exposure period, six animals were sacrificed and the rest held for an additional 29-wk recovery period. The animals exhibited rhinitis and hyperplastic and metaplastic changes of the nasal epithelium.[33] No evidence of carcinogenicity was seen.

Respiratory-System Effects

Groups of mice and guinea pigs were exposed to low concentrations of acrolein. In mice, a concentration of 1.7 ppm produced a 50% decrease in respiratory rate.[50] Another group was exposed to a mixture of acrolein and formaldehyde. A 50% decrease in respiratory rate was produced at a combination of 1.87-ppm acrolein and 1.42-ppm formaldehyde; hence, the effect is not additive.[49] A group of guinea pigs exposed to acrolein at 0.4-1.0 ppm for 2 h showed definite decreases in respiratory rate.[67] In a study with dogs, it was shown that acrolein is taken up more readily by the upper respiratory system, but the uptake is considerably less than that of formaldehyde.[24]

Cardiovascular-System Effects

Acrolein, in common with many other aldehydes, causes an increase in blood pressure (vasopressor effect). Anesthetized rats were given acrolein at 0.05-5.0 mg/kg by intravenous injection or exposed at

4.4-2,200 ppm for 1 min. With intravenous doses up to 0.25 mg/kg, an increase in blood pressure predominanted as an effect. At higher doses, a decrease in blood pressure predominated. With exposure by inhalation, a vasopressor effect of increasing magnitude was observed within 15 s after the onset of exposure. Within 10 s after exposure ceased, there was a rapid return to normal.[25]

Metabolism/Pharmacokinetics

Acrolein is formed during the degradation of oxidized spermine and spermidine. It is a probable metabolite of allyl alcohol and has been shown to be a metabolite of the antitumor agent cyclophosphamide.[21]

Carcinogenic Potential

Acrolein was not carcinogenic in a 52-wk inhalation study in hamsters.[33] Preliminary results from an NCI-sponsored inhalation study, also in hamsters, confirm this conclusion.[21] Acrolein did not produce sarcomas in a subcutaneous-injection study in mice.[105] When tested by skin application in a promotion-initiation study with croton oil, acrolein had little or no tumor-initiating activity.[82] Similarly, it had no effect on the carcinogenic activity of diethylnitrosamine and had a minimal effect on the activity of benzo[a]pyrene.[33]

Mutagenic Potential

Acrolein showed some mutagenic activity in the Ames test. Mutagenic effects were observed in Salmonella typhimurium strains TA 1538 and TA 98 (strains that detect frameshift mutations), whereas no activity was seen in strains TA 1535 and TA 100 (strains that detect base-pair substitutions).[9] In another test of the Ames type, acrolein did not induce point mutations in eight strains of histidine-dependent mutants of S. typhimurium. The authors indicated that this test might not be able to identify weak mutagens.[7] Acrolein had mutagenic activity in the fruit fly Drosophila melanogaster[78] and in a DNA-polymerase-deficient Escherichia coli.[10] However, no activity was seen in strains of E. coli capable of detecting forward and reverse mutations.[27,28] No activity was seen in yeast Saccharomyces cerevisiae[45] or in a dominant-lethal assay in mice.[29]

Embryotoxic and Teratogenic Potential

Acrolein did not exhibit embryotoxicity in an inhalation study in rats. Male and female rats were continuously exposed at 0.55 ppm and allowed to mate after the fourth day of exposure. No significant differences could be observed at this low concentration between control and test animals in number of pregnant animals, number of fetuses, or mean fetal weight.[12] The fetuses were not examined for malformations; therefore, no information on teratogenic potential is available from this study. No evidence of teratogenicity was observed in embryos from acrolein-treated chicken eggs.[51]

ACROLEIN IN HUMANS

Acrolein is predominantly an ocular and respiratory irritant. Its toxicity can involve the senses, reflexes, nervous system, and respiratory system, alter biochemical reactions, and affect the composition of blood at 0.2-6.0 ppm.[17] Its conjugated unsaturated bonds at the 1,2-position result in eye irritation (threshold, 0.2 ppm[17]) 2.5 times greater than that from formaldehyde.[2] At higher concentrations (0.5-1.0 ppm), this difference increases to 4-5 times.[1] Such irritancy is particularly important, because acrolein, as a partially oxidized organic emission, is a major contributor to the irritant quality of cigarette smoke[117] and photochemical smog.[17] Acrolein in Los Angeles smog was thought to be responsible for 35-75% as much eye irritation as formaldehyde (the major aldehyde in auto exhaust).[2] The occupational threshold limit value for acrolein (0.1 ppm) is low enough to minimize irritation in exposed persons: 0.25 ppm is considered to be moderately irritating, 0.5 ppm is thought to be a practical working concentration, and 1 ppm causes marked irritation of the eyes and nose with lacrimation in less than 5 min.

The principal site of attack of acrolein is the mucous membranes of the upper respiratory tract; high concentrations can produce pulmonary edema. Table 8-8 lists human responses to various concentrations of acrolein. The results of several studies concerning eye irritation are found in Table 8-9.

Descriptions of acrolein toxicity are largely in the form of ocular-irritation studies. However, the conclusions drawn from these studies should be tempered, because the relationships between atmospheric acrolein concentrations and indexes of eye irritation are nonlinear and subject to a large element of variability.[72,79] Ocular exposure to 0.5% acrolein for 5 min[95] caused discomfort with stinging in 30-60 s and lacrimation, pain, and eyelid flickering and heaviness in 3-4 min. Attesting to the variability of ocular response, a chronaximetric study of the eyes of three persons showed reduced chronaxy in two subjects at 0.64 ppm, but prolonged chronaxy in the third. The optical-chronaxy reflex threshold was determined to be about 0.7 ppm. Similarly, sensitivity to light was increased or decreased and then gradually returned to normal. The optical-chronaxy method of eye-irritation detection may not be sensitive enough, inasmuch as other studies[17,43,87,117] have indicated lower eye-irritation thresholds measured by the more subjective methods of blink response, lacrimation, or pain response.

TABLE 8-8

Thresholds of Response after Exposure to Acrolein

Acrolein Concentration, ppm	Response
0.2[a]	Eye-irritation threshold[17]
0.33–0.40[a]	Odor threshold[5,6,87]
0.40–1.0	Prolonged deep respiration[87]
0.62	Respiratory-response threshold[6]
0.73	Chronaximetric-response threshold[6,77]
0.8	Severe mucosal irritation[6]
1.0	Immediately detectable[4]
5.5	Intense irritation[4]
$\geq$10.0 (vapor, estimated)	Lethal in a short time[4]
24.0	Unbearable[4]

[a] In a simulated-smog study, the acrolein eye-irritation threshold (without olfaction) for a 30-s exposure was 1.27 ppm, and the odor threshold was 0.08–0.29 ppm.[15] The eye-irritation threshold was the same whether determined by increasing concentrations over a constant period or by increasing the duration of exposure over a series of concentrations (M. Jones, H. Buchberg, K. Lindh, and K. Wilson, unpublished data).

TABLE 8-9

Ocular Response to Airborne Acrolein[a]

Ocular Exposure Concentration, ppm	Duration of Exposure	Effect
0.8	10 min	Extremely irritating; only just tolerable
1-2	5 min	87% of test panel reported irritation
1	5 min	82% of test panel reported irritation
0.5	5 min	35% of test panel reported irritation
0.5	5 min	19% of test panel reported irritation
0.5	12 min	91% of test panel reported irritation
1.8	30 s	(Odor)
	1 min	Slight irritation
	2 min	Distinct irritation (and slight nasal irritation)
	4 min	Profuse lacrimation; practically intolerable
5.5	5 s	Moderate irritation (and odor and moderate nasal irritation)
	20 s	Painful irritation (and painful nasal irritation)
	60 s	Marked lacrimation; practically intolerabl
21.8	--	Intolerable
30.25	5 min	Moderate irritation
4	5 min	Severe irritation
0.06	--	Irritation 0.471 on scale of 0-2
1.3-1.6	--	Irritation 1.182 on scale of 0-2
2.0-2.3	--	Irritation 1.476 on scale of 0-2

[a] Reprinted with permission from Kane.[48]

Inhalation of acrolein at 0.22-0.75 ppm[87] has generally resulted in a depressed respiratory rate due to its anesthetic effect,[77,117] although 1 ppm has been found to be tolerated with no significant respiratory change.[43] Higher inhaled concentrations result in respiratory irritation. The TC_{LO} for irritation of the upper respiratory tract is 1 ppm,[11] although nasal irritation occurs at lower concentrations.[21,43,87]

In liquid form, acrolein causes severe skin irritation.[5] Dermal application of a 1% solution produced a positive patch test.[21]

The effects of acrolein on biochemical functions have not been thoroughly studied.[21,87] At 0.22 ppm or higher, prolonged inhalation caused a reduction in lung lactic dehydrogenase. Acrolein is highly reactive with thiol groups (this is related to its lacrimatory effect), and it can rapidly conjugate with glutathione and cysteine. Acrolein is also a potent in vitro inhibitor of human polymorphonuclear leukocyte chemotaxis (EC_{50}, 15 μm), but has no effect on leukocyte integrity or glucose metabolism. Chemotaxis is assumed to be inhibited by the reaction of acrolein with essential thiol groups of cellular proteins involved in chemotaxis. A decrease in cholinesterase activity and an alteration of liver enzyme activity have also been noted.[87]

Industrially, acrolein is expected to cause serious intoxication only rarely, because of human intolerance of its irritating effects. The inhalation LC_{LO} (lowest lethal concentration) for humans has been estimated at 153 ppm for a 10-min exposure.[104] Two cases of occupational poisoning (one fatal) have been reported.[84] It is speculated that the greatest occupational danger of acrolein poisoning is associated with the welding of fat and oil cauldrons.

The EPA has determined that, for protection of human health from the toxic properties of acrolein ingested through water and contaminated aquatic organisms and through contaminated aquatic organisms alone, the ambient-water criteria are 0.320 and 0.780 mg/L, respectively.[112]

OTHER ALDEHYDES

Tables 8-2, 8-3, and 8-4 list toxicity information on other common aldehydes. The eye and respiratory tract irritation caused by formaldehyde, acrolein, and, to some extent, acetaldehyde is also caused by propionaldehyde,[96,100] butyraldehyde (Sim and Pattle;[96] Smyth et al.;[100,101] Du Pont Company, unpublished data), and chloral (Du Pont Company, unpublished data). Chloral is unique, in that its inhalation toxicity puts it in the highly toxic category (Du Pont Company, unpublished data). However, it is widely used as a sedative. Chloral has also been shown to be mutagenic in the Ames test (Minnich et al.;[65] Du Pont Company, unpublished data) and has shown some embryotoxic properties.[110] None of the other aldehydes have toxicity values that would be inconsistent with the values typical of their class.

BENZALDEHYDE

No information is available on the human health effects of benzaldehyde. Increased concentrations were found in the blood of New Orleans residents during 1970-1975.[57]

BUTYRALDEHYDE AND ISOBUTYRALDEHYDE

These aldehydes are not human irritants. The estimated inhalation TC_{Lo} of butyraldehyde[104] is 580 mg/m^3. Exposure at 200 ppm for 30 min results in no irritant effects.[6] However, isobutyraldehyde at the same concentration causes nausea.[6] Butyraldehyde has been implicated[11] as an etiologic factor in the cancer epidemiologic study discussed in the section on acetaldehyde.

CHLORAL HYDRATE

Chloral hydrate is converted to trichloroethanol and trichloroacetic acid in man. Some of the alcohol derivative is excreted as the glucuronide. Bromal hydrate is metabolized differently and is more toxic.[31]

CHLOROACETALDEHYDE

Chloroacetaldehyde is somewhat more irritating to the eye, nose, and throat than is formaldehyde. Contact with a 40% aqueous solution produces serious eye injury and skin corrosion.[4] Dilute aqueous solutions of 0.1% are capable of causing marked skin irritation. The carcinogenic activity of vinyl chloride has been attributed to its metabolic activation in the liver to 2-chloroacetaldehyde. 2-Chloroacetaldehyde increases the revertant power of S. typhimurium mutant strains; this suggests that the presence of an oxidase in the microsomal fraction of human liver is responsible for converting vinyl chloride to mutagenic metabolites.[8]

CROTONALDEHYDE (β-METHYL ACROLEIN)

Crotonaldehyde, whose threshold limit value of 2 ppm is based on animal studies,[4] produces symptoms similar to those produced by acrolein. A strong odor is detectable at 15 ppm, and exposure at 45 ppm results in disagreeable, conjunctival irritation. An increased cancer incidence in workers exposed to crotonaldehyde and other agents is discussed in the section on acetaldehyde.[11]

FURFURAL

There are conflicting reports on the toxicity of furfural. One report indicating only mild effects contrasts with another describing numbness of the buccal membranes and tongue, loss of taste, and respiratory distress.[84] The latter report indicated that 1.9-14 ppm caused bloodshot eyes, lacrimation, throat irritation, headache, and possibly damage to eyesight. Furfural is metabolized by conversion of the aldehyde group to an acid and conjugation with glycine.[31] Other than an occasional allergic skin manifestation, no injury from occupational exposure to furfural has been reported.[31] The inhalation TC_{Lo} has been estimated[104] at 600 µg/m^3. As a result of the primary irritation induced by furfural, a threshold limit value of 5 ppm has been established.[4]

Feron[32] conducted an intratracheal-instillation study in Syrian golden hamsters with furfural, benzo[a]pyrene, and benzo[a]pyrene plus furfural. Furfural alone produced no evidence of carcinogenic activity, but the results suggested a cocarcinogenic effect of furfural on the respiratory tract of hamsters. In comparison with treatment with benzo[a]pyrene alone, intratracheal instillation of benzo[a]pyrene plus furfural resulted in earlier development of metaplastic changes of the tracheobronchial epithelium, a shorter latent period for tracheobronchial tumors, and a few more bronchial and peripheral squamous-cell carcinomas.

GLUTARALDEHYDE

Glutaraldehyde is a strong nasal irritant and a mild optic or dermal irritant. Occasional dermal contact can cause an allergic response leading to contact dermatitis. Activated glutaraldehyde (pH, 7.5-8.0) is a stronger irritant and affects the upper respiratory tract. Occupationally, a 2% aqueous solution (0.38 ppm) in an operator's breathing zone produces severe eye, nose, and throat irritation and headache. This has led to the establishment of a threshold limit value (ceiling) of 0.2 ppm.[4]

Glutaraldehyde at 5-10% has been effective in reducing hyperhidrosis.[20 36 37 47 86] Sensitization to glutaraldehyde occurs much less frequently than sensitization to formaldehyde, and cross-reaction of formaldehyde-sensitive subjects does not seem to occur.[47] Strong staining of the skin limits the usefulness of glutaraldehyde. As with formaldehyde, the blockage of sweating can be reversed by stripping the stratum corneum with tape.[36 75 86]

GLYOXAL

The antitumor activity of glyoxal has been mentioned.[87] Concentrations of 0.5 mM or higher inhibit human fibroblast cell division and synthesis of DNA, RNA, and proteins.[8]

4-HYDROXYPENTENAL

This compound inhibits mitosis in human kidney cells and is 4 times as active an inhibitor as kethoxal.[87]

KETOALDEHYDES

Ketoaldehydes complexed with hydrogen sulfite and an amino group were less harmful than free aldehydes in tumor therapy. These complexes showed significant antitumor activity in in vitro adenocarcinomas cultured from breast and colon tissues and in epidermal carcinoma.[87]

MALONALDEHYDE

Reaction of malonaldehyde (50 µg/ml) with DNA from human fibroblasts reduces hypochromia, changes the "temperature-absorption curve," and increases resistance to degradation by DNase.[87] Malonaldehyde is a product of peroxidative fat metabolism and is formed in the tissues of animals whose diet is deficient in antioxidants. Shamberger et al.[93] applied malonaldehyde once to the shaved backs of female Swiss mice. Daily treatment with 0.1% croton oil produced tumors in 52% of the mice at 30 wk. Malonaldehyde concentrations in mouse skin increased after application of benzo[a]pyrene, 7,12-dimethylbenz[a]anthracene, and 3-methylcholanthrene. A weak link has also been established from epidemiologic data between beef fat in the diet, malonaldehyde content in beef, and the incidence of large bowel cancer.[91] See the discussion in Schauenstein et al.[87] (Chapter 5) for additional discussion of the metabolism of malonaldehyde.

PROPIONALDEHYDE AND AMINOPROPIONALDEHYDE

Inhalation of propionaldehyde at 134 ppm for 30 min, or 0.1-6.0 µg/m^3, has been demonstrated to cause mild irritation of mucosal surfaces.[6,21] β-Aminopropionaldehyde is a natural component of human serum. A decrease in its oxidation leads to its accumulation, with an additional accumulation of spermidine and a decrease in malonaldehyde. Increases in propionaldehyde and spermidine cause an increase in RNA synthesis.[87] Propionaldehyde is a skin carcinogen whose structure resembles that of malondialdehyde.[92]

SYNAPALDEHYDE (CINNAMALDEHYDE)

Cinnamaldehyde is a natural ingredient in the essential oils of cinnamon leaves and bark, hyacinth, and myrrh. It is used primarily as a fragrance in soaps, creams, lotions, and perfumes.

Cinnamaldehyde is oxidized to cinnamic acid, which is further degraded to benzoic acid;[118] much of it shows up in the urine of rats as hippuric acid.[109]

When administered by intubation to rats and guinea pigs, cinnamic aldehyde had an LD_{50} of 2,220 and 1,160 mg/kg, respectively. The toxicity in rats was expressed as depression, diarrhea, and a scrawny appearance.[46] The intraperitoneal LD_{50} of cinnamaldehyde in mice was 2,318 mg/kg, and its oral LD_{50} in rats was 3,350 mg/kg.

In a study of the toxicity of synthetic and natural products, the LD_{100}s in mice were 6,000 and 12,000 mg/kg, respectively. In chronic-toxicity studies, liver lipid content was reportedly increased by 20% in the first generation and 22% in the second generation.[103]

Published reports indicate that cinnamaldehyde is a skin irritant and strong sensitizer. Sensitization reactions have been produced in guinea pigs after challenge with 0.5% cinnamaldehyde by the method of Buehler.[107] Kligman[52][53] tested cinnamaldehyde on human subjects at 3% and 8% in petrolatum. The lower concentration produced no irritation, but the 8% concentration was severely irritating. Studies completed by the North American Contact Dermatitis Research Group indicated that cinnamaldehyde may be a frequent cause of allergic reactions to perfumes.[89] This Group and Schorr[90] reported positive reactions to cinnamaldehyde in more than 3% of those tested.

When tested in rabbits, cinnamaldehyde converted resting EEG patterns to arousal patterns in the gallamine-paralyzed preparation with the intact brain. A centrally originating deactivation was produced through direct and indirect excitatory action on the brainstem reticular formation.[41] It also produced positive inotropic and chronotropic effects in isolated guinea pig heart preparations and hypotensive effects in anesthetized dogs and guinea pigs secondary to its peripheral vasodilatation.[42]

Cinnamaldehyde was shown to be weakly mutagenic, with a tendency to produce nondisjunction in tests using late embryos and young larvae of <u>Drosophila melanogaster</u>.[114] The incidence of primary lung tumors in both male and female A-strain mice was not increased over control values after intraperitoneal injection at 4.00 or 0.8 g/kg over an 8-wk period.[106]

Epidemiologic evidence from a study in Buckinghamshire in Oxfordshire, England, suggested increased nasal and sinus cancers in woodworkers of the furniture industry.[1] Other investigators have not found a relationship between occupation and nasopharyngeal cancer in retrospective surveys.[18][56][62][94] Nasopharyngeal cancers are apparently found in a wide variety of occupations, and Buell,[16] using occupation as an indicator of economic status, found a twofold excess of nasopharyngeal cancers among those of lower socioeconomic status.

As a result of these epidemiologic findings, a number of constituents of wood are being tested for carcinogenic activity, including the lignin constituents, methoxy-substituted cinnamaldehydes, and cinnamalcohols. The latter would yield respective aldehydes in the course of metabolic oxidation by alcohol dehydrogenases. Preliminary data indicate that 3,4,5-

trimethoxycinnamaldehyde is a potent carcinogen and may be involved in the carcinogenic action of some woods and their products.[88] Cinnamaldehyde has not been shown to be carcinogenic, but glycidaldehyde (2,3-epoxypropionaldehyde) has been shown to be carcinogenic in mice and rats.[113]

HeLa cells in permanent culture suffered irreversible damage from exposure to 50 mM glyceraldehyde, with high toxicity at concentrations lower by a factor of 10-100.[87] The HeLa cells produced 4-hydroxy-2-oxobutanal as a bacteria-inhibiting factor.[87]

VALERALDEHYDE AND ISOVALERALDEHYDE (2-METHYLBUTYRALDEHYDE)

Human data on valeraldehyde are not available. A threshold limit value of 50 ppm is based on animal data.[4] Several chemists engaged in distilling isovaleraldehyde[31] developed chest discomfort, nausea, emesis, and headaches. Although exposures were not measured, the odor was pronounced, and ambient concentrations may have been high. All symptoms were reversed within a few days without further consequence.

MISCELLANEOUS

Eye Irritation from Oxidation Products of Paraffinic, Olefinic, and Aromatic Hydrocarbons and Aldehydes

Photochemical auto smog consists largely (ca. 15%) of unburned and partially oxidized organic materials, including aliphatic, olefinic, and aromatic hydrocarbons, and aldehydes. Gaseous NO_x is also present as an oxidant.[17]

Chemical reactivity increases in the order of paraffinic hydrocarbons < ethylene, toluene, propionaldehyde < 1-butene, 1,3-butene, 1,3,5-trimethylbenzene.[3] In measurements of human eye irritation caused by oxidation products of these compounds combined with NO_x, 1,3-butadiene proved to be the most potent (eye irritation index, 20 on a 1-30 scale). Less-saturated or shorter-chain olefins produced less eye irritation, and isolated oxidized olefins at concentrations of less than 1 ppm produced no irritation. Of the aromatic hydrocarbons, 45% oxidation of mesitylene to aliphatic aldehydes produced only slight eye irritation (index, 6 on a 1-30 scale). In contrast with the strong irritating effect of oxidized olefinic hydrocarbon mixtures, olefinic and aromatic hydrocarbons together, when oxidized, produced only slight irritation. Oxidation of 3.5-ppm propionaldehyde produced 0.08-ppm formaldehyde, which resulted in moderate eye irritation (index, 8 on a 1-30 scale). Lower propionaldehyde concentrations resulted in only slight irritation.[3] In combination with acetaldehyde, oxidation resulted in eye irritation of 0.2-2.1 on a scale with an upper limit of 5.[2] For constituents of photochemical smog, variations in eye-irritation threshold appear to be due to the amount of unsaturated hydrocarbon present in the smog. The thresholds of saturated hydrocarbons oxidized to saturated aldehydes are lower than expected. This suggests the presence of unsaturated aldehyde (e.g., acrolein) resulting from the oxidation of unsaturated precursors.[15]

Naturally Occurring Aldehydes

In addition to the presence of propionaldehyde as a natural constituent of human blood, several other aldehydes are also natural body constituents. Some of these higher, naturally occurring aldehydes and their functions are as follows:[87]

Indol-3-ylacetaldehyde 5-Hydroxyindol-3-ylacetaldehyde	Metabolic products of tryptophan
Pyridoxal phosphate and pyridoxal	Coenzymes and catalysts
Retinal and dehydroretinal	Vitamins A_1 and A_2, respectively; parts of the light-sensitive optic pigments (rhodopsin)
Collagenaldehyde	Part of the collagen cross-linkage reaction mechanism

These and other aldehydes (formaldehyde, acetaldehyde, butyraldehyde, isobutyraldehyde, and crotonaldehyde) are oxidized to the acid form by formyl hydrate dehydrogenase, which is in human blood serum.[87]

REFERENCES

1. Acheson, E. D., R. H. Cowdell, E. Hadfield, and R. G. Macbeth. Nasal cancer in woodworkers in the furniture industry. Br. Med. J. 2:587-596, 1968.
2. Altshuller, A. P. Assessment of the contribution of chemical species to the eye irritation potential of photochemical smog. J. Air Pollut. Control Assoc. 28:594-598, 1978.
3. Altshuller, A. P., D. L. Klosterman, P. W. Leach, I. J. Hindawi, and J. E. Sigsby, Jr. Products and biological effects from irradiation of nitrogen oxides with hydrocarbons or aldehydes under dynamic conditions. Int. J. Air Wat. Pollut. 10:81-98, 1966.
4. American Conference of Governmental Industrial Hygienists Inc. Documentation of the Threshold Limit Values. Fourth Edition. 1980. Cincinnati: American Conference of Governmental Industrial Hygienists Inc., 1980. 486 pp.
5. American Industrial Hygiene Association. Acrolein. Hygiene Guide Series. Detroit: American Industrial Hygiene Association, 1963. 3 pp.
6. American Industrial Hygiene Association. Community air quality guides. Aldehydes. J. Am. Ind. Hyg. Assoc. 29:505-512, 1968.

7. Andersen, K. J., E. G. Leighty, and M. T. Takahashi. Evaluation of herbicides for possible mutagenic properties. J. Agr. Food Chem. 20:649-656, 1972.
8. Auerbach, C., M. Moutschen-Dahmen, and J. Moutschen. Genetic and cytogenetical effects of formaldehyde and related compounds. Mutat. Res. 39:317-362, 1977.
9. Bignami, M., G. Cardamone, P. Comba, V. A. Ortali, G. Morpurgo, and A. Carere. Relationship between chemical structure and mutagenic activity in some pesticides. The use of Salmonella typhimurium and Aspergillus nidulans. Mutat. Res. 46:243-244, 1977.
10. Bilimoria, M. H. The detection of mutagenic activity of chemicals and tobacco smoke in a bacterial system. Mutat. Res. 31:328, 1975.
11. Bittersohl, G. Epidemiological research on cancer risk by aldol and aliphatic aldehydes. Environ. Qual. Saf. 4:235-238, 1975.
12. Bouley, G., A. Dubreuil, J. Godin, M. Biosset, and C. Boudene. Phenomena of adaptation in rats continuously exposed to low concentrations of acrolein. Ann. Occup. Hyg. 19:27-32, 1976.
13. Boyland, E. Experiments on the chemotherapy of cancer. 4. Further experiments with aldehydes and their derivatives. Biochem. J. 34:1196-1201, 1940.
14. Boytsov, A. N., Y. S. Rotenberg, V. G. Mulenkova. Toxicological evaluation of chloral in the process of its liberation during spraying and pouring of polyurethane foams. Gig. Tr. Prof. Zabol. 14:26-29, 1970.
15. Buchberg, H., M. H. Jones, K. G. Lindh, and K. W. Wilson. air pollution studies with simulated atmospheres. Report No. 61-44. Los Angeles: University of California Department of Engineering, 1961. 168 pp.
16. Buell, P. Nasopharynx cancer in Chinese of California. Br. J. Cancer 19:459-470, 1965.
17. Campbell, K. I. Effects of gaseous air pollution on body systems. Vet. Toxicol. 16:73-81, 1974.
18. Ch'in, K. Y., and C. Szutu. Lymphoepithelioma. Pathological study of 97 cases. Chin. Med. J. 3 (Suppl.):94-119, 1940.
19. Commoner, B. Reliability of Bacterial Mutagenesis Techniques to Distinguish Carcinogenic and Noncarcinogenic Chemicals. U.S. Environmental Protection Agency Report No. 600/1-76/022. PB-259 934. Springfield, Va.: National Technical Information Service, 1976. 114 pp.
20. Cullen, S. I. Management of hyperhidrosis. Postgrad. Med. 52:77-79, 1972.
21. Dirken, P. Acrolein. Ambient Water Quality Criteria. Syracuse: Syracuse Research Corp., for U.S. Environmental Protection Agency, 15 May 1979. (2nd draft)
22. Dixon, M. Reaction of lachrymators with enzymes and proteins, pp. 39-49. In Biochemical Society Symposia No. 2. Cambridge, England: Cambridge University Press, 1948.

23. Egle, J. L., Jr. Effects of inhaled acetaldehyde and propionaldehyde on blood pressure and heart rate. Toxicol. Appl. Pharmacol. 23:131-135, 1972.
24. Egle, J. L., Jr. Retention of inhaled formaldehyde, propionaldehyde, and acrolein in the dog. Arch. Environ. Health 25:119-124, 1972.
25. Egle, J. L., Jr., and P. M. Hudgins. Dose-dependent sympathomimetic and cardioinhibitory effects of acrolein and formaldehyde in the anesthetized rat. Toxicol. Appl. Pharmacol. 28:358-366, 1974.
26. Egle, J. L., Jr., P. M. Hudgins, and F. M. Lai. Cardiovascular effects of intravenous acetaldehyde and propionaldehyde in the anesthetized rat. Toxicol. Appl. Pharmacol. 24:636-644, 1973.
27. Ellenberger, J., and G. R. Mohn. Comparative mutagenicity testing of cyclophosphamide and some of its metabolites. Mutat. Res. 38:120-121, 1976.
28. Ellenberger, J., and G. R. Mohn. Mutagenic activity of major mammalian metabolites of cyclophosphamide toward several genes of _Escherichia_ _coli_. J. Toxicol. Environ. Health 3:637-650, 1977.
29. Epstein, S. S., E. Arnold, J. Andrea, W. Bass, and Y. Bishop. Detection of chemical mutagens by the dominant lethal assay in the mouse. Toxicol. Appl. Pharmacol. 23:288-325, 1972.
30. Eriksson, C. J. P., and H. W. Sippel. The distribution and metabolism of acetaldehyde in rats during ethanol oxidation--I. The distribution of acetaldehyde in liver, brain, blood and breath. Biochem. Pharmacol. 26:241-247, 1977.
31. Fassett, D. W. Aldehydes and acetals, pp. 1959-1989. In D. W. Fassett and D. D. Irish, Eds. Toxicology. Vol. 2. In F. A. Patty, Ed. Industrial Hygiene and Toxicology. 2nd ed. New York: John Wiley & Sons, Inc., 1963.
31a. Feron, V. J. Effects of exposure to acetaldehyde in Syrian hamsters simultaneously treated with benzo(a)pyrene or diethylnitrosamine. Prog. Exp. Tumor Res. 24:162-176, 1979.
32. Feron, V. J. Respiratory tract tumors in hamsters after intratracheal instillations of benzo[a]pyrene alone and with furfural. Cancer Res. 32:28-36, 1972.
33. Feron, V. J., and A. Kruysse. Effects of exposure to acrolein vapor in hamsters simultaneously treated with benzo[a]pyrene or diethylnitrosamine. J. Toxicol. Environ. Health 3:379-394, 1977.
34. Feron, V. J., A. Kruysse, H. P. Til, and H. R. Immel. Repeated exposure to acrolein vapour: Subacute studies in hamsters, rats and rabbits. Toxicology 9:47-57, 1978.
35. Gage, J. C. Subacute inhalation toxicity of 109 industrial chemicals. Brit. J. Ind. Med. 27:1-18, 1970.
36. Gordon, B. I., and H. I Maibach. Eccrine anhidrosis due to glutaraldehyde, formaldehyde and iontophoresis. J. Invest. Dermatol. 53:436-439, 1969.
37. Gordon, H. H. Hyperhidrosis: Treatment with glutaraldehyde. Cutis 9:375-378, 1972.

38. Guillerm, R., R. Badre, and B. Vignon. Inhibitory effects of tobacco smoke on the ciliary activity of the respiratory epithelium and nature of the responsible constituents. Bull. Acad. Nat. Med. Paris 145:416-423, 13-20 Jun 1961. (in French)
39. Gillerm, R., J. Hee, M. Bourdin, H. Burnet, and G. Siou. Contribution a la détermination de la valeur limite de concentration de l'acroléine. Cah. Notes Document. 77:527-535, 1974.
40. Gusev, M. I., A. I. Svechnikova, I. S. Dronov, M. D. Grebenskova, and A. I. Golovina. Determination of the daily average maximum permissible concentration of acrolein in the atmosphere. Gig. Sanit. 31:8-13, 1966. (in Russian)
41. Harada, M., Y. Fujii, and J. Kamiya. Pharmacological studies on Chinese cinnamon. III. Electroencephalographic studies of cinnamaldehyde in the rabbit. Chem. Pharm. Bull. (Tokyo) 24:1784-1788, 1976.
42. Harada, M., and S. Yano. Pharmacological studies on Chinese cinnamon. II. Effects of cinnamaldehyde on the cardiovascular and digestive systems. Chem. Pharm. Bull. (Tokyo) 23:941, 1975.
43. Hine, C. H., F. Meyers, F. Ivanhoe, S. Walker, and G. H. Takahashi. Simple tests of respiratory function and study of sensory response in human subjects exposed to respiratory tract irritants, pp. 28-30. In Proceedings of the Afternoon Sessions of The First Day of the Air Pollution Medical Research Conference, 4 December 1961, Los Angeles. Berkeley: California State Department of Health, 1961.
44. Iwanoff, N. Experimentelle Studien über den Einfluss technisch und hygienisch wichtiger Gase und Dämpfe auf den Organismus. Teil XVI, XVII, XVIII. Über einige praktisch wichtige Aldehyde (Formaldehyd, Acetaldehyd, Akrolein). Arch. Hyg. 73:307-340, 1911.
45. Izard, C. Mutagenic effects of acrolein and its two epoxides: glycidol and glycidal, in Saccharomyces cerevisiae. C. R. Acad. Sci., Sér. D. 276:3037-3040, 1973. (in French)
46. Jenner, P. M., E. C. Hagan, J. M Taylor, E. L. Cook, and O G. Fitzhugh. Food flavourings and compounds of related structure. I. Acute oral toxicity. Food Cosmet. Toxicol. 2:237, 1964.
47. Juhlin, L., and H. Hansson. Topical glutaraldehyde for plantar hyperhidrosis. Arch. Dermatol. 97:327-330, 1968.
48. Kane, L. E. Sensory Irritation from Photochemical Oxidant Mixtures and Their Components: Development of an Animal Model System. Dissertation. Pittsburgh: University of Pittsburgh, Graduate School of Public Health, 1977. 124 pp.
49. Kane, L. E., and Y. Alarie. Evaluation of sensory irritation from acrolein-formaldehyde mixtures. Am. Ind. Hyg. Assoc. J. 39:270-274, 1978.
50. Kane, L. E. and Y. Alarie. Sensory irritation to formaldehyde and acrolein during single and repeated exposures in mice. Am. Ind. Hyg. Assoc. J. 38:509-522, 1977.

51. Kankaanpää, J., E. Elovaara, K. Hemminki, and H. Vainio. Embryotoxicity of acrolein, acrylonitrile and acrylamide in developing chick embryos. Toxicol. Lett. 4:93-96, 1979.
52. Kligman, A. M. Report to Research Institute for Fragrance Materials, 10 October 1973.
53. Kligman, A. M. Report to Research Institute for Fragrance Materials, 22 August 1974.
54. Kruysse, A., V. J. Feron, and H. P. Til. Repeated exposure to acetaldehyde vapor. Studies in Syrian golden hamsters. Arch. Environ. Health 30:449-452, 1975.
55. Kryatov. I. A. Hygienic evaluation of sodium p-chlorobenzenesulfonate and chloral as water pollutant. Hyg. Sanit. 35:333-338, 1970.
56. Laing, D. Nasopharyngeal carcinoma in the Chinese in Hong Kong. Trans. Amer. Acad. Ophthalmol. Otolaryngol. 71:934-950, 1967.
57. Laseter, J., and B. J. Dowty. Association of biorefractories in drinking water and body burden in people, pp. 547-556. In H. F. Draybill, C. J. Daive, J. C. Harshbarger, and R. G. Tardiff. Aquatic Pollutants and Biological Effects with Emphasis on Neoplasia. Ann. N. Y. Acad. Sci. 298, 1977.
58. Lieber, C. S. Metabolism of ethanol, pp. 1-29. In C. S. Lieber, Ed. Metabolic Aspects of Alcohol. Baltimore: University Park Press, 1977.
59. Lindros, K. O. Acetaldehyde--its metabolism and role in the actions of alcohol. Res. Adv. Alcohol Drug Prob. 4:111-176, 1978.
60. Lyon, J. P., L. J. Jenkins, Jr., R. A. Jones, R. A. Coon, and J. Siegel. Repeated and continuous exposure of laboratory animals to acrolein. Toxicol. Appl. Pharmacol. 17:726-732, 1970.
61. Manufacturing Chemists' Association, Inc. Properties and Essential Information for Safe Handling and Use of Acetaldehyde. Chemical Safety Data Sheet SD-43. Washington, D.C.: Manufacturing Chemists' Association, Inc., adopted 1952.
62. Martin, H., and S. Quan. Racial incidence (Chinese) of nasopharyngeal cancer. Ann. Otol. Rhinol. Laryngol. 60:168-174, 1951.
63. Matsuzaki, S., and C. S. Lieber. Metabolism and toxicity of acetaldehyde in the liver. Saishin Igaku 31:2099-2107, 1976.
64. McLaughlin, R. S. Chemical burns of the human cornea. Am. J. Ophthalmol. 29:1355-1362, 1946.
65. Minnich, V., M. E. Smith, D. Thompson, and S. Kornfeld. Detection of mutagenic activity in human urine using mutant strains of Salmonella typhimurium. Cancer 38:1253-1258, 1976.
66. Murphy, S. D., H. V. Davis, and V. L. Zaratzian. Biochemical effects in rats from irritating air contaminants. Toxicol. Appl. Pharmacol. 6:520-528, 1964.
67. Murphy, S. D., D. A. Klingshirn, and C. E. Ulrich. Respiratory response of guinea pigs during acrolein inhalation and its modification by drugs. J. Pharmacol. Exp. Therap. 141:79-83, 1963.

68. Nakahara, W., and K. Mori. Experimental production of liver cirrhosis by furfural feeding. GANN 35:208-231, 1941.
69. Obe, G., and B. Beek. Mutagenic activity of aldehydes. Drug Alcohol Depend. 4(1-2):91-94, 1979.
70. Omel'yanets, N. I., N. V. Mironets, N. V. Martyshchenko, I. A. Gubareva, L. F. Piven, and S. N. Starchenko. Experimental substantiation of the maximum permissible concentrations of acetone and acetaldehyde in reclaimed potable water. Biol. Aviakosm. Med. 1978:67-70, 1978. (in Russian)
71. Opdyke, D. L. J. Inhibition of sensitization reactions induced by certain aldehydes. Food Cosmet. Toxicol. 14:197-198, 1976.
72. Orcutt, J. A., and J. R. Taylor. Correlation of Eye Irritation with Concentrations of Ambient Air Contaminants by Probit Analysis of the Quantal Response of Human Panels. Preliminary Results of APCD-APF Project 60. Analysis Paper No. 36. Los Angeles: County of Los Angeles Air Pollution Control District, 28 November 1960.
73. O'Shea, K. S., and M. H. Kaufman. The teratogenic effect of acetaldehyde: implications for the study of the fetal alcohol syndrome. J. Anat. 128:65-76, 1979.
74. Pace, D. M., and A. Elliott. Studies on the effects of acetaldehyde on tissue cells cultivated in vitro. Cancer Res. 20:868-875, 1966.
75. Papa, C. M., and A. M. Kligman. Mechanism of eccrine anhidrosis. I. High level blockage. J. Invest. Dermatol. 47:1-9, 1966.
76. Pavlova, L. P. Toxicological characteristics of trichloroacetaldehyde. Tr. Azerb. Nauchno-Issled. Inst. Gig. Tr. Prof. Zabol. 1975:99-105, 1975. (in Russian)
77. Plotnikova, M. M. Acrolein as an atmospheric air pollutant. Gig. Sanit. 22:10-15, 1957. (in Russian)
78. Rapoport, I. A. Mutations under the effect of unsaturated aldehydes. Dokl. Akad. Nauk S.S.S.R. 61:713-715, 1948. (in Russian)
79. Renzetti, N. A., and R. J. Bryan. Atmospheric sampling for aldehydes and eye irritation in Los Angeles smog--1960. J. Air Pollut. Control Assoc. 11:421-424, 427, 1961.
80. Riley, J. F., and A. B. Wallace. Stimulation and inhibition of connective tissue in mice following injection of benzpyrene. Br. J. Exper. Pathol. 22:24-28, 1941.
81. Ristow, H., and G. Obe. Acetaldehyde induces cross-links in DNA and causes sister-chromatid exchanges in human cells. Mutat. Res. 58:115-119, 1978.
82. Salaman, M. H., and F. J. C. Roe. Further test for tumour-initiating activity: N,N-di-(2-chloroethyl)-p-aminophenylbutyric acid (CB1348) as an initiator of skin tumour formation in the mouse. Br. J. Cancer 10:363-378, 1956.
83. Salem, H., and H. Cullumbine. Inhalation toxicities of some aldehydes. Toxicol. Appl. Pharmacol. 2:183-187, 1960.
84. Santodonato, J., J. E. Hoecker, D. Orzel, and W. Meylan. Information Profiles on Potential Occupational Hazards--Classes

of Chemicals. Syracuse, New York: Syracuse Research Corporation, Center for Chemical Hazard Assessment, for the National Institute for Occupational Safety and Health, 1978. 14 pp.

85. Sasaki, Y., and R. Endo. Mutagenicity of aldehydes in *Salmonella*. Mutat. Res. 54:251-252, 1978.

86. Sato, K., and R. L. Dobson. Mechanism of the antiperspirant effect of topical glutaraldehyde. Arch. Dermatol. 100:564-569, 1969.

87. Schauenstein, E., H. Esterbauer, and H. Zollner. Aldehydes in Biological Systems. Their Natural Occurrence and Biological Activities. London: Pion Limited, 1977. 205 pp.

88. Schoental, R. Chapter 12. Carcinogens in plants and microorganisms, pp. 626-689. In Searle, C. E., Ed. ACS Monograph No. 173: Chemical Carcinogens. Washington, D.C.: American Chemical Society, 1976.

89. Schorr, W. F. Allergic skin disease caused by cosmetics. Am. Fam. Physician 12:90-95, September 1975.

90. Schorr, W. F. Cinnamic aldehyde allergy. Contact Dermatitis 1:108, 1975.

91. Shamberger, R. J. Antioxidants and cancer. VII. Presence of malonaldehyde in beef and other meats and its epidemiological significance, pp. 36-43. In D. D. Hemphill, Ed. Trace Substances in Environmental Health. XI. Proceedings of University of Missouri's 11th Annual Conference on Trace Substances in Environmental Health, June 7-9, 1977, Columbia, Missouri. Columbia, Mo.: Curators of the University of Missouri, 1977.

92. Shamberger, R. J. Increase of peroxidation in carcinogenesis. J. Nat. Cancer Inst. 48:1491-1497, 1972.

93. Shamberger, R. J., T. L. Andreone, and C. E. Willis. Antioxidants and cancer. IV. Initiating activity of malonaldehyde as a carcinogen. J. Nat. Cancer Inst. 53:1771-1773, 1974.

94. Shanmugaratnam, K., and J. Higginson. Pp. 130-134. In C. S. Muir and K. Shanmugaratnam, Eds. Cancer of the Nasopharynx. Copenhagen: Munksgaard, 1967.

95. Shimizu, K., M. Harada, M. Miuata, S. Ishikawa, and I. Mizoguchi. Effect of photochemical smog on the human eye. J. Clin. Ophthalmol. 30:407-418, 1976.

96. Sim, V. M., and R. E. Pattle. Effect of possible smog irritants on human subjects. J. Am. Med. Assoc. 165:1908-1913, 1957.

97. Skog, E. A toxicological investigation of lower aliphatic aldehydes. I. Toxicity of formaldehyde, acetaldehyde, propionaldehyde and butyraldehyde; as well as of acrolein and crotonaldehyde. Acta Pharmacol. 6:299-318, 1950.

98. Smyth, H.F., Jr., and C. P. Carpenter. The place of the range finding test in the industrial toxicology laboratory. J. Ind. Hyg. Toxicol. 26:269-273, 1944.

99. Smyth, H. F., Jr., C. P. Carpenter, and C. S. Weil. Range-finding toxicity data, list III. J. Ind. Hyg. Toxicol. 31:60-62, 1949.
100. Smyth, H. F., Jr., C. P. Carpenter, and C. S. Weil. Range-finding toxicity data: List IV. Arch. Ind. Hyg. Occup. Med. 4:119-122, 1951.
101. Smyth, H. F., Jr., C. P. Carpenter, C. S. Weil, and U. C. Pozzani. Range-finding toxicity data. List V. Arch. Ind. Hyg. Occup. Med. 10:61-68, 1954.
102. Smyth, H. F., Jr., C. P. Carpenter, C. S. Weil, U. C. Pozzani, and J. A. Streigel. Range-finding toxicity data. List VI. Am. Ind. Hyg. Assoc. J. 23:95-108, 1962.
103. Sporn, A., I. Dinu, and V. Stanciu. The toxicity of cinnamaldehyde. Igiena 14:339-346, 1965. (in Rumanian; English abstract in Chem. Abstr. 64:10293a, 1966)
104. Stanford Research Institute. Profiles on Occupational Hazards for Criteria Document Priorities, pp. [5-15]. Arlington, Virginia: Stanford Research Institute, for U.S. Department of Health, Education, and Welfare, National Institute for Occupational Safety and Health, March 1977.
105. Steiner, P. E., R. Steele, and F. C. Koch. The possible carcinogenicity of overcooked meats, heated cholesterol, acrolein, and heated sesame oil. Cancer Res. 3:100-107, 1943.
106. Stoner, G. D., M. B. Shimkin, A. J. Kniazeff, J. H. Weisburger, E. K. Weisburger, and G. B. Gori. Test for carcinogenicity of food additives and chemotherapeutic agents by the pulmonary tumor response in strain A mice. Cancer Res. 33:3069-3085, 1973.
107. Suskind, R. R., and V. A. Majeti. Occupational and environmental allergic problems of the skin. J. Dermatol. 3:3, 1976.
108. Teschke, R., Y. Hasumura, and C. S. Lieber. Hepatic pathways of ethanol and acetaldehyde metabolism and their role in the pathogenesis of alcohol-induced liver injury. Nutr. Metab. 21 (Suppl. 1):144-147, 1977.
109. Teuchy, H., J. Quatacker, G. Wolf, and C. F. Van Sumere. Quantitative investigation of the hippuric acid formation in the rat after administration of some possible aromatic and hydroaromatic precursors. Arch. Int. Physiol. Biochim. 79:573, 1971.
110. Tittmar, H. G. Some effects of ethanol, presented during the pre-natal period, on the development of rats. Br. J. Alcohol Alcohol. 12:71-83, 1977.
111. U.S. Department of Health, Education, and Welfare, Public Health Service, Center for Disease Control, National Institute for Occupational Safety and Health. National Occupational Hazard Survey, conducted 1972-1974. Computerized data file. 1980.
112. U.S. Environmental Protection Agency, Office of Water Regulations and Standards. Ambient Water Quality Criteria for Acrolein. U.S. Environmental Protection Agency Report No. EPA 440/5-80-016. Washington, D.C.: U.S. Government Printing Office, 1980. 100 pp.

113. Van Duuren, B. L., L. Langseth, L. Orris, M. Baden, and M. Kuschner. Carcinogenicity of epoxides, lactones, and peroxy compounds. V. Subcutaneous injection in rats. J. Nat. Cancer Inst. 39:1213-1216, 1967.

114. Venkatasetty, R. Genetic variation induced by radiation and chemical agents in *Drosophila melanogaster*. Dissertation. Bowling Green, Kentucky: Bowling Green State University, 1971. 138 pp.

115. Watanabe, F., and S. Sugimoto. Study on the carcinogenicity of aldehyde. 3rd report. Four cases of sarcomas of rats appearing in the areas of repeated subcutaneous injections of acetaldehyde. GANN 47:599-601, 1956. (in Japanese)

116. Watanabe, T., and D. M. Aviado. Functional and biochemical effects on the lung following inhalation of cigarette smoke and constituents. II. Skatole, acrolein, and acetaldehyde. Toxicol. Appl. Pharmacol. 30:201-209, 1974.

117. Weber-Tschopp, A., T. Fischer, R. Gierer, and E. Grandjean. Experimentally induced irritating effects of acrolein on humans. Int. Arch. Occup. Environ. Health 40:117-130, 1977.

118. Williams. R. T. Detoxication Mechanisms. The Metabolism and Detoxication of Drugs, Toxic Substances and Other Organic Compounds. 2nd ed. New York: John Wiley & Sons, Inc., 1959. 796 pp.

CHAPTER 9

EFFECTS OF ALDEHYDES ON VEGETATION

For over 80 yr, formaldehyde was assumed to play an important role in plant metabolism as the first product of photosynthesis. According to the hypothesis of the German chemist von Baeyer, carbon dioxide absorbed from the air was dissociated by green plants into carbon monoxide that was reduced to formaldehyde, which in turn polymerized to a carbohydrate.[10] However, experimental evidence never supported this hypothesis. Researchers were unable to distill formaldehyde from huge quantities of leaves or to enhance sugar production in leaves by adding formaldehyde. As the concept of the essentiality of formaldehyde in plants faded, the phytotoxic nature of the compound began to emerge. Once Benson and Calvin had demonstrated that 3-phosphoglyceric acid was the first product of photosynthesis, further interest in formaldehyde was focused on its phytotoxicity.

The greatest incentive for the investigation of aldehydes as a class of compounds was probably the occurrence of photochemical smog in California and other highly populated areas of the United States in 1945. Although hydrocarbons and oxides of nitrogen were suspected of being the principal reactants in smog,[8] the specific pollutants responsible for plant damage had not been identified. Several groups of investigators experimentally subjected intact plants or plant parts to known doses of artificially generated aldehydes and then described symptom development or measured the impairment of some physiological process.

In addition to these investigations related to ambient air quality,[1-4,7-9,13,19] two plant studies were prompted by reports of the emission of formaldehyde vapors in confined areas under special conditions.[17,23] In effect, growth chambers or seeding magazines made of wood or particleboard were found to release formaldehyde that proved to be injurious to seeds or seedlings stored in them. These case histories--one in the United States and the other in Australia--were reported 25 yr apart.

Finally, information on the response of plants to aldehydes has been uncovered as a result of the use of aldehyde-containing compounds in specific plant practices, such as postharvest treatment of fruit and the collection of maple syrup.

It is the purpose of this chapter to assemble experimental data from these diverse sources related to the effect of aldehydes on vegetation. The material is critically reviewed with the intent of arriving at a definitive statement regarding the phytotoxicity of aldehydes.

There is wisdom in the maxim that "those who cannot remember the past are condemned to repeat it." Before presenting information on aldehydes, a rather "new" group of pollutants, we should consider two models that have been painstakingly derived from studies of more thoroughly investigated pollutants, such as ozone, sulfur dioxide, and hydrogen fluoride.

The first is a conceptual model of factors that influence the effects of air pollutants on vegetation (Figure 9-1). This model was adapted from an evaluation of the phytotoxicity of ozone and photochemical oxidants.[16] The model shows that one must understand many factors before one can predict the response of a plant species to a specific pollutant. Those factors include genetic variability, stage of plant development, climatic and edaphic factors, interactions among pollutants, interactions among pathogens and insects, and pollutant dosage. Plant responses are classified as visible or subtle effects.

In a second model, plant responses are classified according to the degree and type of effect produced at each level of biologic organization, and an attempt is made to relate effects at the cellular level with those anticipated at the level of the intact plant or plant community (Table 9-1). This model was used by the National Research Council[15] to evaluate the effects of fluoride on vegetation.

ALDEHYDE IN AMBIENT AIR AND PLANT INJURY

The only report correlating aldehyde concentrations in ambient air with plant injury was published by Brennan et al.[1] In New Jersey, foliar symptoms in Snowstorm petunias (Petunia hybrida Vilma "Snowstorm") were similar to those reported by Taylor et al. in the field in California.[21] Leaves that were rapidly expanding in size appeared watersoaked between the veins; and after several hours in sunlight, the upper leaf surfaces developed characteristic necrotic bands, and the lower leaf surfaces, a glazed appearance. The youngest leaves were marked only slightly, if at all, at the apex; and the oldest leaves escaped injury. (According to Stephens et al.,[20] similar symptoms were experimentally induced in petunias with irradiated ozone-olefin mixtures, irradiated nitrogen dioxide and hydrocarbons, irradiated aldehydes, or peroxyacetylnitrate, PAN, which was common to all irradiated nitrogen oxide mixtures.) The appearance of symptoms could be correlated with increased concentrations of aldehyde in ambient air on either of the previous two days--concentrations generally exceeding 0.20 ppm for 2 h or 0.30 ppm for 1 h by the bisulfite test. Inasmuch as the oxidant concentration in ambient air measured by a Mast sensor was lower than normal, the researchers assumed that neither peroxyacetylnitrate nor ozone was responsible for the injury to petunias. It was not established whether there was a causal relationship, rather than correlation, between aldehydes and plant damage.

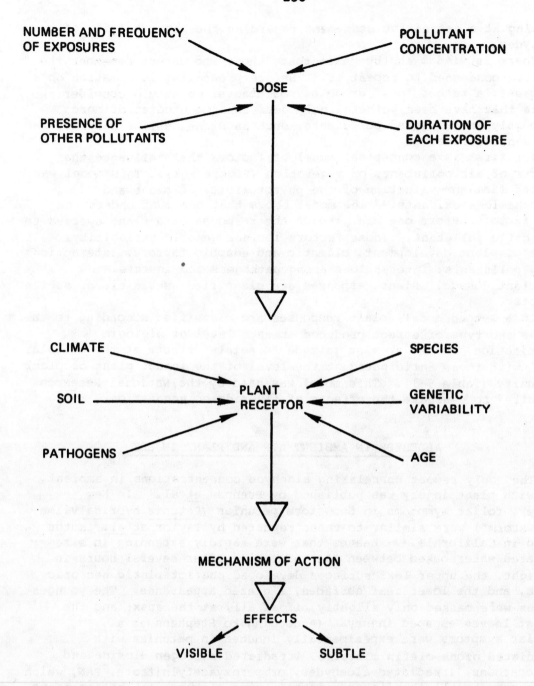

FIGURE 9-1 Conceptual model of factors involved in air-pollution effects on vegetation. Adapted from National Research Council.[16]

TABLE 9-1

Nature of Pollutant-Induced Effects in Plants at
Four Levels of Biologic Organization[a]

Cell	Tissue or Organ	Organism	Ecosystem
(Altered cell milieu?)			
Effects on enzymes and metabolites	Decreased assimilation Altered respiration		
Modification of cell organelles and metabolism	Altered growth and development	Modified growth	
Pathway disruption	Chlorotic lesions	Reduced reproduction	
Cellular modification	Necrotic lesions	Decreased fitness for environment	Changes in composition and structure of plant community
Disruption and death of cell	Death or abscission of leaf	Death of plant	Desolation

[a] Adapted from National Research Council.[15]

In a later paper, the New Jersey investigators reported that 6-14 petunia-damaging episodes related to aldehydes occurred each year in the state over a 4-yr period. The white petunias were generally sensitive, the mixed white were intermediate, and the red, pink, purple, and blue tended to be resistant.[2] In 1978, Lewis and Brennan noted that the injury syndrome observed on petunias in the field could be reproduced with a mixture of ozone and sulfur dioxide in experimental fumigations.[12]

EXPERIMENTAL FUMIGATIONS

VISIBLE PLANT INJURY CAUSED BY ALDEHYDES

In an effort to simulate the plant injury observed in California as a result of so-called smog in the mid-1940s, Haagen-Smit et al.[8] exposed five plant species that appeared to be the most sensitive in the field (spinach, endives, sugar beets, oats, and alfalfa) to a variety of organic and inorganic compounds in a fumigation chamber at concentrations generally less than 1 ppm. Several aldehydes were among the compounds tested. Formaldehyde had little effect on the test plants: exposure at 2 ppm for 2 h did not visibly affect any of the species, and exposure at 0.7 ppm for 5 h produced a symptom only in alfalfa (Table 9-2), and it was atypical. A 4-h exposure to trichloroacetaldehyde at 0.8 ppm caused smoglike symptoms on alfalfa--speckled necrosis and marginal wilting of the leaves--but did not damage the other species. Exposure to the unsaturated aldehyde, acrolein, at 0.1 ppm for 9 h also produced symptoms on alfalfa resembling natural smog damage, but there was no suggestion of damage to the other species. Higher doses of acrolein (0.6 ppm for 3 h or 1.2 ppm for 4.5 h) produced numerous sunken pits on both surfaces of spinach, endives, and beets, but the injury was unlike that observed in the field. Having failed to reproduce typical smog symptoms on four of the five sensitive plant species, the group of investigators in California concluded that aldehydes were not responsible for plant damage in the Los Angeles area.

In 1960, Darley et al.[3] had occasion to evaluate acrolein effects on pinto beans as they were testing the phytotoxicity and eye-irritation severity of varius ozone-hydrocarbon mixtures. They reported damage to bean plants from exposures to acrolein at 2.0 ppm for 70 min that, "while not severe, was definite and indistinguishable from the underside bronzing typical of oxidant damage." It should be noted that from 1940 to 1960 the term "oxidant damage" was used to describe under-surface leaf injury that was later proved to be caused by PAN.

A more recent study by N. Masaru and K. Fukaya (personal communication) indicated a greater phytotoxicity of acrolein than previously reported. Experiments in Japan revealed that bean leaves exposed to acrolein at 0.5 ppm developed brown foliar lesions after 4 h, and morning-glories developed similar symptoms after 6-7 h. Damage was more severe if the plants were fumigated in wet, rather than dry,

TABLE 9-2

Effects of Fumigation with Various Aldehyde Gases on Plants[a]

Fumigant	Duration of Exposure, h	Concentration, ppm (vol/vol)	Injury[b] Spinach	Endive	Beet	Oats	Alfalfa
Formaldehyde	2	2.0	0	0	0	0	0
	5	0.7	0	0	0	0	A
Acrolein	4.5	1.2	A	A	A	0	T
	3	0.6	A	A	A	0	T
	9	0.1	0	0	0	0	T
Trichloro-acetaldehyde	4	0.8	0	0	0	—	T

[a] Reprinted with permission from Haagen-Smit et al.[8]

[b] 0, no injury; A, injury that is atypical of smog damage in Los Angeles area; T, injury typical of smog damage.

conditions. Radish leaves did not respond until the acrolein exposure was increased to 6-7 h at 1.5 ppm, and neither geraniums nor tomato plants were affected even at the greater exposure. Thus, Masaru and Fukaya demonstrated two principles that have been apparent when air pollutants have been studied more extensively: species respond differently to a given exposure to pollutant, and environmental factors affect plant response.

VISIBLE PLANT INJURY CAUSED BY IRRADIATED ALDEHYDES

While the California group was considering an ozone-olefin reaction as the probable source of eye irritation and plant damage, Stephens et al.[19] recognized that aldehydes were products of such reactions. They irradiated selected aldehydes in static systems with 48 Blacklite fluorescent tubes that emitted radiation of wavelength less than 3000 Å. They then passed the aldehydes and their reaction products over petunias and pinto bean leaves (8 or 14 d old) for 1-1.5 h (Table 9-3). The concentrations of aldehydes were 4.5-9.0 ppm at the start of the fumigation and decreased to 1.5-4.5 ppm by the end of the fumigation. Formaldehyde and acetaldehyde and their reaction products caused little or no injury to the plants. Propionaldehyde and butyraldehyde and their reaction products produced a glazing of the lower surface of petunia leaves and of trifoliate and 8-d old primary leaves of pinto beans, but did not injure 14-d old primary leaves of the beans. Irradiated ozone-olefin mixtures or irradiated nitrogen dioxide-hydrocarbon mixtures caused a similar response.

Speculating that small concentrations of nitrogen oxides may have been present in the Stephens et al. fumigation, Hindawi and Altshuller[9] investigated the phytotoxicity of irradiated formaldehyde and propionaldehyde in the presence of low and high concentrations of NO_x (Table 9-4). The species that they tested (petunias, tobacco, and pinto beans) were not affected by a combination of formaldehyde at 6.1 ppm and NO_x at 0.9 ppm for 4 h, despite the generation of oxidant at 0.65 ppm. They speculated that the oxidant was not ozone, but a nontoxic compound. The next higher homologue, propionaldehyde, proved more toxic, causing injury to plants at 0.52 ppm in the presence of NO_x at 0.5 ppm. On the basis of symptom type and species of plant affected, the researchers identified the same five classes of injury that they had observed in the same three plant species exposed to irradiated automobile exhaust. In discussing the aldehydes, they expressed an opinion that irradiated acetaldehyde did not cause significant damage, inasmuch as Stephens et al. had observed no phytotoxicity with a mixture of irradiated cis-2-butene and ozone, despite a high yield of acetaldehyde. They also assumed that acrolein did not cause significant damage to petunias, pinto beans, and tobacco leaves, because there was no phytotoxicity in their own experiments with irradiated mixtures of 1,3-butadiene and nitrogen oxide, although acrolein at 1.0 ppm was formed as a product. It cannot be assumed,

TABLE 9-3

Plant Damage Caused by Four Irradiated Aldehydes

Concentration of Aldehydes, ppm[b]				Duration of Fumigation, h	Plant Injury		
A	B	C	D		14-d-old Pinto	8-d-old Pinto	Petunia
				Butyraldehyde			
17.7	15.8	8.5	---	1.5	0	Severe	Severe
17.9	16.0	8.3	2.8	1	0	Severe	Severe
				Propionaldehyde			
14.6	13.8	6.9	3.8	1.25	0	Severe	Severe
12.5	11.8	6.3	2.5	1	0	Severe	Severe
				Acetaldehyde			
17.5	16.5	8.2	4.5	1	0	0	Light
22.0	21.4	9.0	3.6	1	0	0	0
11.2	10.6	5.1	1.7	1.5	0	Light	Light
				Formaldehyde			
16.0	12.0	6.5	1.5	0.25	Atypical	0	0
14.4	8.9	4.5	1.6	1	Atypical	0	0

[a]Reprinted with permission from Stephens et al.[19]

[b]A, in cell before irradiation; B, in cell after 2 h of irradiation with black lights in static system; C, at beginning of plant fumigation, after circulation through plant box; D, at end of fumigation.

TABLE 9-4

Pinto Bean, Tobacco, and Petunia Damage Caused by Irradiated Mixtures of Aldehydes and Nitrogen Oxides[a]

Gas Concentration, ppm				Average Oxidants[c]	Damage[b] Average Index	Class 1	Class 2	Class 3	Class 4	Class 5
Propion-aldehyde	Form-aldehyde	NO_x	N_2 Peak							
3.4	---	0.1	0.15	0.1	2.0	xx	xxx	xx	xx	xxx
3.5	---	0.9	0.7	0.3	2.6	xx	xxx	xx	xx	xxx
0.52	---	0.5	0.5	0.05	1.5	tr	xxx	x	x	xxx
---	5.6	0.05	0.1	0.25	0.0	0	0	0	0	0
---	6.1	0.9	0.6	0.65	0.0	0	0	0	0	0

[a] Reprinted with permission from Hindawi and Altshuller.[9]

[b] tr, trace; x, little; xx, moderate; xxx, heavy. Class 1: young leaves, upper surface. Class 2: tobacco, young leaves, lower surface; pinto bean, young and trifoliate leaves, lower surface. Class 3: tobacco and pinto bean, young leaves, upper surface. Class 4: tobacco and pinto bean, old leaves, upper surface. Class 5: petunia, old leaves, upper surface.

[c] Readings from a Mast oxidant instrument were averaged over a 4-h exposure period with correction for nitrogen dioxide. Multiplication of these values by 1.5 usually resulted in concentrations in reasonable agreement with those obtained by colorimetric analyses.

however, that the phytotoxicity of an aldehyde in a complex mixture is the same as when it occurs singly.

SUBTLE PLANT INJURY CAUSED BY ALDEHYDES

In addition to the investigations involving visible effects of aldehydes on plants, there has been some experimental work on their physiologic effects. Among the processes examined have been photosynthesis, respiration, transpiration, and pollen germination.

Photosynthesis and Respiration

Researchers in Canada[4] evaluated the effect of formaldehyde on photosynthesis of an alga, Euglena gracilis. When they passed air containing formaldehyde at 0.075 ppm through a 5-ml sample of euglena in bicarbonate buffer for 1 h, the rates of photosynthesis and respiration of the cells were slightly but not statistically significantly reduced (Table 9-5). In fasted cells (those suspended in buffer for 4.5 h before aldehyde exposure) the researchers noted that formaldehyde might even have a beneficial effect. Propionaldehyde at 0.100 ppm decreased the rates of photosynthesis and respiration of euglena; again, fasting of the cells offered protection against the toxic effects (Table 9-5).

The notion that formaldehyde may be beneficial to algae is consistent with the results of studies by Doman et al.[5] and Krall and Tolbert,[11] who demonstrated that [^{14}C]formaldehyde was absorbed by leaves of kidney bean and barley plants and that in light it was fixed rapidly in products similar to those formed from carbon dioxide.

Transpiration

The effect of aldehydes on transpiration was evaluated by Fries et al.[7] of Sweden. Having observed in a prior investigation that ethereal oils in gaseous form reduced transpiration rates in leaves, they proceeded to try to determine whether specific aldehydes were responsible. They enclosed wheat seedlings in a cuvette through which air containing a specific aliphatic aldehyde was passed for 1 h at a constant flow rate, temperature, and relative humidity. All six aliphatic aldehydes tested (trans-2-hexenal, pentanal, hexanal, heptanal, octanal, and nonanal) at 1.0 µM (24 ppm) caused a decrease in transpiration rate. Because the aldehyde treatment caused a reduction in transpiration rate even greater than that caused by complete darkness, there was some question whether the change was due entirely to stomatal closure. Irrespective of the mechanism, Fries et al. concluded that volatile aldehydes may play a role in the control of transpiration of plants under field conditions. It would be

TABLE 9-5

Effect of Exposure to Formaldehyde (at 0.075 ppm for 1 h) and
Propionaldehyde (at 0.100 ppm for 1 h) on Rates of Photosynthesis
and Respiration of Euglena gracilis[a]

	Rate[b]			
	Formaldehyde		Propionaldehyde	
	Control	Exposure	Control	Exposure
Unfasted cells				
Photosynthesis	5.25	4.54	6.09	4.71
Respiration	2.26	1.83	2.18	1.85
Fasted cells				
Photosynthesis	4.22	4.61	5.25	5.45
Respiration	1.47	1.54	1.75	1.66

[a]Reprinted with permission from deKoning and Jegier.[4]

[b]For photosynthesis, micromoles of oxygen given off by 6.3×10^6 cells in 10 min; for respiration, micromoles of oxygen absorbed by 6.3×10^6 cells in 10 min.

important to know whether the effect persisted after the removal of the pollutant.

Pollen Germination

Pollen germination has proved sensitive to various air pollutants, such as ozone.[6] The implication is that the inhibition of pollen germination will be reflected as an adverse effect on reproductive capacity of a species. In 1976, Masaru et al.[13] reported on their examination of the effects of formaldehyde, acrolein, sulfur dioxide, nitrogen dioxide, and ozone on lily pollen. They sowed pollen grains on culture medium, placed the medium in a fumigation chamber with pollutants at various concentrations, and measured pollen tube length after 24 h (Table 9-6). A 5-h exposure to formaldehyde at 0.37 ppm resulted in a significant reduction in pollen-tube length, whereas a 1- or 2-h exposure was innocuous. When formaldehyde was increased to 2.4 ppm, a 1-h exposure caused a decrease in tube length. The investigators observed that, with respect to pollen, the activity of formaldehyde was comparable with that of nitrogen dioxide. Acrolein proved to be more injurious to pollen than any of the other pollutants tested. At 0.40 ppm, acrolein caused a 40% decrease in pollen-tube elongation after 2 h; at 1.70 ppm, it completely prevented extension of the pollen tube. Having previously observed that exposure to acrolein at 0.50 ppm for 6-7 h caused acute foliar injury to lily, Masaru et al.[13] concluded that lily pollen was as sensitive as foliage to aldehyde treatment. Masaru et al. also tested combinations of pollutants on lily pollen. Pollen grains exposed to sulfur dioxide at 0.69 ppm for 30 min or to nitrogen dioxide at 0.15 ppm for 30 or 60 min showed little inhibition of tube elongation; if they were then exposed to formaldehyde at 0.26 ppm, significant inhibition occurred (Table 9-7).

PLANT EXPOSURES TO ALDEHYDES UNDER SPECIAL CONDITIONS

WOODEN CONTAINERS

Two reports of aldehyde damage to seedlings arose from similar circumstances in the United States and Australia some 25 yr apart. While culturing oat seedlings in a growth chamber made of ponderosa pine and hardboard (Masonite Tempered Presdwood), Weintraub and Price[23] observed a marked retardation of seedling growth. Other species—including wheat, corn, sorghum, barley, tomato, bean, lettuce, and radish—were similarly affected. Hypothesizing that a toxic agent was liberated from the box, they confined seeds under a bell jar with small pieces of well-seasoned pine board or with hardboard and again observed the inhibitory action. Nine other species of wood were associated with the same effects. In an attempt to identify the volatile compound, they placed vials of various compounds in a desiccator containing oat seeds. The most inhibitory compounds

TABLE 9-6

Pollen-Tube Length of _Lilium longiflorum_ after Exposure of Pollen Grains to Various Pollutants[a]

Pollutant Gas	Pollutant Concentration, ppm	Pollen-Tube Length, % of control, after Exposure Lasting:		
		1 h	2 h	5 h
Sulfur dioxide	0.40	96.2	92.4	0.0
	0.71	45.0	21.2	0.0
	2.50	32.2	0.0	0.0
Nitrogen dioxide	0.57	89.2	85.7	80.0
	1.70	77.5	60.3	17.2
	2.00	31.7	0.0	0.0
Ozone	0.28	81.8	80.0	72.7
	2.09	88.2	95.5	80.0
Formaldehyde	0.37	100.0	100.0	27.7
	1.40	86.5	67.3	0.0
	2.40	62.5	41.6	0.0
Acrolein	0.40	90.0	40.0	0.0
	1.40	44.4	0.0	0.0
	1.70	0.0	0.0	0.0

[a] Reprinted with permission from Masaru et al.[13]

TABLE 9-7

Pollen-Tube Length of _Lilium longiflorum_ after Exposure to a Single Pollutant or Two Pollutants in Sequence[a]

First Exposure			Second Exposure—to Formaldehyde		Pollen-Tube Length, % of Control
Pollutant	Concentration, ppm	Duration, min	Concentration, ppm	Duration, min	
Sulfur dioxide	0.69	30	--	--	95.2
		60	--	--	58.7
		30	0.26	30	39.6
		60	0.26	60	31.7
Nitrogen dioxide	0.15	30	--	--	91.2
		60	--	--	80.0
		30	0.26	30	32.8
		60	0.26	60	37.1
Ozone	1.85	30	--	--	101.4
		60	--	--	108.8
		30	0.34	30	88.2
		60	0.34	60	80.8

[a] Reprinted with permission from Masaru et al.[13]

proved to be acrolein, crotonaldehyde, hydrogen peroxide, crotonic acid, and acrylic acid. Concentrations of the compounds were not determined.

About 25 yr later, wheat-breeders experienced a similar problem in using a magazine planting device made of bonded particleboard. If seeds were stored in such a magazine for 1-3 d, the germination percentage was reduced; if stored for 3 wk, the seeds completely fail to germinate.[17] The authors suspected that free formaldehyde released from the bonding resin of the particleboard was responsible for the problem. They conducted a series of experiments with wheat in contact with or at various distances from particleboard bonded with urea-formaldehyde. After 1 d in a seeding magazine, the emergence rate for seedlings started to decline; after 1.5 d, there was only 3% emergence. If wheat seeds were stored in a paper bag at various heights above the particleboard for 30 mo, all seeds within 7.5 cm of the board were adversely affected. If the particleboard was cured for 5 d at 40°C, the volatile substance was no longer released, and seeds were not affected. The authors recommended that bonded particleboard not be used in the construction of seeding magazines.

POSTHARVEST TREATMENTS

The literature of plant pathology contains information on two uses of aldehydes that provide additional data on plant effects. Acetaldehyde vapor at concentrations far in excess of ambient exposures has been used to prevent postharvest decay of strawberries caused by Botrytis cinerea and Rhizopus stolonifer.[18] Exposure to 1% acetaldehyde vapor prevented decay and had no adverse effect on quality (as indicated by total solids and pH) of the berries. Exposure to 4% acetaldehyde produced objectionable results, decreasing the quality and injuring the caps of the berries.

MAPLE-SYRUP COLLECTION

Paraformaldehyde pills have been used on tapholes drilled into sugar maple trees to increase or prolong the yield of sap. Apparently, paraformaldehyde temporarily inhibits the growth of microorganisms in the taphole that would normally restrict sap flow. Walters and Shigo[22] studied the long-range effect of such treatment. They treated some 200 mature sugar maple trees with a 250-mg paraformaldehyde pill for 2 mo and harvested selected trees over a 35-mo period. They found a higher incidence of discolored or decayed wood in the treated trees than in the controls. Paraformaldehyde altered the vascular and ray systems that play an important role in vessel plugging of trees and thereby facilitated invasion by wood-decaying fungi.

DISCUSSION

To what extent does the information on aldehydes satisfy the two models (Figure 9-1 and Table 9-1) and approximate the impact of this class of pollutants on vegetation? Inspection of Figure 9-1 reveals that a cluster of factors related to the plant receptor and another cluster related to pollutant dosage determine the nature and degree of plant response likely to be elicited by a given pollutant.

With regard to the factors influencing the plant receptor, investigations involving air pollutants, such as ozone and fluoride, have emphasized that genetic makeup is foremost in determining the sensitivity or tolerance of a plant. Well over 100 plant species, as well as many groups of cultivars of some 20 species have been tested for their reactions to ozone and fluoride, and lists of plants that are highly sensitive, of intermediate or slight sensitivity, or resistant to each pollutant have been compiled. In contrast, only 15 species have been tested for sensitivity to aldehydes; the only results on intraspecific variations were those related to petunias in the ambient-air study conducted in New Jersey.

In addition to the genetic component, the stage of plant development influences the response of a plant receptor. Most frequently, it is the vegetative parts, rather than the fruit or floral parts, that exhibit toxicity symptoms, although there are exceptions, such as peach fruit injury due to fluoride. The age of the tissue is critical. For example, plants at an age associated with nearly complete expansion of leaves are at their peak of ozone sensitivity, but are past their peak of PAN sensitivity. The stage of maximal sensitivity to aldehydes has not been determined, although there is a clue in the greater susceptibility of young bean leaves than of old leaves. Many species must be tested to determine the part of their life cycle when injury is most likely.

The sensitivity of a plant receptor is also influenced by many climatic and edaphic factors. Some of the climatic factors that have been important with respect to more thoroughly investigated air pollutants are temperature, relative humidity, light quality and intensity, photoperiod, and rate of air movement. None of these has been systematically evaluated in aldehyde fumigations. Masaru et al., however, did observe that wet leaves were injured more than dry leaves. Among the edaphic factors that influence the growth and development of the plant receptor--and hence the response to a pollutant--are soil moisture, aeration, and nutrients. None of these has been evaluated in aldehyde studies.

Finally, biotic factors have been found to alter a plant receptor. Research with ozone has demonstrated that the presence of a pathogen in a plant may increase or decrease ozone phytotoxicity. Information of this nature on aldehydes is lacking.

Turning from the factors that act directly on the receptor, it is necessary to consider the factors that are involved in the "dose" component. According to Figure 9-1, pollutant concentration, duration of exposure, and number of exposures are important. Obviously, a high dose of pollutant is more apt to be injurious than a low dose.

Although it is not self-evident, it may also be true for some aldehydes, as it is for ozone, that a given dose applied over a short period produces a greater plant reponse than the same dose applied over a long period. An acute exposure may, in fact, evoke a response different from that to a chronic exposure, depending on the mechanism of action of the pollutant and the mechanism of resistance of the plant. In aldehyde research, excluding the work with particleboard containers, exposures have been short (1-6 h). Concentrations of aldehydes used with higher plants generally have ranged from 0.2 to 2.0 ppm. Because the investigators used analytic techniques of varied sensitivity and precision for measuring aldehydes (Table 9-8), it is futile to attempt to compare their results. (Methods for aldehyde determination are presented in Chapter 6). Indeed, whether the concentrations used in experimental work are realistic, with respect to those occurring in ambient air, will not be known until there is a refined standard method for use in experimental and ambient atmospheres. In a sense, history would be repeating itself, in that methods for ozone (oxidant) determination have progressed through a series of "acceptable" techniques since the toxicity of ozone was first recognized. Even assuming the availability of better analytic techniques, one must recognize that aldehydes are present in complex mixtures with other pollutants that may also be phytotoxic and interact with the aldehydes. In 1966, Menser and Heggestad[14] established that administration of mixtures of sulfur dioxide (0.50 ppm) and ozone (0.03 ppm) for 2 h caused 23% foliar injury on tobacco, whereas administration of the gases separately produced no injury. This type of experimentation has not been done with aldehydes, except that of Masaru et al.,[13] who observed that exposure to sulfur dioxide or nitrogen dioxide, followed by exposure to formaldehyde, resulted in greater inhibition of pollen germination than did exposoure to either pollutant singly. Exposure to formaldehyde after ozone exposure appears to decrease pollen-tube length, but the differences were not significant.

Inspection of the second model reveals the need for assessing air-pollution effects on plants at many levels of biologic organization, including cell, tissue, organism, and ecosystem. Ideally, one would know whether any of the structural or functional alterations initiated at the cellular level are expressed at any of the higher levels. For example, are the changes in rates of photosynthesis and respiration responsible for foliar lesions in an intact leaf? Does the presence of a necrotic or chlorotic lesion have an important effect on the plant _in toto_? Is growth or yield reduced? Does injury to individual plants constitute a threat to the ecosystem? These questions cannot yet be answered with respect to the aldehydes--the available information is too sparse.

TABLE 9-8

Analytic Methods for Aldehydes Used in Plant Studies

Compound	Method	Reference
Aldehydes (including acrolein)	Gravimetric precipitation of dimedons or 2,4-dinitrophenyl hydrazones	8
Acrolein	Absorption in buffered semicarbazide hydrochloride solution and reading on spectrophotometer	3
Acrolein	Absorption in 0.1 N hydroxylamine hydrochloride solution and measurement by m-aminophenol method	13
Aldehydes	Bisulfite addition	1
Aldehydes	Long-path infrared cell	21
Formaldehyde	Chromotropic acid	13
Formaldehyde, propionaldehyde	3-Methyl-2-benzothiazolone hydrazone test	4

REFERENCES

1. Brennan, E., I. A. Leone, and R. H. Daines. Atmospheric aldehydes related to petunia leaf damage. Science 143:818-820, 1964.
2. Brennan, E., I. A. Leone, R. H. Daines, and A. Mitlehner. Polluted petunias. Florists' Rev. 139(3599):29, 75-76, 1966.
3. Darley, E. F., J. T. Middleton, and M. J. Garber. Plant damage and eye irritation from ozone-hydrocarbon reactions. Agric. Food Chem. 8:483-485, 1960.
4. deKoning, H., and Z. Jegier. Effect of aldehydes on photosynthesis and respiration of Euglena gracilis. Arch. Environ. Health 20:720-722, 1976.
5. Doman, N. G., A. K. Romanova, and Z. A. Terent'eva. Conversion of some volatile organic substances absorbed by leaves from the atmosphere. Doklady Akad. Nauk S.S.S.R. 138:702, 1961. (in Russian)
6. Feder, W. A. Reduction in tobacco pollen germination and tube elongation, induced by low levels of ozone. Science 160:1122, 1968.
7. Fries, N., K. Flodin, J. Bjurman, and J. Parsby. Influence of volatile aldehydes and terpenoids on the transpiration of wheat. Naturwissenschaften (Berlin) 61:452-453, 1974.
8. Haagen-Smit, A. J., E. F. Darley, M. Zaitlin, H. Hull, and W. M. Noble. Investigation on injury to plants from air pollution in the Los Angeles area. Plant Physiol. 27:18-34, 1952.
9. Hindawi, I. J., and A. P. Altshuller. Plant damage caused by irradiation of aldehydes. Science 146:540-542, 1964.
10. Kostychev, S. P. Chemical Plant Physiology. Philadelphia: P. Blakiston's Sons and Co., 1931. 354 pp.
11. Krall, A. R., and N. E. Tolbert. A comparison of the light dependent metabolism of carbon monoxide by barley leaves with that of formaldehyde, formate and carbon dioxide. Plant Physiol. 32:321-326, 1957.
12. Lewis, E., and E. Brennan. Ozone and sulfur dioxide mixtures cause a PAN-type injury to petunia. Phytopathology 68:1011-1014, 1978.
13. Masaru, N., F. Syozo, and K. Saburo. Effects of exposure to various injurious gases on germination of lily pollen. Environ. Pollut. 11:181-187, 1976.
14. Menser, H. A., and H. E. Heggestad. Ozone and sulfur dioxide synergism: Injury to tobacco plants. Science 153:424-425, 1966.
15. National Research Council, Committee on Biologic Effects of Atmospheric Pollutants. Effects of fluoride on vegetation, pp. 77-133. In Fluorides. Washington, D.C.: National Academy of Sciences, 1971.
16. National Research Council, Committee on Medical and Biologic Effects of Environmental Pollutants. Plants and microorganisms, pp. 437-585. In Ozone and Other Photochemical Oxidants. Washington, D.C.: National Academy of Sciences, 1977.
17. O'Brien, L., and R. A. Orth. An unexpected cause of poor germination and emergence of wheat. Crop Sci. 18:510-512, 1978.

18. Prasad, K., and G. J. Stadelbacher. Effect of acetaldehyde vapor on postharvest decay and market quality of fresh strawberries. Phytopathology 64:948-951, 1974.
19. Stephens, E. R., E. F. Darley, and O. C. Taylor. Air Pollution Research at University of California, Riverside. An Interim Report on the Cooperative Program between William E. Scott and Associates, Perkasie, Pa., and the University of California, Riverside. Prepared for meeting of Research Advisory Committee Smoke and Fumes Committee, American Petroleum Institute, held at Riverside, California, October 1-2, 1959.
20. Stephens, E. R., E.F. Darley, O. C. Taylor, and W. E. Scott. Photochemical reaction products in air pollution. Int. J. Air Water Pollut. 4:79-100, 1961.
21. Taylor, O. C., E. R. Stephens, E. F. Darley, and E. A. Cardiff. Effect of air-borne oxidants on leaves of the pinto bean and petunia. Proc. Amer. Soc. Hort. Sci. 75:434-444, 1960.
22. Walters, R. S., and A. L. Shigo. Discolouration and decay associated with paraformaldehyde-treated tapholes in sugar maple. Can. J. Forest Res. 8:54-60, 1978.
23. Weintraub, R. L., and L. Price. Inhibition of plant growth by emanations from oils, varnishes, and woods. Smithson. Misc. Collect. 107(17):1-13, 1948.

CHAPTER 10

EFFECTS OF ALDEHYDES ON AQUATIC ORGANISMS

This chapter presents an overview of current knowledge of the effects of aldehydes on aquatic organisms. It addresses 39 aldehydes, which were considered to be potentially more hazardous to aquatic life than other aldehydes by having been identified in water, having been produced or consumed in the United States in amounts of at least 1 million pounds per year, having been used as pesticides, or being currently considered by EPA as high-priority water pollutants. These aldehydes are listed in Table 10-1.

Of the 39 aldehydes, 36 have been identified in water, including industrial and sewage-treatment plant discharges, surface waters, and drinking water (see Chapter 5); the 36 comprise 10 of the 11 high-production or high-consumption aldehydes and all the pesticide aldehydes except metaldehyde. Only two aldehydes--acrolein and endrin aldehyde--are currently considered by EPA to be high-priority water pollutants. The reason for so classifying endrin aldehyde is unclear; it may be a transformation product of the pesticide endrin.

TOXICITY TO AQUATIC ORGANISMS

Very little is known about the toxicity of most of these potentially hazardous aldehydes to aquatic organisms. Eighteen have been evaluated for toxicity, but only one has been evaluated for chronic toxicity and bioconcentration potential. Median tolerance limits (e.g., LC_{50}s and EC_{50}s) have been reported for seven; the remaining 11 have been evaluated only for selective piscicidal activity.

ACROLEIN

Of all the aldehydes that have been evaluated for toxicity to aquatic organisms, acrolein is the most toxic. For fish, the LC_{50} for exposure periods of 24-144 h ranges from 0.046 to 0.24 ppm (Table 10-2). Aquatic invertebrates appear to be as sensitive as fish. Butler[7] estimated the 48-h LC_{50} for a marine shrimp (<u>Penaeus azteca</u>) to be 0.1 ppm. He also estimated that exposure at 0.055 ppm for 96 h would reduce the growth rate of oysters (<u>Crassostrea</u>

TABLE 10-1

Aldehydes Potentially Hazardous to Aquatic Organisms

Aldehyde	Identified in Water[a]	High Production or Consumption	Pesticide	EPA Priority Pollutant
Acetaldehyde[b]	X (ESD)	X	X	
Acrolein[b]	X (ESD)	X	X	X
Anisaldehyde[b]	X (ES)	X		
Benzaldehyde[b]	X (ESD)	X	X	
Butyraldehyde[b]	X (ESD)	X		
Capraldehyde	X (S)			
Caproaldehyde	X (SD)			
Caprylaldehyde	X (S)			
Chloral[b]	X (D)	X		
Cinnamaldehyde	X (SD)			
Citronellal		X		
Crotonaldehyde[b]	X (D)			
Di-*tert*-butylhydroxy-4-benzaldehyde	X (S)			
Dichlorobenzaldehyde[b]	X (S)			
Dimethylbenzaldehyde	X (SD)			
Enanthaldehyde[b]	X (SD)			
Endrin aldehyde				X
2-Ethylbutyraldehyde	X (D)			
2-Ethylcaproaldehyde	X (S)			
Formaldehyde[b]	X (E)	X		
Furaldehyde[b]	X (ED)			
Isobutyraldehyde[b]	X (D)			
Isopropionaldehyde	X (D)			
Isovaleraldehyde	X (ED)			
Mesitaldehyde[b]	X (S)			
Methacrolein[b]	X (D)			
Metaldehyde			X	
2-Methylpropionaldehyde	X (ESD)			
3-Methylvaleraldehyde	X (D)			
Nonylaldehyde	X (S)			
Paraldehyde	X (ESD)			
Propionaldehyde[b]	X (D)	X		
Salicylaldehyde[b]	X (E)	X		
Sorbaldehyde	X (E)			
Syringaldehyde	X (E)	X		
Undecylaldehyde	X (S)			
Valeraldehyde	X (D)			
Vanillin[b]	X (ES)			
Veratraldehyde[b]	X (E)			

[a] E = industrial or sewage treatment plant effluent; S = surface water; D = drinking water.

[b] Evaluated for toxicity.

TABLE 10-2

Acute Toxicity of Acrolein to Fish

Species	LC_{50}, ppm	Exposure Time, h	Reference
Oncorhynchus tschawytscha (king salmon)	0.08	24	5
Salmo gairdnerii (rainbow trout)	0.065	24	5
Salmo trutta (brown trout)	0.046	24	6
Lepomis macrochirus (bluegill sunfish)	0.10	96	19
Micropterus salmoides (largemouth bass)	0.16	96	19
Amia calva (bowfin)	0.062	24	19
Pimephales promelas (fathead minnow)	0.084	144	20
Gambusia affinis (mosquito fish)	0.061	48	19
Fundulis similis (longnose killifish)	0.24	48	7

virginica) by 50%. For the water flea (Daphnia magna), Macek and co-workers[20] reported a 48-h LC_{50} of 0.057 mg/L.

Studies by the Shell Development Company[10] have shown acrolein to be lethal to various aquatic flora--such as Hydrodictyon sp., Spirogyra sp., Potomogeton sp., Zannichellia sp., Cladophera sp., and Ceratophyllum sp.--at concentrations ranging from 1.5 to 7.5 ppm.

Macek and co-workers[20] evaluated acrolein for chronic effects in the water flea (D. magna) and the fathead minnow (Pimephales promelas) with the flow-through exposure technique, which provides for continuous replacement of the test solutions in the exposure tanks. They also based their toxicity estimates on measured acrolein concentrations. The water flea was exposed at five concentrations, from 0.0032 to 0.043 mg/L, for 64 d (three generations). Although the compound had no statistically significant effect on fecundity at the concentrations tested, it significantly reduced survival at 0.034 and 0.043 mg/L.

In the chronic test with minnows, the test concentrations ranged from 0.0046 to 0.042 mg/L, and the test was begun with 27-d-old fish. None of the tested concentrations affected the growth, survival, or reproductive capacity of these fish; however, at 0.042 mg/L, the compound significantly reduced the survival of their offspring.

The EPA has determined that acrolein has acute and chronic toxic effects on freshwater aquatic organisms at concentrations as low as 0.068 and 0.021 mg/L, respectively, and acute toxic effects on marine organisms down to 0.055 mg/L.[32] There are no data on chronic toxicity in sensitive marine organisms. Toxic effects would occur at lower concentrations among species that are more sensitive than those tested.

FORMALDEHYDE

Kitchens and co-workers[18] reviewed all the available published information on formaldehyde as an environmental pollutant. They discussed the structure and chemical and physical properties of formaldehyde, its production and uses, the sources of environmental formaldehyde, monitoring and analytic methods, and its human health and environmental effects. Much of the aquatic toxicologic information presented by Kitchens and associates was from a review by Schnick[30] of the toxicity of formalin.

Formalin has been evaluated for acute toxicity with a variety of fish, amphibians, invertebrates, and algae. Schnick's review[30] presented LC_{50}s for 20 freshwater and marine fish. A comparison of the 24-h LC_{50}s (the most common reported) showed striped bass (Morone saxitalis) to be the most sensitive of the fish tested. The 24-h LC_{50} of formalin for that species was 10-30 μl/L (3.7-11.1 mg/L as formaldehyde). Young fish were more sensitive than older fish. For the other species tested, the 24-h LC_{50}s ranged from about 50 to 120 mg/L as formaldehyde.

Formalin is probably the most widely used agent for treating fish for ectoparasitic infections and fish eggs for fungal infections.[33] Treatment is usually very short, but frequently repetitive. The recommended concentration for treating ectoparasitic infections is formaldehyde at about 160-250 mg/L applied in the water for 1 h/d for up to 3 d. With fish reared in ponds, formalin is added to the pond to achieve a formaldehyde concentration of 5-9 mg/L and permitted to dissipate naturally. To treat eggs with fungal infections, much higher concentrations are used (about 620 mg/L); however, the exposure period is reduced to 15 min/d. These treatment schedules indicate that fish and fish eggs can tolerate concentrations considerably higher than the 24-h LC_{50}s, but for only short periods.

In bullfrog tadpoles exposed to formalin at 275-325 µl/L (about 100-120 mg/L as formaldehyde) for 48 h, 20-30% mortality has been reported;[14] however, 100% mortality has been observed in bullfrog tadpoles exposed for 72 h to formalin at as low as 40 µl/L (15 mg/L as formaldehyde) and in tadpoles of the leopard frog (Rana pipiens) and toad (Bufo sp.) exposed at 30 and 50 µl/L (11 and 18.5 mg/L as formaldehyde), respectively.[4,14,17] At a concentration of 100 µl/L (37 mg/L as formaldehyde), formalin was not toxic to larvae of the salamander, Amblystoma tigrinum in 72 h.[14] In toxicity tests with the freshwater invertebrate Daphnia magna, mortality occurred at formalin concentrations as low as 13.5 µl/L (5 mg/L as formaldehyde).[26] In a review by McKee and Wolf,[23] the median threshold concentration for formaldehyde was reported to be 2 mg/L for Daphnia sp. (2-d exposure). Helms[14] reported observing no effect in crayfish (Procambarus blandingi) exposed to formalin at up to 100 µl/L for up to 72 h.

Gellman[12] estimated the toxic concentration of formaldehyde for aerobic aquatic microorganisms to be between 130 and 175 mg/L, and Hermann[15] found that 740 mg/L inhibited their oxygen utilization by 50%.

Helms[14] observed no effect in the aquatic algae Aphanothece sp., Oscillatoria sp., and Rhizoclonium sp. exposed for 7 d to formalin at up to 100 µl/L (37 mg/L as formaldehyde). However, cultures of Scenedesmus sp., Sirogonium sp., Spyrogyra sp., and Stigeoclonium sp. did not survive at concentrations of 15 µl/L (5.6 mg/L as formaldehyde) or higher. Euglena gracilis, exposed to formaldehyde at 0.075 ppm for 1 h, showed reduced photosynthesis and respiration,[16] but the reduction in photosynthesis was not statistically significant.

OTHER ALDEHYDES

Table 10-3 presents acute-toxicity estimates reported for acetaldehyde in two species of fish, a shrimp, and two species of algae. Unpublished studies performed by the Dow Chemical Company (R.J. Moolenau, personal communication) on acetaldehyde showed 70 ppm to be lethal to fathead minnows (Pimephales promelas) in 96 h; however, exposure for 96 h at 60 ppm caused no toxic effect. Acetaldehyde thus appears to be acutely lethal over a very narrow concentration range.

For furaldehyde, Middlebrooks and co-workers,[24] using the harlequin fish (Rasbora heteromorpha), determined the 24- and 48-h

TABLE 10-3

Acute Toxicity of Acetaldehyde to Acquatic Organism

Species	Statistic	Concentration, mg/L	Reference
Lagodon rhomboides (pinfish)	24-h LC_{50}	70	11
Lepomis macrochirus (bluegill sunfish)	96-h LC_{50}	53	8
Crangon crangon (shrimp)	24-h LC_{50}	>100	29
Nitzchia linearis (alga)	5-d EC_{50} (growth)	237	1
Navicula seminulum (alga)	EC_{50} (growth)	239	28

LC_{50}s to be 31 and 23 ppm, respectively. Mattson and co-workers[22] reported a 96-h LC_{50} of 32 ppm for the fathead minnow (P. promelas). With the bluegill sunfish (L. macrochirus), Turnbull and co-workers[31] determined the 24- and 48-h LC_{50}s of furaldehyde to be 32 and 24 mg/L, respectively. In very turbid water, the 24-h LC_{50} for bluegills has been reported as 44 mg/L, and the 48- and 96-h LC_{50}s both have been determined to be 24 mg/L.[34] The lowest reported 96-h LC_{50} of furaldehyde is 1.2 ppm,[35] for bluegills.

Dawson and co-workers[9] tested crotonaldehyde and propionaldehyde with bluegill sunfish (L. macrochirus) and tidewater silversides (Menidia beryllina). The 96-h LC_{50}s of crotonaldehyde for the bluegill and silversides were 3.5 and 1.3 mg/L, respectively. The 96-h LC_{50}s of propionaldehyde were 130 mg/L for the bluegill and 100 mg/L for the silversides.

According to Mattson and co-workers,[22] the 96-h LC_{50} of vanillin for fathead minnows is 112-121 mg/L, on the basis of two tests. Palmer and Maloney[27] determined the toxicity of vanillin to six species of algae with concentrations up to 2 mg/L. This concentration slightly inhibited the growth of Gomphonema sp., but had no effect on the other species. In a search for a chemical agent that would selectively kill the Oregon squawfish (Ptychocheilus oregonensis), MacPhee and Ruelle[21] screened nearly 1,900 compounds, including five aldehydes, for toxicity to the squawfish, steelhead trout (Salmo gairdnerii), chinook salmon (Oncorhynchus tschawytscha), and coho salmon (Oncorhynchus kisutch). Each compound was evaluated at only one concentration. Some of the compounds were tested with only some of the species. Compounds that caused death, loss of equilibrium, or other signs of distress were considered toxic. At 10 ppm, anisaldehyde was toxic to all the species. At 1 ppm, polymeric butyraldehyde was toxic to chinook salmon, but not to the other species. At 10 ppm, chloral (as chloral hydrate) had no effect on any of the species; however, at the same concentration, mesitaldehyde was toxic to the squawfish, steelhead trout, and coho salmon (the chinook salmon was not used as a test species). At 2.5 ppm, none of the species was affected by salicylaldehyde.

Applegate and co-workers[2] performed a similar study on about 4,400 chemical compounds (including 13 aldehydes) to find one that would selectively affect the marine lamprey (Petromyzon marinus). Other fish included in the study were the rainbow trout (Salmo gairdnerii) and the bluegill sunfish (Lepomis macrochirus). With some of the compounds, tests with the trout and bluegill were deleted. The test concentrations were 0.1, 1.0, and 5.0 ppm, and the maximal exposure time was 24 h. At 5 ppm, acrolein had no effect on any of the species. This result is not in agreement with those presented earlier. Other aldehydes that had no effect on any of the species at 5 ppm were anisaldehyde, benzaldehyde, butyraldehyde (polymer), chloral, dichlorobenzaldehyde, enanthaldehyde, isobutyraldehyde, metacrolein, propionaldehyde, salicylaldehyde, and veratraldehyde. Of the aldehydes tested, only mesitaldehyde was toxic to the lamprey at 5 ppm; it was not tested with the other fish species.

Neither of these studies showed chloral to have any effect at up to 10 ppm (10 mg/L). That is surprising, because the recommended concentration for producing anesthesia in fish is about 2-3 mg/L,[3] at which concentration narcosis usually occurs in less than 5 min.

BIOCONCENTRATION

Bioconcentration is the process by which a chemical becomes more concentrated in an organism than it is in the environment of the organism.[25] Chemicals that bioconcentrate are generally considered to be more hazardous than those which do not, because, at sublethal concentrations, they may eventually produce toxic effects as the body burden increases or may cause a progressive increase in the body burden of organisms at higher trophic levels as the compound is transferred through food webs or chains.

The propensity of a chemical to bioaccumulate can be determined experimentally by exposing an organism to it and determining the concentrations of the compound in the tissues and in the exposure medium. The ratio of these two concentrations is called the bioconcentration factor (BCF). Of the 39 aldehydes addressed in this chapter, acrolein is the only one for which a BCF has been experimentally derived. The value was 344, and it was determined with the bluegill sunfish (S. Petrocelli, personal communication).

An indirect method of determining the propensity of a chemical to accumulate in tissues is to determine its octanol-water partition coefficient, which is an experimentally derived ratio of the concentrations of a compound in N-octanol and in water after N-octanol is mixed with water that contains the compound. The logarithms of the octanol-water partition coefficients for the 36 aldehydes are shown in Table 10-4. The log P values were calculated by the method of Hansch and Leo,[13] except that for acrolein, which was determined experimentally (Petrocelli, personal communication). Hydration will reduce the calculated log P values of the aliphatic aldehydes by 0.38 and will increase the values of the aromatic aldehydes by 0.39. Of the log P values shown, capraldehyde, caprylaldehyde, 3,5-di-tert-butyl- 4-hydroxybenzaldehyde, mesitaldehyde, nonylaldehyde, and undecylaldehyde have values over 3.0. These may bioconcentrate appreciably in aquatic organisms.

TABLE 10-4

Logarithms of Octanol-Water Partition Coefficients for 36 Aldehydes[a]

Aldehyde	Log P
Acetaldehyde	-0.21
Acrolein	0.90
Anisaldehyde	1.54
Benzaldehyde	1.48
Butyraldehyde	0.87
Capraldehyde	4.11
Caproaldehyde	1.95
Caprylaldehyde	3.03
Chloral	0.51
Cinnamaldehyde	1.92
Crotonaldehyde	0.55
3,5-Di-tert-butyl-4-hydroxybenzaldehyde	4.75
Dichlorobenzaldehyde	2.00
Dimethylbenzaldehyde	2.82
Enanthaldehyde	2.49
2-Ethylbutyraldehyde	1.73
2-Ethylcaproaldehyde	2.81
Formaldehyde	-0.87
Furaldehyde	0.88
Isobutyraldehyde	0.65
Isopropionaldehyde	1.82
Isovaleraldehyde	1.28
Mesitaldehyde	3.48
Methacrolein	0.33
2-Methylpropionaldehyde	0.65
3-Methylvaleraldehyde	1.82
Nonylaldehyde	3.57
Paraldehyde	1.15
Propionaldehyde	0.33
Salicylaldehyde	1.89
Sorbaldehyde	1.08
Syringaldehyde	2.15
Undecylaldehyde	4.65
Valeraldehyde	1.41
Vanillin	0.89
Veratraldehyde	1.61

[a]Calculated by the method of Hansch and Leo,[13] except the value for acrolein, which was experimentally derived (S. Petrocelli, personal communication).

REFERENCES

1. Academy of Natural Sciences. The Sensitivity of Aquatic Life to Certain Chemicals Commonly Found in Industrial Wastes. Final Report. Washington, D.C.: U.S. Public Health Service, 1960. 89 pp.
2. Applegate, V. C., J. H. Howell, A. E. Hall, Jr., and M. A. Smith. Toxicity of 4,346 Chemicals to Larval Lampreys and Fishes. U.S. Department of the Interior Fish and Wildlife Service Special Scientific Report--Fisheries No. 207. Washington, D.C.: U.S. Department of the Interior, 1957. 157 pp.
3. Bell, G. R. A Guide to the Properties, Characteristics, and Uses of Some General Anaesthetics for Fish. Fisheries Research Board of Canada Bulletin No. 148. Ottawa: Roger Duhamel, F.R.S.C., Queen's Printer and Controller of Stationery, 1964. 4 pp.
4. Bennett, G. W. Management of Lakes and Ponds. 2nd ed. New York: Van Nostrand Reinhold Company, 1971. 375 pp.
5. Bond, C. E., R. H. Lewis, and J. L. Fryer. Toxicity of various herbicidal materials to fishes, pp. 96-101. In C. M. Tarzwell, Compiler. Biological Problems in Water Pollution. Transactions of the 1959 Seminar. Robert A. Taft Sanitary Engineering Center Technical Report W60-3. Cincinnati, Ohio: U.S. Department of Health, Education, and Welfare, Public Health Service, Robert A. Taft Sanitary Engineering Center, 1960.
6. Burdick, G. E., H. J. Dean, and E. J. Harris. Toxicity of aqualin to fingerling brown trout and bluegills. N.Y. Fish Game J. 11:106-114, 1964.
7. Butler, P. A. Effects of herbicides on estuarine fauna. Proc. South. Weed Conf. 18:576-580, 1965.
8. Cairns, J., Jr., and A. Scheier. A comparison of the toxicity of some common industrial waste components tested individually and combined. Prog. Fish Cult. 30:3-8, 1968.
9. Dawson, G. W., A. L. Jennings, D. Drozdowski, and E. Rider. The acute toxicity of 47 industrial chemicals to fresh and salt water fishes. J. Hazardous Mater. 1:303-318, 1977.
10. Ferguson, F. F., C. S Richards, and J. R. Palmer. Control of Australorbis glabratus by acrolein in Puerto Rico. Public Health Rep. 76:461-468, 1961.
11. Garrett, J. T. Toxicity investigations on aquatic and marine life. Public Works J. 88:95-96, 1957.
12. Gellman, I. Studies on the biochemical oxidation of sewage, industrial wastes, and organic compounds. Thesis. Water Pollut. Abstr. 27:1859, pp. 307-313, 1954.
13. Hansch, C., and A. Leo. Chapter 4. The fragment method of calculating partition coefficients. In Substituent Constants for Correlation Analysis in Chemistry and Biology. New York: Wiley-Interscience, 1979.
14. Helms, D. R. Use of formalin for selective control of tadpoles in the presence of fishes. Prog. Fish Cult. 29:43-47, 1967.
15. Hermann, E. R. Toxicity index for industrial wastes. Ind. Eng. Chem. 51:84A-87A, 1959.

16. Hill, R. D. The use of acrolein, acrylaldehyde:2-propenal in the treatment of submerged weeds in farm ponds. Ohio Agric. Exp. Stn., 1960. 3 pp.
17. Kemp, H. T., J. P. Abrams, and R. C. Overbeck. Water Quality Criteria Data Book. Vol. 3. Effects of Chemicals on Aquatic Life. Selected Data from the Literature through 1968. U.S. Environmental Protection Agency Project No. 18050 GWV05/71. Washington, D.C.: U.S. Government Printing Office, 1971. 528 pp.
18. Kitchens, J. E., R. E. Casner, G. S. Edwards, W. E. Harward III, and B. J. Macri. Investigation of Selected Potential Environmental Contaminants: Formaldehyde. U.S. Environmental Protection Agency Report No. EPA-560/2-76-009. Washington, D.C.: U.S. Environmental Protection Agency, Office of Toxic Substances, 1976. 217 pp.
19. Louder, D. E., and F. G. McCoy. Preliminary investigations of the use of aqualin for collecting fishes, pp. 240-242. In Proceedings of the 16th Annual Conference, Southeastern Association of Game and Fish Commissioners. New Orleans: Southeastern Association of Game and Fish Commissioners, 1962.
20. Macek, K. J., M. A. Lindberg, S. Sauter, K. S. Buxton, and P. A. Costa. Toxicity of Four Pesticides to Water Fleas and Fathead Minnows. Acute and Chronic Toxicity of Acrolein, Heptachlor, Endosulfan, and Trifluralin to the Water Flea (_Daphnia magna_) and the Fathead Minnow (_Pimephales promelas_). U.S. Environmental Protection Agency Report No. EPA-600/3-76-099, 1976. Washington, D.C.: U.S. Government Printing Office, 1976. 68 pp.
21. MacPhee, C., and R. Ruelle. Lethal effects of 1888 chemicals upon four species of fish from western North America. Univer. Idaho For. Wildl. Range, Exp. Stn., Bull. No. 3, 1969. 112 pp.
22. Mattson, V. R., J. W. Arthur, and C. T. Walbridge. Acute Toxicity of Selected Organic Compounds to Fathead Minnows. U.S. Environmental Protection Agency Report No. EPA-600/e-76-097. Washington, D.C.: U.S. Government Printing Office, 1976. 12 pp.
23. McKee, J. E., and H. W. Wolf, Eds. Water Quality Criteria. 2nd ed. The Resources Agency of California, State Water Quality Control Board Publication No. 3-A. Sacramento: State Water Quality Control Board, 1963. 548 pp.
24. Middlebrooks, E. J., M. J. Gaspar, R. D. Gaspar, J. H. Reynolds, and D. B. Porcella. Effects of Temperature on the Toxicity to the Aquatic Biota of Waste Discharges. A Compilation of the Literature. PRWG Report 105-1. Washington, D.C.: U.S. Department of the Interior, 1973. 170 pp.
25. National Research Council, Environmental Studies Board. Principles for Evaluating Chemicals in the Environment. Washington, D.C.: National Academy of Sciences, 1975. 454 pp.
26. Nazareuko, J. V. Effect of formaldehyde on aquatic organisms. Tr. Vses. Gidrobiol. Ova. 10:170, 1960. (in Russian)
27. Palmer, C. M., and T. E. Maloney. Preliminary screening for potential algicides. Ohio J. Sci. 55:1-8, 1955.
28. Patrick, R., J. Cairns, Jr., and A. Scheier. The relative sensitivity of diatoms, snails, and fish to twenty common

constituents of industrial wastes. Prog. Fish Cult. 30:137-140, 1968.

29. Portmann, J. E., and K. W. Wilson. Toxicity of 140 Substances to the Brown Shrimp and Other Marine Animals. Shellfish Information Leaflet No. 22. Burnham-on-Couch, Essex, England: Ministry of Agriculture, Fish, and Food; Fisheries Laboratory, 1971. 12 pp.

30. Schnick, R. A. Formalin as a therapeutant in fish culture. U.S. Fish and Wildlife Service Report FWS/LR-74/09. Springfield, Virginia: National Technical Information Service, Publication No. PB-237 198, 1973. 175 pp.

31. Turnbull, H., J. G. DeMann, and R. F. Weston. Toxicity of various refinery materials to fresh water fish. Ind. Eng. Chem. 46:324-333, 1954.

32. U.S. Environmental Protection Agency, Office of Water Regulations and Standards. Ambient Water Quality Criteria for Acrolein. U.S. Environmental Protection Agency Report No. EPA 440/5-80-016. Washington, D.C.: U.S. Government Printing Office, 1980. 100 pp.

33. U.S. Fish and Wildlife Service, Bureau of Sport Fisheries and Wildlife. Fish Disease Manual. Region 3. 1971. 183 pp. Available from National Fisheries Center, Kearneysville, W. Va.

34. Wallen, I. E., W. C. Greer, and R. Lasater. Toxicity to *Gambusia affinis* of certain pure chemicals in turbid waters. Sewage Ind. Wastes 29:695-711, 1957.

35. Wilber, C. G. The Biological Aspects of Water Pollution. Springfield, Illinois: Charles C Thomas, 1969. 296 pp.

APPENDIX

Properties, Uses, and Synonyms of Selected Aldehydes

KEY TO TABLES A-1 AND A-2

Aldehyde	Entry in Tables A-1 and A-2
p-Acetaldehyde	ACETALDEHYDE, Paraldehyde
Acetaldol	BUTANAL, 3-Hydroxy-
Acetic aldehyde	ACETALDEHYDE
Acetylformaldehyde	PROPANAL, 2-Oxo-
Acetylformyl	PROPANAL, 2-Oxo-
Acrolein	2-PROPENAL
Acrylaldehyde	2-PROPENAL
Acrylic aldehyde	2-PROPENAL
AgriStrep	STREPTOMYCIN sulfate
Aldehyde B	PROPANAL, α-Methyl-4-(1-methylethyl)benzene-
Aldehyde C-7	n-HEPTANAL
Aldehyde C-8	OCTANAL
Aldehyde C-10	DECANAL
Aldehyde C-12	DECANAL, Do-
Aldehyde M.N.A.	UNDECANAL, 2-Methyl-
Aldesan	1,5-PENTANEDIAL
Aldol	BUTANAL, 3-Hydroxy-
Allyl aldehyde	2-PROPENAL
Amylcinnamaldehyde	HEPTANAL, 2-(Phenylmethylene)-
α-Amylcinnamaldehyde	HEPTANAL, 2-(Phenylmethylene)-
α-Amyl-β-phenylacrolein	HEPTANAL, 2-(Phenylmethylene)-
Anhydrous chloral	ACETALDEHYDE, Trichloro-
m-Anisaldehyde	BENZALDEHYDE, 3-Methoxy-
o-Anisaldehyde	BENZALDEHYDE, 2-Methoxy-
p-Anisaldehyde	BENZALDEHYDE, 4-Methoxy-
2-Anisaldehyde	BENZALDEHYDE, 2-Methoxy-
4-Anisaldehyde	BENZALDEHYDE, 4-Methoxy-
o-Anisic aldehyde	BENZALDEHYDE, 2-Methoxy-
p-Anisic aldehyde	BENZALDEHYDE, 4-Methoxy-
Antifoam-LF	OCTANAL
Aqualin	2-PROPENAL
Artificial almond oil	BENZALDEHYDE
Aubepine	BENZALDEHYDE, 4-Methoxy-
Benzaldehyde-2,4-disulfonic acid	BENZALDEHYDE, 4-Formyl-1,3-benzenedisulfonic acid
Benzaldehyde FFC	BENZALDEHYDE
Benzaldehyde-o-sulfonic acid sodium salt	BENZALDEHYDE, 2-Formylbenzenesulfonic acid sodium salt
Benzeneacetaldehyde	ACETALDEHYDE, Benzene
Benzenecarbonal	BENZALDEHYDE
Benzenecarboxaldehyde	BENZALDEHYDE
p-Benzenedicarboxaldehyde	1,4-BENZENEDICARBOXALDEHYDE
Benzoic aldehyde	BENZALDEHYDE

Benzylacetaldehyde	PROPANAL, Benzene-
Benzylideneacetaldehyde	2-PROPENAL, 3-Phenyl-
Biformal	ETHANEDIAL
Biformyl	ETHANEDIAL
Bourbonal	BENZALDEHYDE, 3-Ethoxy-4-hydroxy-
Butal	BUTANAL
Butaldehyde	BUTANAL
n-Butanal	BUTANAL
Butanaldehyde	BUTANAL
trans-2-Butenal	2-BUTENAL
Butyl aldehyde	BUTANAL
n-Butyl aldehyde	BUTANAL
p-tert-Butylbenzaldehyde	BENZALDEHYDE, 4-(1,1-Dimethylethyl)-
4-tert-Butylbenzaldehyde	BENZALDEHYDE, 4-(1,1-Dimethylethyl)-
t-Butylcarboxaldehyde	PROPANAL, 2,2-Dimethyl-
4-tert-Butylcyclohexanecarboxaldehyde	CYCLOHEXANECARBOXALDEHYDE, 4-(1,1-Dimethylethyl)-
t-Butylformaldehyde	PROPANAL, 2,2-Dimethyl-
4-tert-Butylhexahydrobenzaldehyde	CYCLOHEXANECARBOXALDEHYDE, 4-(1,1-Dimethylethyl)-
p-tert-Butyl-α-methylhydrocinnamaldehyde	PROPANAL, 4-(1,1-Dimethylethyl)-α-methylbenzene-
Butyral	BUTANAL
Butyraldehyde	BUTANAL
n-Butyraldehyde	BUTANAL
Butyric aldehyde	BUTANAL
Butyrylaldehyde	BUTANAL
BVF	FORMALDEHYDE
Capraldehyde	DECANAL
Capric aldehyde	DECANAL
Caprinaldehyde	DECANAL
Caprinic aldehyde	DECANAL
Caproaldehyde	HEXANAL
n-Caproaldehyde	HEXANAL
Caproic aldehyde	HEXANAL
Capronaldehyde	HEXANAL
Caprylaldehyde	OCTANAL
n-Caprylaldehyde	OCTANAL
Caprylic aldehyde	OCTANAL
Carbomethene	ETHENONE
Cassia aldehyde	2-PROPENAL, 3-Phenyl-
Chloral	ACETALDEHYDE, Trichloro-
o-Chlorobenzaldehyde	BENZALDEHYDE, 2-Chloro-
p-Chlorobenzaldehyde	BENZALDEHYDE, 4-Chloro-

4-((2-Chloroethyl)ethylamino)-o-tolualdehyde	BENZALDEHYDE, 4-((2-Chloroethyl)ethylamino)-2-methyl-
p-((2-Chloroethyl)methylamino)-benzaldehyde	BENZALDEHYDE, 4((2-Chloroethyl)methylamino)-
Cinnamal	2-PROPENAL, 3-Phenyl-
Cinnamaldehyde	2-PROPENAL, 3-Phenyl-
Cinnamic aldehyde	2-PROPENAL, 3-Phenyl-
Cinnamyl aldehyde	2-PROPENAL, 3-Phenyl-
Citral	2,6-OCTADIENAL, 3,7-Dimethyl-
Citronellal hydrate	OCTANAL, 7-Hydroxy-3,7-dimethyl-
Citronelloxyacetaldehyde	ACETALDEHYDE, 3,7-Dimethyl-6-octenyloxy-
Coniferaldehyde	2-PROPENAL, 3-(4-Hydroxy-3-methoxyphenyl)-
p-Coniferaldehyde	2-PROPENAL, 3-(4-Hydroxy-3-methoxyphenyl)-
Coniferyl aldehyde	2-PROPENAL, 3-(4-Hydroxy-3-methoxyphenyl)-
Crategine	BENZALDEHYDE, 4-Methoxy-
Crotonal	2-BUTENAL
Crotonaldehyde	2-BUTENAL
Crotonic aldehyde	2-BUTENAL
Crotylaldehyde	2-BUTENAL
Cumaldehyde	BENZALDEHYDE, 4-(1-Methylethyl)-
Cumene aldehyde	ACETALDEHYDE, α-Methylbenzene-
Cumic aldehyde	BENZALDEHYDE, 4-(1-Methylethyl)-
Cuminal	BENZALDEHYDE, 4-(1-Methylethyl)-
p-Cuminaldehyde	BENZALDEHYDE, 4-(1-Methylethyl)-
Cuminic aldehyde	BENZALDEHYDE, 4-(1-Methylethyl)-
Cuminyl aldehyde	BENZALDEHYDE, 4-(1-Methylethyl)-
Cyclalia	OCTANAL, 7-Hydroxy-3,7-dimethyl-
Cyclamal	PROPANAL, α-Methyl-4-(1-methylethyl)benzene-
Cyclamen aldehyde	PROPANAL, α-Methyl-4-(1-methylethyl)benzene-
3-Cyclohexen-1-aldehyde	3-CYCLOHEXENE-1-CARBOXALDEHYDE
Cyclohexene-4-carboxaldehyde	3-CYCLOHEXENE-1-CARBOXALDEHYDE
Cyclosia	OCTANAL, 7-Hydroxy-3,7-dimethyl-
p-Cymene-7-carboxaldehyde	ACETALDEHYDE, 4-(1-Methylethyl)benzene-
Decaldehyde	DECANAL
n-Decaldehyde	DECANAL
n-Decanal	DECANAL
Decanaldehyde	DECANAL
Decyl aldehyde	DECANAL
n-Decyl aldehyde	DECANAL
Decylic aldehyde	DECANAL
o-2-Deoxy-2-(methylamino)-α-1-glucopyranosyl-(1-4)-N,N'-bis(aminoiminomethyl)D-streptamine	STREPTOMYCIN
Diethylacetaldehyde	BUTANAL, 2-Ethyl-

p-(Diethylamino)benzaldehyde	BENZALDEHYDE, 4-(Diethylamino)-
Diformyl	ETHANEDIAL
1,4-Diformylbenzene	1,4-BENZENEDICARBOXALDEHYDE
Dihydrocinnamaldehyde	PROPANAL, Benzene-
(1,3-Dihydro-1,3,3-trimethyl-2H-indol-2-ylidene)-acetaldehyde	ACETALDEHYDE, 1,3,3-Trimethyl-Δ-(2,α)-indoline-
3,4-Dihydroxybenzaldehyde methylene ketal	1,3-BENZODIOXOLE-5-CARBOXALDEHYDE
3,4-Dimethoxybenzenecarbonal	BENZALDEHYDE, 3,4-Dimethoxy-
3,5-Dimethoxy-4-hydroxybenzaldehyde	BENZALDEHYDE, 4-Hydroxy-3,5-dimethoxy-
3,5-Dimethoxy-4-hydroxybenzene carbonal	BENZALDEHYDE, 4-Hydroxy-3,5-dimethoxy-
Dimethyl acetaldehyde	PROPANAL, 2-Methyl-
α,4-Dimethylbenzeneacetaldehyde	ACETALDEHYDE, α,4-Dimethylbenzene-
3,4-Dimethylenedioxybenzaldehyde	1,3-BENZODIOXOLE-5-CARBOXALDEHYDE
2,6-Dimethyl-5-hepten-1-al	HEPTENAL, 2,6-Dimethyl-5-
3,7-Dimethyl-7-hydroxyoctanal	OCTANAL, 7-Hydroxy-3,7-dimethyl-
3,7-Dimethyl-6-octenal	CITRONELLAL (d isomer)
3,7-Dimethyl-6-octenyl-oxy-acetaldehyde	ACETALDEHYDE, 3,7-Dimethyl-6-octenyl-oxy-
6,10-Dimethyl-3-oxa-9-undecanal	ACETALDEHYDE, 3,7-Dimethyl-6-octenyl-oxy-
α,α-Dimethylpropanal	PROPANAL, 2,2-Dimethyl-
α,α-Dimethylpropionaldehyde	PROPANAL, 2,2-Dimethyl-
2,2-Dimethylpropionaldehyde	PROPANAL, 2,2-Dimethyl-
2,4-Disulfobenzaldehyde	BENZALDEHYDE, 4-Formyl-1,3-benzenedisulfonic acid
n-Dodecanal	DECANAL, Do-
1-Dodecanal	DECANAL, Do-
Dodecanaldehyde	DECANAL, Do-
Dodecyl aldehyde	DECANAL, Do-
n-Dodecyl aldehyde	DECANAL, Do-
Dodecylic aldehyde	DECANAL, Do-
Enanthal	n-HEPTANAL
Enanthaldehyde	n-HEPTANAL
Enanthic aldehyde	n-HEPTANAL
Enanthole	n-HEPTANAL
Epihydrinaldehyde	OXIRANECARBOXALDEHYDE
2,3-Epoxypropanal	OXIRANECARBOXALDEHYDE
2,3-Epoxypropionaldehyde	OXIRANECARBOXALDEHYDE
Ethanal	ACETALDEHYDE
Ethanedione	ETHANEDIAL
1,2-Ethanedione	ETHANEDIAL

Ethavan	BENZALDEHYDE, 3-Ethoxy-4-hydroxy-
Ethovan	BENZALDEHYDE, 3-Ethoxy-4-hydroxy-
4-Ethoxy-m-anisaldehyde	BENZALDEHYDE, 4-Ethoxy-3-methoxy-
Ethoxybenzaldehyde	BENZALDEHYDE, 4-Ethoxy-
p-Ethoxybenzaldehyde	BENZALDEHYDE, 4-Ethoxy-
β-Ethoxy-α-ketobutyraldehyde	BUTANAL, 3-Ethoxy-2-oxo-
Ethyl aldehyde	ACETALDEHYDE
α-Ethylbutyraldehyde	BUTANAL, 2-Ethyl-
2-Ethylbutyraldehyde	BUTANAL, 2-Ethyl-
2-Ethylbutyric aldehyde	BUTANAL, 2-Ethyl
Ethylprotal	BENZALDEHYDE, 3-Ethoxy-4-hydroxy-
Ethylvanillin	BENZALDEHYDE, 3-Ethoxy-4-hydroxy-
Fannoform	FORMALDEHYDE
Ferulaldehyde	2-PROPENAL, 3-(4-Hydroxy-3-methoxyphenyl)-
Fisher's aldehyde	ACETALDEHYDE, 1,3,3-Trimethyl-Δ-(2,α)-indoline-
Fixol	OCTANAL, 7-Hydroxy-3,7-dimethyl-
Flomine	HEPTANAL, 2-(Phenylmethylene)-
Flo-Mor	FORMALDEHYDE, Para-
Formaldehyde solution	FORMALDEHYDE
Formaldehyde trimer	FORMALDEHYDE, 1,3,5-Trioxane
Formalin	FORMALDEHYDE
Formalith	FORMALDEHYDE
Formic aldehyde	FORMALDEHYDE
Formol	FORMALDEHYDE
p-Formylanisole	BENZALDEHYDE, 4-Methoxy-
p-Formylbenzaldehyde	BENZENEDICARBOXALDEHYDE, 1,4-
4-Formylbenzaldehyde	1,4-BENZENEDICARBOXALDEHYDE
α-Formylbenzene acetic acid	ACETALDEHYDE, α-Formylbenzene acetic acid
4-Formyl-m-benzenedisulfonic acid	BENZALDEHYDE, 4-Formyl-1,3-benzenedisulfonic acid
o-Formylbenzenesulfonic acid	BENZALDEHYDE, 2-Formylbenzenesulfonic acid
o-Formylbenzenesulfonic acid sodium salt	BENZALDEHYDE, 2-Formylbenzenesulfonic acid sodium salt
5-Formyl-1,3-benzodioxole	1,3-BENZODIOXOLE-5-CARBOXALDEHYDE
2-Formylbutane	BUTANAL (dl), 2-Methyl-
1-Formyl-3-cyclohexene	3-CYCLOHEXENE-1-CARBOXALDEHYDE
4-Formylcyclohexene	3-CYCLOHEXENE-1-CARBOXALDEHYDE
p-Formyl-N,N-diethylaniline	BENZALDEHYDE, 4-(Diethylamino)-

2-Formylfuran	2-FURANCARBOXCALDEHYDE
5-Formylguaiacol	BENZALDEHYDE, 3-Hydroxy-4-methoxy-
6-Formylguaiacol	BENZALDEHYDE, 2-Hydroxy-3-methoxy-
α-Formylisobutylbenzene	BUTANAL, 3-Methyl-2-phenyl-
4-Formyl-2-methoxyphenol	BENZALDEHYDE, 4-Hydroxy-3-methoxy-
6-Formyl-2-methoxyphenol	BENZALDEHYDE, 2-Hydroxy-3-methoxy-
2-(Formylmethylene)-1,3,3-trimethylindoline	ACETALDEHYDE, 1,3,3-Trimethyl-Δ-(2,α)indoline-
3-Formyl-2-methylindole	INDOLE-3-CARBOXALDEHYDE, 2-Methyl-1H-
1-Formyl-2-naphthol	2-Hydroxy-1-NAPHTHALENECARBOXALDEHYDE
2-Formylpentane	PENTANAL, 2-Methyl-
3-Formylpentane	BUTANAL, 2-Ethyl-
m-Formylphenol	BENZALDEHYDE, 3-Hydroxy-
o-Formylphenol	BENZALDEHYDE, 2-Hydroxy-
p-Formylphenol	BENZALDEHYDE, 4-Hydroxy-
2-Formylphenol	BENZALDEHYDE, 2-Hydroxy-
3-Formylphenol	BENZALDEHYDE, 3-Hydroxy-
α-Formylphenylacetic acid	ACETALDEHYDE, α-Formylbenzene acetic acid
2-Formylpyridine	2-PYRIDINECARBOXALDEHYDE
p-Formyltoluene	BENZALDEHYDE, 4-Methyl-
2-Formyltoluene	BENZALDEHYDE, 2-Methyl-
3-Formyltoluene	BENZALDEHYDE, 3-Methyl-
Fural	2-FURANCARBOXALDEHYDE
Furaldehyde	2-FURANCARBOXALDEHYDE
α-Furaldehyde	2-FURANCARBOXALDEHYDE
2-Furaldehyde	2-FURANCARBOXALDEHYDE
Furale	2-FURANCARBOXALDEHYDE
2-Furanaldehyde	2-FURANCARBOXALDEHYDE
Furancarbonal	2-FURANCARBOXALDEHYDE
2-Furancarbonal	2-FURANCARBOXALDEHYDE
Furfural	2-FURANCARBOXALDEHYDE
2-Furfural	2-FURANCARBOXALDEHYDE
Furfuraldehyde	2-FURANCARBOXALDEHYDE
2-Furfuraldehyde	2-FURANCARBOXALDEHYDE
Furfurole	2-FURANCARBOXALDEHYDE
Furfurylaldehyde	2-FURANCARBOXALDEHYDE
Furole	2-FURANCARBOXALDEHYDE
α-Furole	2-FURANCARBOXALDEHYDE
2-Furylaldehyde	2-FURANCARBOXALDEHYDE
Fyde	FORMALDEHYDE
Gallaldehyde 3,5-dimethyl ether	BENZALDEHYDE, 4-Hydroxy-3,5-dimethoxy-

Geliotropin	1,3-BENZODIOXOLE-5-CARBOXALDEHYDE
Geranial	2,6-OCTADIENAL, 3,7-Dimethyl-
Geranialdehyde	2,6-OCTADIENAL, 3,7-Dimethyl-
Glutaral	1,5-PENTANEDIAL
Glutaraldehyde	1,5-PENTANEDIAL
Glutardialdehyde	1,5-PENTANEDIAL
Glutaric dialdehyde	1,5-PENTANEDIAL
Glycidal	OXIRANECARBOXALDEHYDE
Glycidaldehyde	OXIRANECARBOXALDEHYDE
Glyoxal	ETHANEDIAL
Glyoxal aldehyde	ETHANEDIAL
Glyoxylaldehyde	ETHANEDIAL
Heliotropin	1,3-BENZODIOXOLE-5-CARBOXALDEHYDE
Heliotropine	1,3-BENZODIOXOLE-5-CARBOXALDEHYDE
Hendecanal	UNDECANAL
Hendecanaldehyde	UNDECANAL
Heptaldehyde	n-HEPTANAL
n-Heptaldehyde	n-HEPTANAL
Heptanal	n-HEPTANAL
Heptanaldehyde	n-HEPTANAL
n-Heptylaldehyde	n-HEPTANAL
2,4-Hexadien-1-al	2,4-HEXADIENAL
Hexaldehyde	HEXANAL
n-Hexanal	HEXANAL
(E)-2-Hexenal	HEXENAL, 2-
trans-Hex-2-enal	HEXENAL, 2-
2-trans-Hexenal	HEXENAL, 2-
trans-2-Hexenal	HEXENAL, 2-
trans-2-Hexen-1-al	HEXENAL, 2-
α-n-Hexylcinnamaldehyde	OCTANAL, 2-(Phenylmethylene)-
Hexyl cinnamic aldehyde	OCTANAL, 2-(Phenylmethylene)-
Hexylenic aldehyde	HEXENAL
α-n-Hexyl-β-phenylacrolein	OCTANAL, 2-(Phenylmethylene)-
Hospex	1,5-PENTANEDIAL
Hyacinthal	ACETALDEHYDE, α-Methylbenzene-
Hyacinthin	ACETALDEHYDE, Benzene
Hydratropa aldehyde	ACETALDEHYDE, α-Methylbenzene-
Hydratropaldehyde	ACETALDEHYDE, α-Methylbenzene-
Hydratropic aldehyde	ACETALDEHYDE, α-Methylbenzene-
Hydrocinnamaldehyde	PROPANAL, Benzene-
Hydrocinnamic aldehyde	PROPANAL, Benzene-
2-Hydroxy-m-anisaldehyde	BENZALDEHYDE, 2-Hydroxy-3-methoxy-
3-Hydroxy-p-anisaldehyde	BENZALDEHYDE, 3-Hydroxy-4-methoxy-

4-Hydroxy-m-anisaldehyde	BENZALDEHYDE, 4-Hydroxy-3-methoxy-
m-Hydroxybenzaldehyde	BENZALDEHYDE, 3-Hydroxy-
o-Hydroxybenzaldehyde	BENZALDEHYDE, 2-Hydroxy-
p-Hydroxybenzaldehyde	BENZALDEHYDE, 4-Hydroxy-
β-Hydroxybutanal	BUTANAL, 3-Hydroxy-
Hydroxycitronellal	OCTANAL, 7-Hydroxy-3,7-dimethyl-
7-Hydroxycitronellal	OCTANAL, 7-Hydroxy-3,7-dimethyl-
4-Hydroxy-3,5-dimethoxycinnamaldehyde	2-PROPENAL, 3-(4-Hydroxy-3,5-dimethoxyphenyl)-
4-Hydroxy-3-ethoxybenzaldehyde	BENZALDEHYDE, 3-Ethoxy-4-hydroxy-
p-Hydroxy-m-methoxybenzaldehyde	BENZALDEHYDE, 4-Hydroxy-3-methoxy-
4-Hydroxy-3-methoxycinnamaldehyde	2-PROPENAL, 3-(4-Hydroxy-3-methoxyphenyl)-
4-(4-Hydroxy-4-methylpentyl)-Δ^3-tetrahydrobenzaldehyde	3-CYCLOHEXENE-1-CARBOXALDEHYDE, 4-(4-Hydroxy-4-methylpentyl)-
2-Hydroxynaphthaldehyde	NAPHTHALENECARBOXALDEHYDE, 2-Hydroxy-1-
2-Hydroxy-α-naphthaldehyde	NAPHTHALENECARBOXALDEHYDE, 2-Hydroxy-1-
2-Hydroxy-1-naphthaldehyde	NAPHTHALENECARBOXALDEHYDE, 2-Hydroxy-1-
2-Hydroxy-1-naphthylaldehyde	NAPHTHALENECARBOXALDEHYDE, 2-Hydroxy-1-
Isobutanal	PROPANAL, 2-Methyl-
Isobutenal	2-PROPENAL, 2-Methyl-
Isobutyraldehyde	PROPANAL, 2-Methyl-
Isodihydrolavandulyl aldehyde	HEXENAL, 5-Methyl-2-(1-methylethylidene)4-
Isopentanal	BUTANAL, 3-Methyl-
Isophthalaldehyde	BENZENEDICARBOXALDEHYDE, 1,3-
Isopropyl aldehyde	PROPANAL, 2-Methyl-
p-Isopropylbenzaldehyde	BENZALDEHYDE, 4-(1-Methylethyl)-
4-Isopropylbenzaldehyde	BENZALDEHYDE, 4-(1-Methylethyl)-
Isopropyl formaldehyde	PROPANAL, 2-Methyl-
2-Isopropylidene-5-methyl-4-hexenal	HEXENAL, 5-Methyl-2-(1-methylethylidene)4-
p-Isopropyl-α-methylhydrocinnamaldehyde	PROPANAL, α-Methyl-4-(1-methylethyl)benzene-
(p-Isopropylphenyl)acetaldehyde	ACETALDEHYDE, 4-(1-Methylethyl)-benzene-
3-(4-Isopropylphenyl)-2-methylpropanal	PROPANAL, α-Methyl-4-(1-methylethyl)benzene-
p-Isopropylphenyl-α-methylpropyl aldehyde	PROPANAL, α-Methyl-4-(1-methylethyl)benzene-
Isovaleral	BUTANAL, 3-Methyl-

Isovaleraldehyde	BUTANAL, 3-Methyl-
Isovaleric aldehyde	BUTANAL, 3-Methyl-
Isovanillin	BENZALDEHYDE, 3-Hydroxy-4-methoxy-
Ivalon	FORMALDEHYDE
Jasminaldehyde	HEPTANAL, 2-(Phenylmethylene)-
Ketene	ETHENONE
Kethoxal	BUTANAL, 3-Ethoxy-2-oxo-
α-Ketopropionaldehyde	PROPANAL, 2-Oxo-
2-Ketopropionaldehyde	PROPANAL, 2-Oxo-
Lauraldehyde	DECANAL, Do-
n-Lauraldehyde	DECANAL, Do-
Lauric aldehyde	DECANAL, Do-
Laurine	OCTANAL, 7-Hydroxy-3,7-dimethyl-
Leaf aldehyde	HEXENAL, 2-
Lilial	PROPANAL, 4-(1,1-Dimethylethyl)-α-methylbenzene-
Lilyal	PROPANAL, 4-(1,1-Dimethylethyl)-α-methylbenzene-
Lilyl	OCTANAL, 7-Hydroxy-3,7-dimethyl-
Lioxin	BENZALDEHYDE, 4-Hydroxy-3-methoxy-
Lyral	3-CYCLOHEXENE-1-CARBOXALDEHYDE, 4-(4-Hydroxy-4-methylpentyl)-
Lysoform	FORMALDEHYDE
Malonaldehyde	PROPANEDIAL
Malondialdehyde	PROPANEDIAL
Malonic dialdehyde	PROPANEDIAL
Malonyldialdehyde	PROPANEDIAL
Metaformaldehyde	FORMALDEHYDE, 1,3,5-Trioxane
Methacrolein	2-PROPENAL, 2-Methyl-
2-Methacrolein	2-PROPENAL, 2-Methyl-
Methacrylaldehyde	2-PROPENAL, 2-Methyl-
Methacrylic aldehyde	2-PROPENAL, 2-Methyl-
Methanal	FORMALDEHYDE
Methional	PROPANAL, 3-(Methylthio)-
m-Methoxybenzaldehyde	BENZALDEHYDE, 3-Methoxy-
o-Methoxybenzaldehyde	BENZALDEHYDE, 2-Methoxy-
p-Methoxybenzaldehyde	BENZALDEHYDE, 4-Methoxy-

2-Methoxybenzenecarboxaldehyde	BENZALDEHYDE, 2-Methoxy-
o-Methoxycinnamaldehyde	2-PROPENAL, 3-(2-Methoxyphenyl)-
2-Methoxycinnamaldehyde	2-PROPENAL, 3-(2-Methoxyphenyl)-
o-Methoxycinnamic aldehyde	2-PROPENAL, 3-(2-Methoxyphenyl)-
2-Methoxy-4-formylphenol	BENZALDEHYDE, 4-Hydroxy-3-methoxy-
3-Methoxy-2-hydroxybenzaldehyde	BENZALDEHYDE, 2-Hydroxy-3-methoxy-
3-Methoxy-4-hydroxybenzaldehyde	BENZALDEHYDE, 4-Hydroxy-3-methoxy-
3-Methoxysalicylaldehyde	BENZALDEHYDE, 2-Hydroxy-3-methoxy-
Methylacetaldehyde	PROPANAL
α-Methylacrolein	2-PROPENAL, 2-Methyl-
β-Methylacrolein	2-BUTENAL
2-Methylacrolein	2-PROPENAL, 2-Methyl-
Methylacrylaldehyde	2-PROPENAL, 2-Methyl-
Methyl aldehyde	FORMALDEHYDE
m-Methylbenzaldehyde	BENZALDEHYDE, 3-Methyl-
o-Methylbenzaldehyde	BENZALDEHYDE, 2-Methyl-
p-Methylbenzaldehyde	BENZALDEHYDE, 4-Methyl-
α-Methylbutanal	BUTANAL (dl), 2-Methyl-
β-Methylbutanal	BUTANAL, 3-Methyl-
2-Methylbutanal-4	BUTANAL, 3-Methyl-
E-2-Methyl-2-butenal	2-BUTENAL (E), 2-Methyl-
trans-2-Methyl-2-butenal	2-BUTENAL (E), 2-Methyl-
α-Methyl-p-(tert-butyl)-hydrocinammaldehyde	PROPANAL, 4-(1,1-Dimethylethyl)-α-methylbenzene-
α-Methylbutyraldehyde	BUTANAL (dl), 2-Methyl-
2-Methylbutyraldehyde	BUTANAL (dl), 2-Methyl-
3-Methylbutyraldehyde	BUTANAL, 3-Methyl-
α-Methylbutyric aldehyde	BUTANAL (dl), 2-Methyl-
2-Methylbutyric aldehyde	BUTANAL (dl), 2-Methyl-
α-Methylcinnamaldehyde	2-PROPENAL, 2-Methyl-3-phenyl-
α-Methylcinnamic aldehyde	2-PROPENAL, 2-Methyl-3-phenyl-
2-Methylcrotonaldehyde	2-BUTENAL (E), 2-Methyl-
3,4-(Methylenedioxy)benzaldehyde	1,3-BENZODIOXOLE-5-CARBOXALDEHYDE
Methylene oxide	FORMALDEHYDE
Methylethylacetaldehyde	BUTANAL (dl), 2-Methyl-
Methylformaldehyde	ACETALDEHYDE
2-Methyl-3-formylindole	INDOLE-3-CARBOXALDEHYDE, 2-Methyl-1H-
Methylglyoxal	PROPANAL, 2-Oxo-
p-Methylhydratropaldehyde	ACETALDEHYDE, α,4-Dimethylbenzene-
p-Methylhydratropicaldehyde	ACETALDEHYDE, α,4-Dimethylbenzene-
2-Methylindole-3-carboxaldehyde	INDOLE-3-CARBOXALDEHYDE, 2-Methyl-1H-
1-Methyl-4-isohexylcyclohexane-1-carboxaldehyde	BENZALDEHYDE, 1-Methyl-4-isohexylhexahydro-

α-Methyl-p-isopropylhydro- cinnamaldehyde	PROPANAL, α-Methyl-4-(1-methyl- ethyl)benzene-
2-Methyl-3-(p-isopropylphenyl) propionaldehyde	PROPANAL, α-Methyl-4-(1-methyl- ethyl)benzene-
β-(Methylmercapto)propionaldehyde	PROPANAL, 3-(Methylthio)-
3-(Methylmercapto)propionaldehyde	PROPANAL, 3-(Methylthio)-
1-Methyl-4(4-methylpentyl)cyclo- hexane-1-carboxaldehyde	BENZALDEHYDE, 1-Methyl-4-iso- hexylhexahydro-
Methylnonylacetaldehyde	UNDECANAL, 2-Methyl-
Methyl-n-nonyl acetaldehyde	UNDECANAL, 2-Methyl-
Methyl nonyl acetic aldehyde	UNDECANAL, 2-Methyl-
α-Methylpentanal	PENTANAL, 2-Methyl-
p-Methylphenylacetaldehyde	ACETALDEHYDE, 4-Methylbenzene-
(4-Methylphenyl)acetaldehyde	ACETALDEHYDE, 4-Methylbenzene-
2-Methyl-3-phenylacrolein	2-PROPENAL, 2-Methyl-3-phenyl-
2-Methyl-3-phenylacrylaldehyde	2-PROPENAL, 2-Methyl-3-phenyl-
2-(p-Methylphenyl)propionaldehyde	ACETALDEHYDE, α,4-Dimethyl- benzene-
2-Methylpropenal	2-PROPENAL, 2-Methyl
α-Methylpropionaldehyde	PROPANAL, 2-Methyl-
2-Methylpropionaldehyde	PROPANAL, 2-Methyl-
β-(Methylthio)propionaldehyde	PROPANAL, 3-(Methylthio)-
3-Methylthiopropionaldehyde	PROPANAL, 3-(Methylthio)-
α-Methyl-α-toluic aldehyde	ACETALDEHYDE, α-Methylbenzene-
2-Methyl-1-undecanal	UNDECANAL, 2-Methyl-
2-Methylvaleraldehyde	PENTANAL, 2-Methyl-
Methylvanillin	BENZALDEHYDE, 3,4-Dimethoxy-
Morbicid	FORMALDEHYDE
Muguet synthetic	OCTANAL, 7-Hydroxy-3,7-dimethyl-
Muguettine	OCTANAL, 7-Hydroxy-3,7-dimethyl-
Myristaldehyde	DECANAL, Tetra-
Myristylaldehyde	DECANAL, Tetra-
β-Naphthol-1-aldehyde	2-Hydroxy-1-NAPHTHALENECARBOX- ALDEHYDE
2-Naphthol-1-carboxaldehyde	2-Hydroxy-1-NAPHTHALENECARBOX- ALDEHYDE
Neopentanal	PROPANAL, 2,2-Dimethyl-
o-Nicotinaldehyde	2-PYRIDINECARBOXALDEHYDE
m-Nitrobenzaldehyde	BENZALDEHYDE, 3-Nitro-
Nonaldehyde	NONANAL
n-Nonaldehyde	NONANAL
Nonanoic aldehyde	NONANAL
n-Nonylaldehyde	NONANAL
Nonylic aldehyde	NONANAL
NSC 8819	2-PROPENAL

Obepin	BENZALDEHYDE, 4-Methoxy-
Octaldehyde	OCTANAL
n-Octaldehyde	OCTANAL
n-Octanal	OCTANAL
Octanaldehyde	OCTANAL
Octanoic aldehyde	OCTANAL
n-Octylal	OCTANAL
Octylaldehyde	OCTANAL
Oenanthal	n-HEPTANAL
Oenanthaldehyde	n-HEPTANAL
Oenanthic aldehyde	n-HEPTANAL
Oenanthol	n-HEPTANAL
Oenanthole	n-HEPTANAL
Oxal	ETHANEDIAL
Oxalaldehyde	ETHANEDIAL
Oxomethane	FORMALDEHYDE
p-Oxybenzaldehyde	BENZALDEHYDE, 4-Hydroxy-
Oxymethylene	FORMALDEHYDE
Paraform	FORMALDEHYDE, Para-
Paraldehyde	ACETALDEHYDE, Paraldehyde
Pelargonaldehyde	NONANAL
Pelargonic aldehyde	NONANAL
1,3-Pentadiene-1-carboxaldehyde	2,4-HEXADIENAL
n-Pentanal	PENTANAL
1,5-Pentanedione	1,5-PENTANEDIAL
α-Pentylcinnamaldehyde	HEPTANAL, 2-(Phenylmethylene)-
Phenylacetaldehyde	ACETALDEHYDE, Benzene
Phenylacetic aldehyde	ACETALDEHYDE, Benzene
β-Phenylacrolein	2-PROPENAL, 3-Phenyl-
3-Phenylacrolein	2-PROPENAL, 3-Phenyl-
Phenylethanal	ACETALDEHYDE, Benzene
Phenylformaldehyde	BENZALDEHYDE
Phenylmethanal	BENZALDEHYDE
1-Phenyl-1-octene-2-carboxaldehyde	OCTANAL, 2-(Phenylmethylene)-
2-Phenylpropanal	ACETALDEHYDE, α-Methylbenzene-
3-Phenylpropanal	PROPANAL, Benzene-
3-Phenyl-1-propanal	PROPANAL, Benzene-
3-Phenylpropenal	2-PROPENAL, 3-Phenyl-
α-Phenylpropionaldehyde	ACETALDEHYDE, α-Methylbenzene-
β-Phenylpropionaldehyde	PROPANAL, Benzene-
2-Phenylpropionaldehyde	ACETALDEHYDE, α-Methylbenzene-
3-Phenylpropionaldehyde	PROPANAL, Benzene-
3-Phenylpropyl aldehyde	PROPANAL, Benzene-
Phixia	OCTANAL, 7-Hydroxy-3,7-dimethyl-
m-Phthalaldehyde	BENZENEDICARBOXALDEHYDE, 1,3-
p-Phthalaldehyde	BENZENEDICARBOXALDEHYDE, 1,4-
Picolinal	2-PYRIDINECARBOXALDEHYDE
Picolinaldehyde	2-PYRIDINECARBOXALDEHYDE

2-Picolinaldehyde	2-PYRIDINECARBOXALDEHYDE
2-Picolinealdehyde	2-PYRIDINECARBOXALDEHYDE
Picolinic aldehyde	2-PYRIDINECARBOXALDEHYDE
Piperonal	1,3-BENZODIOXOLE-5-CARBOXALDEHYDE
Piperonaldehyde	1,3-BENZODIOXOLE-5-CARBOXALDEHYDE
Piperonylaldehyde	1,3-BENZODIOXOLE-5-CARBOXALDEHYDE
Pivalaldehyde	PROPANAL, 2,2-Dimethyl-
Pivalic aldehyde	PROPANAL, 2,2-Dimethyl-
Propaldehyde	PROPANAL
1,3-Propanedialdehyde	PROPANEDIAL
1,3-Propanedione	PROPANEDIAL
Propenal	2-PROPENAL
Prop-2-en-1-al	2-PROPENAL
2-Propen-1-one	2-PROPENAL
Propional	PROPANAL
Propionaldehyde	PROPANAL
Propionic aldehyde	PROPANAL
Propylaldehyde	PROPANAL
Propylic aldehyde	PROPANAL
Protocatechualdehyde dimethyl ether	BENZALDEHYDE, 3,4-Dimethoxy-
Protocatechuic aldehyde dimethyl ether	BENZALDEHYDE, 3,4-Dimethoxy-
Protocatechuic aldehyde ethyl ether	BENZALDEHYDE, 3-Ethoxy-4-hydroxy-
Protocatechuic aldehyde methylene ether	1,3-BENZODIOXOLE-5-CARBOXALDEHYDE
2-Pyridaldehyde	2-PYRIDINECARBOXALDEHYDE
Pyridine-2-aldehyde	2-PYRIDINECARBOXALDEHYDE
2-Pyridylcarboxaldehyde	2-PYRIDINECARBOXALDEHYDE
Pyromucic aldehyde	2-FURANCARBOXALDEHYDE
Pyroracemic aldehyde	PROPANAL, 2-Oxo-
Pyruvaldehyde	PROPANAL, 2-Oxo-
Pyruvic aldehyde	PROPANAL, 2-Oxo-
Quantrovanil	See BENZALDEHYDE, 3-Ethoxy-4-hydroxy-
d-Rhodinal	CITRONELLAL (d isomer)
Salicylal	BENZALDEHYDE, 2-Hydroxy-
Salicylaldehyde	BENZALDEHYDE, 2-Hydroxy-
Salicylic aldehyde	BENZALDEHYDE, 2-Hydroxy-
Sesquisulfate	STREPTOMYCIN sulfate

Sinapaldehyde	2-PROPENAL, 3-(4-Hydroxy-3,5-dimethoxyphenyl)-
Sinapic aldehyde	2-PROPENAL, 3-(4-Hydroxy-3,5-dimethoxyphenyl)-
Sinapyl	2-PROPENAL, 3-(4-Hydroxy-3,5-dimethoxyphenyl)-
Sodium o-benzaldehyde sulfonate	BENZALDEHYDE, 2-Formylbenzenesulfonic acid sodium salt
Sodium benzaldehyde-2-sulfonate	BENZALDEHYDE, 2-Formylbenzenesulfonic acid sodium salt
Sodium o-formylbenzenesulfonate	BENZALDEHYDE, 2-Formylbenzenesulfonic acid sodium salt
Sodium 2-formylbenzenesulfonate	BENZALDEHYDE, 2-Formylbenzenesulfonic acid sodium salt
Sonacide	1,5-PENTANEDIAL
Sorbaldehyde	2,4-HEXADIENAL
Sorbic aldehyde	2,4-HEXADIENAL
Streptobrettin	STREPTOMYCIN sulfate
Streptomycin A	STREPTOMYCIN
Streptorex	STREPTOMYCIN sulfate
o-Sulfobenzaldehyde	BENZALDEHYDE, 2-Formylbenzenesulfonic acid
2-Sulfobenzaldehyde	BENZALDEHYDE, 2-Formylbenzenesulfonic acid
2-Sulfobenzaldehyde sodium salt	BENZALDEHYDE, 2-Formylbenzenesulfonic acid sodium salt
Superlysoform	FORMALDEHYDE
Syringaldehyde	BENZALDEHYDE, 4-Hydroxy-3,5-dimethoxy-
Syringic aldehyde	BENZALDEHYDE, 4-Hydroxy-3,5-dimethoxy-
Syringylaldehyde	BENZALDEHYDE, 4-Hydroxy-3,5-dimethoxy-
Terephthalaldehyde	BENZENEDICARBOXALDEHYDE, 1,4-
Terephthalic aldehyde	BENZENEDICARBOXALDEHYDE, 1,4-
Tetradecylaldehyde	DECANAL, Tetra-
Δ^3-Tetrahydrobenzaldehyde	3-CYCLOHEXENE-1-CARBOXALDEHYDE
1,2,3,6-Tetrahydrobenzaldehyde	3-CYCLOHEXENE-1-CARBOXALDEHYDE
Tiglaldehyde	2-BUTENAL (E), 2-Methyl-
Tiglic acid aldehyde	2-BUTENAL (E), 2-Methyl-
Tiglic aldehyde	2-BUTENAL (E), 2-Methyl-
m-Tolualdehyde	BENZALDEHYDE, 3-Methyl-
o-Tolualdehyde	BENZALDEHYDE, 2-Methyl-
p-Tolualdehyde	BENZALDEHYDE, 4-Methyl-
α-Tolualdehyde	ACETALDEHYDE, Benzene
2-Tolualdehyde	BENZALDEHYDE, 2-Methyl-
4-Tolualdehyde	BENZALDEHYDE, 4-Methyl-
o-Toluic aldehyde	BENZALDEHYDE, 2-Methyl-
α-Toluic aldehyde	ACETALDEHYDE, Benzene
o-Toluylaldehyde	BENZALDEHYDE, 2-Methyl-
p-Toluyl aldehyde	BENZALDEHYDE, 4-Methyl-

p-Tolylacetaldehyde	ACETALDEHYDE, 4-Methylbenzene-
o-Tolylaldehyde	BENZALDEHYDE, 2-Methyl-
p-Tolylaldehyde	BENZALDEHYDE, 4-Methyl-
Trichloroethanal	ACETALDEHYDE, Trichloro-
Triformol	FORMALDEHYDE, 1,3,5-Trioxane
3,4,5-Trimethoxycinnamaldehyde	2-PROPENAL, 3-(3,4,5-Tri-methoxyphenyl)-
Trimethylacetaldehyde	PROPANAL, 2,2-Dimethyl-
1,3,3-Trimethyl-2-(formyl-methylene)indolene	ACETALDEHYDE, 1,3,3-trimethyl-Δ-(2,α)indoline-
2,4,6-Trimethyl-1,3,5-trioxane	ACETALDEHYDE, Paraldehyde
Trioxan	FORMALDEHYDE, 1,3,5-Trioxane
Trioxane	FORMALDEHYDE, 1,3,5-Trioxane
s-Trioxane	FORMALDEHYDE, 1,3,5-Trioxane
sym-Trioxane	FORMALDEHYDE, 1,3,5-Trioxane
Trioxin	FORMALDEHYDE, 1,3,5-Trioxane
Trioxymethylene	FORMALDEHYDE, 1,3,5-Trioxane
Tylan	STREPTOMYCIN, Tylosin
Tylon	STREPTOMYCIN, Tylosin
Tylosin	STREPTOMYCIN, Tylosin
n-Undecanal	UNDECANAL
Undecenoic aldehyde	UNDECENAL, 9-
Undecyl aldehyde	UNDECANAL
9-Undecylene aldehyde	UNDECENAL, 9-
10-Undecylene aldehyde	UNDECENAL, 10-
Undecylenic aldehyde	UNDECENAL, 10-
n-Undecylic aldehyde	UNDECANAL
Valeral	PENTANAL
Valeraldehyde	PENTANAL
n-Valeraldehyde	PENTANAL
Valerianic aldehyde	PENTANAL
Valeric acid aldehyde	PENTANAL
Valeric aldehyde	PENTANAL
Valerylaldehyde	PENTANAL
Vanillal	BENZALDEHYDE, 3-Ethoxy-4-hydroxy-
Vanillic aldehyde	BENZALDEHYDE, 4-Hydroxy-3-methoxy-
Vanillin	BENZALDEHYDE, 4-Hydroxy-3-methoxy-
o-Vanillin	BENZALDEHYDE, 2-Hydroxy-3-methoxy-
Vanillin ethyl ether	BENZALDEHYDE, 4-Ethoxy-3-methoxy-
Vanillin methyl ether	BENZALDEHYDE, 3,4-Dimethoxy-

Vanirom	BENZALDEHYDE, 3-Ethoxy-4-hydroxy-
Veratral	BENZALDEHYDE, 3,4-Dimethoxy-
Veratraldehyde	BENZALDEHYDE, 3,4-Dimethoxy-
Veratric aldehyde	BENZALDEHYDE, 3,4-Dimethoxy-
Veratryl aldehyde	BENZALDEHYDE, 3,4-Dimethoxy-
Vetstrep	STREPTOMYCIN sulfate

TABLE A-1. Physical and Chemical Properties of Selected Aldehydes

Name and Formula	Mol. Wt.	Density, relative to H_2O or g/ml	Vapor Density, (air=1)	Vapor Pressure, mm Hg (°C)	Melting Pt./ Boiling Pt., °C	Flash Point, °F	Solub. in H_2O	Conversion Factors 1 mg/L (ppm)[a]	1 ppm (mg/m^3)[b]
ACETALDEHYDE CH_3CHO	44.05	0.7834 (18/4°C)	1.52	740(20)	-121/ 20.8	-40 (open cup)	Miscible	556	1.8
Benzene acetaldehyde $C_6H_5CH_2CHO$	120.16	1.0272 (20/4°C)	4.1	10(78)	33-34/ 195 (F.P. below -10)	--	Very slightly soluble	204	4.9
α,4-Dimethyl- benzene- acetaldehyde $C_{10}H_{12}O$	148.21	--	--	--	--	--	--	165	6.1
3,7-Dimethyl-6- octenyloxy- acetaldehyde $C_{11}H_{20}O_2$	184.28	--	--	--	--	--	--	133	7.5
α-Formylbenzene acetic acid $C_9H_8O_3$	164.16	--	--	--	--	--	--	149	6.7
α-Methylbenzene- acetaldehyde $C_6H_5CH(CH_3)CHO$	134.18	1.0089 (20/4°C)	--	--	--/ 202-205	--	Insoluble	182	5.5
4-Methylbenzene- acetaldehyde $C_9H_{10}O$	134.18	--	--	--	--	--	--	182	5.5

TABLE A-1. Continued

Name and Formula	Mol. Wt.	Density, relative to H_2O or g/ml	Vapor Density (air=1)	Vapor Pressure, mm Hg (°C)	Melting Pt./ Boiling Pt., °C	Flash Point, °F	Solub. in H_2O	Conversion Factors 1 mg/L (ppm)[a]	1 ppm (mg/m^3)[b]
(ACETALDEHYDE – Contd.)									
4-(1-Methylethyl) benzene acetaldehyde $C_{11}H_{14}O$	162.22	--	--	--	--	--	--	151	6.6
Paraldehyde $(C_2H_4O)_3$	132.16	0.9943 (20/4°C)	--	25.3 (20)	12.6/ 128.0	96	Slightly soluble	--	--
Trichloro- acetaldehyde CCl_3CHO	147.39	1.5121 (20/4°C)	5.1	35 (20)	-57.5/ 97.75/	167	Soluble	166	6.0
1,3,3-Trimethyl-Δ-(2,α)-indoline- acetaldehyde $C_{13}H_{15}NO$	201.26	--	--	--	--	--	--	122	8.2
BENZALDEHYDE C_6H_5CHO	106.13	1.0415 (15/4°C)	3.66	1 (26)	-26/ 178.1 (F.P. -56.9– -55.6)	145	Slightly soluble	231	4.3
2-Chlorobenzaldehyde C_7H_5ClO	140.57	1.2483 (20/4°C)	--	--	12.39/ 211.9	--	Slightly soluble	174	5.7
4-Chlorobenzaldehyde C_7H_5ClO	140.57	1.196 (61/4°C)	--	--	47.5/ 213–214	--	Soluble	174	5.7
4-((2-Chloroethyl) ethylamino)-2- methylbenzaldehyde $C_{12}H_{16}ClNO$	225.72	--	--	--	--	--	--	108	9.2

Compound	MW	Density			mp/bp		Solubility		
4-((2-Chloroethyl)-methylamino)-benzaldehyde C$_{10}$H$_{12}$ClNO	197.67	--	--	--	--	--	--	124	8.1
4-((2-Cyanoethyl)-methylamino)-benzaldehyde C$_{11}$H$_{12}$N$_2$O	188.23	--	--	--	--	--	--	130	7.7
2,6-Dichlorobenzaldehyde C$_7$H$_4$Cl$_2$O	175.02	--	--	--	71/ --	--	Insoluble	140	7.2
4-(Diethylamino)-benzaldehyde C$_{11}$H$_{15}$NO	177.25	--	--	7(174)	41/ --	--	Soluble	138	7.2
2,5-Dimethoxy-benzaldehyde (CH$_3$O)$_2$C$_6$H$_3$CHO	166.18	--	5.7	--	46-49/ --	--	Insoluble	147	6.8
3,4-Dimethoxy-benzaldehyde (CH$_3$O)$_2$CH$_3$CHO	166.18	--	--	10(155)	44/258	--	Slightly soluble	147	6.8
4-(1,1-Dimethyl-ethyl)benzaldehyde C$_{11}$H$_{14}$O	162.23	--	--	--	--	--	--	151	6.6
4-Ethoxybenzaldehyde C$_9$H$_{10}$O$_2$	150.18	1.08 (21/21°C)	--	20(140)	133-134/249	--	--	163	6.1

TABLE A-1. Continued

Name and Formula	Mol. Wt.	Density, relative to H_2O or g/ml	Vapor Density, (air=1)	Vapor Pressure, mm Hg (°C)	Melting Pt./ Boiling Pt., °C	Flash Point, °F	Solub. in H_2O	Conversion Factors 1 mg/L (ppm)[a]	1 ppm (mg/m³)[b]
(BENZALDEHYDE - Contd.)									
3-Ethoxy-4-hydroxy-benzaldehyde $C_9H_{10}O_3$	166.18	--	--	--	77-78/ --	--	Slightly soluble	147	6.8
4-Ethoxy-3-methoxy-benzaldehyde $C_{10}H_{12}O_3$	180.21	--	--	--	64-65/ Sublimes	--	Slightly soluble	136	7.4
4-Formyl-1,3-benzene disulfonic acid $(SO_3H)_2C_6H_3CHO$	266.25	--	--	--	--	--	--	92	10.9
2-Formylbenzene-sulfonic acid $C_7H_6O_4S$	186.19	--	--	--	--	--	--	131	7.6
2-Formylbenzene-sulfonic acid sodium salt $C_7H_5NaO_4S$	208.17	--	--	--	--	--	--	118	8.5
2-Hydroxybenzaldehyde C_6H_4OHCHO	122.13	1.1674 (20/4°C)	4.2	1(33) 25(93)	-7/ 197/ (F.P. -7)	172	Slightly soluble	200	5.0
3-Hydroxybenzaldehyde $C_7H_6O_2$	122.13	--	--	20(161)	108/ 240	--	Slightly soluble	200	5.0
4-Hydroxybenzaldehyde $C_7H_6O_2$	122.13	1.129 (130/4°C)	4.2	1(121) 10(170)	117/ Sublimes	--	Slightly soluble	200	5.0

Name	MW	Density		Solubility (g/100g)	mp/bp (°C)		Solubility in water		
4-hydroxy-3,5-dimethoxybenzaldehyde $C_9H_{10}O_4$	182.18	--	--	--	113/192-193 (14 torr)	--	Slightly soluble	134	7.5
2-Hydroxy-3-methoxybenzaldehyde $C_8H_8O_3$	152.16	--	--	10(128)	44-45/265-266	--	Slightly soluble	161	6.2
3-Hydroxy-4-methoxybenzaldehyde	--	--	--	--	--	--	--	--	--
4-Hydroxy-3-methoxybenzaldehyde $(CH_3O)(OH)C_6H_3CHO$	152.16	1.056^c	5.2	1(107) 15(170)	81-83/285	--	Slightly soluble	161	6.2
2-Methoxybenzaldehyde $C_6H_4(OCH_3)CHO$	136.16	1.1326 (20/4°C)	--	18(125)	37-38/243-244	244	Insoluble	180	5.6
3-Methoxybenzaldehyde $C_6H_4(OCH_3)CHO$	136.16	1.1187 (20/4°C)	--	1(62)	--/230	--	Insoluble	180	5.6
4-Methoxybenzaldehyde $C_6H_4(OCH_3)CHO$	136.16	1.1191 (15/4°C)	--	2(83)	0/249.5	--	Insoluble	180	5.6
2-Methylbenzaldehyde C_8H_8O	120.16	1.0386 (19/4°C)	--	10(94)	--/200	--	Slightly soluble with heat	204	4.9

309

TABLE A-1. Continued

Name and Formula	Mol. Wt.	Density, relative to H_2O or g/ml	Vapor Density, (air=1)	Vapor Pressure, mm Hg (°C)	Melting Pt./ Boiling Pt., °C	Flash Point, °F	Solub. in H_2O	Conversion Factors 1 mg/L (ppm)[a]	1 ppm (mg/m³)[b]
(BENZALDEHYDE - Contd.)									
3-Methylbenzaldehyde C_8H_8O	120.16	1.0189 (21/4°C)	--	17(93-94)	--/199	--	Slightly soluble	204	4.9
4-Methylbenzaldehyde $CH_3C_6H_4CHO$	120.16	1.0194 (17/4°C)	4.1	10(106)	--/204-205	--	Slightly soluble	204	4.9
4-(1-Methylethyl)benz-aldehyde $(CH_3)_2CHC_6H_4CHO$	148.21	0.9755 (20/4°C)	--	10(104)	--/235-236	--	Insoluble	165	6.1
1-Methyl-4-isohexyl-hexahydrobenzaldehyde $C_{14}H_{26}O$	210.36	--	--	--	--	--	--	116	8.6
3-Nitrobenzaldehyde $C_7H_5NO_3$	151.12	1.2792 (20/4°C)	--	23(164)	58/--	--	Slightly soluble	162	6.2
1,3-BENZENEDICARBOX-ALDEHYDE $C_6H_4(CHO)_2$	134.14	--	--	--	89-90/245-248 (771 torr)	--	Slightly soluble	182	5.5
1,4-Benzenedicarbox-aldehyde $C_6H_4(CHO)_2$	134.14	--	--	--	116/247	--	Slightly soluble	182	5.5
1,3-BENZODIOXOLE 5-CARBOXALDEHYDE $C_6H_3(CH_2OO)CHO$	150.14	--	5.2	1(87)	37/263	--	Slightly soluble	163	6.1
BUTANAL $CH_3(CH_2)_2CHO$	72.12	0.8170 (20/4°C)	2.48	92(20)	-99/75.7	15	Soluble	339	3.0

Compound	MW	Density		M.p./B.p. (°C)	Flash point	Solubility in water		
3-Ethoxy-2-oxobutanal $CH_3CH(OC_2H_5)COCHO$	130.14	--	--	--/145	--	Slightly soluble	188	5.3
2-Ethylbutanal $(C_2H_5)_2CHCHO$	100.16	0.8110 (20/4°C)	3.5	13.7 (20)	-89/117-119 (160 torr)	70 (open cup)	Slightly soluble	244	4.1
3-Hydroxybutanal $CH_3CHOHCH_2CHO$	88.12	1.103 (20/4°C)	3.0	21.0 (20)	<0/83(85 decomposes)	150 (open)	Miscible	278	3.6
2-Methylbutanal (dl) $CH_3CH_2CH(CH_3)CHO$	86.14	0.8029 (20/4°C)	3.0	50.0 (ca. 25)	--/92-93	--	Insoluble	284	3.5
3-Methylbutanal $(CH_3)_2CHCH_2CHO$	86.14	0.7977 (20/4°C)	3.0	50.0 (ca. 25)	-51/92.5	--	Slightly soluble	284	3.5
3-Methyl-2-phenyl-butanal $C_{11}H_{14}O$	162.23	--	--	--	--	--	--	151	6.6
trans-2-BUTENAL $CH_3CH=CHCHO$	70.09	0.84942 (25/4°C)	2.4	30.0 (20)	-74/104-105	55	Soluble	349	2.9
2-Methyl-2-butenal (E) $CH_3CH=C(CH_3)CHO$	84.13	0.8710 (20/4°C)	--	119.0 (63-65)	--/116.5-117.5 (738 torr)	--	Soluble	291	3.4
CITRONELLAL(d) $C_9H_{17}CHO$	154.26	0.8573 (20/4°C)	--	--	--/207.8	--	Slightly soluble	159	6.3

TABLE A-1. Continued

Name and Formula	Mol. Wt.	Density, relative to H$_2$O or g/ml	Vapor Density, (air=1)	Vapor Pressure, mm Hg (°C)	Melting Pt./ Boiling Pt., °C	Flash Point, °F	Solub. in H$_2$O	Conversion Factors 1 mg/L (ppm)[a]	1 ppm (mg/m^3)[b]
4-(1,1-Dimethylethyl)-CYCLOHEXANECARBOX-ALDEHYDE $C_{11}H_{20}O$	168.28	--	--	--	--	--	--	145	6.9
3-CYCLOHEXENE-1-CARBOX-ALDEHYDE $C_7H_{10}O$	110.16	0.9709 (20/4°C)	--	--	Freezes -96.1/ 164/	135	Slightly soluble	222	4.5
4-(4-Hydroxy-4-methyl-pentyl)-3-cyclohexene-1-carboxaldehyde $C_{13}H_{22}O_2$	210.32	--	--	--	--	--	--	116	8.6
DECANAL $CH_3(CH_2)_8CHO$	156.27	0.830 (15/4°C)	--	7(81)	ca. -5/ 208-209	--	Insoluble	157	6.4
n-Dodecanal $CH_3(CH_2)_{10}CHO$	184.33	0.8352 (15/4°C)	--	3.5 (100) 100.0 (85)	44.5/ 185 (100 torr)	--	Insoluble	133	7.5
Tetradecanal $C_{14}H_{28}O$	212.38	--	--	--	30/ 166 (24 torr)	--	--	115	8.7
ETHANEDIAL OCHCHO	58.04	1.14 (20°C)	2.0	220.0 (ca. 20)	15/ 50.4	--	Very soluble	422	2.4
ETHENONE $CH_2=CO$	42.04	--	1.45	--	-151/ -56	--	Decomposes	582	1.7

Compound	MW	Density			MP/BP	Flash point	Solubility		
FORMALDEHYDE (gas) HCHO	30.03	0.815 (-20/4°C)	1.075	10(-88)	-92/-21 (F.P. -118)	572	Very soluble	815	1.2
Paraformaldehyde $(HCHO)_x$	$(30)_x$	--	--	--	25(solid)/ Decomposes	158	Soluble to insol.	--	--
1,3,5-Trioxane $C_3H_6O_3$	90.08	1.17 (65°C)	--	--	64/ 114.5(759 torr; sublimes at 115)	113 (open cup)	Very soluble	272	3.7
2-FURANCARBOXALDEHYDE $C_5H_4O_2$	96.09	1.1594 (20/4°C)	3.31	65(90)	-38.7/ 161.7	140(cl. cup)	Soluble at 20°C	255	3.9
n-HEPTANAL $CH_3(CH_2)_5CHO$	114.19	0.8495 (20/4°C)	3.9	3(25)	-43.3/ 152.8	--	Slightly soluble	214	4.7
2-(Phenylmethylene)-heptanal $CH_3(CH_2)_4C(=CHC_6H_5)CHO$	202.30	0.97108 (20°C)	--	--	80/ 174/175 (20 torr)	--	Insoluble	121	8.3
2,6-Dimethyl-5-HEPTENAL $(CH_3)_2C=CH(CH_2)_2CH(CH_3)CHO$	140.23	0.845-0.855 (25/25°C)	--	--	--	144(cl. cup)	--	175	57
2,4-HEXADIENAL $CH_3CH=CHCH=CHCHO$	96.14	0.898 (20°C)	--	--	--/ 173-174 (754 torr)	--	--	255	3.9
HEXANAL $CH_3(CH_2)_4CHO$	100.16	0.81392	3.5	10(20)	-56/ 128	90 (open cup)	Slightly soluble	245	4.1

TABLE A-1. Continued

Name and Formula	Mol. Wt.	Density, relative to H_2O or g/ml	Vapor Density, (air=1)	Vapor Pressure, mm Hg (°C)	Melting Pt./ Boiling Pt., °C	Flash Point, °F	Solub. in H_2O	Conversion Factors 1 mg/L (ppm)[a]	1 ppm (mg/m^3)[b]
2-HEXENAL $CH_3CH_2CH_2CH=CHCHO$	98.15	0.8491 (20/4°C)	--	--	--/ 146-147	--	--	249	4.0
5-Methyl-2-(1-methyl-ethylidene)4-hexenal $C_{10}H_{16}O$	152.24	--	--	--	--	--	--	161	6.2
2-Methyl-1H-INDOLE-3-CARBOXALDEHYDE $C_{10}H_9NO$	159.19	--	--	--	--	--	--	154	6.5
2-Hydroxy-1-NAPHTHALENECARBOX-ALDEHYDE $C_{11}H_8O_2$	172.20	--	--	--	82/ 192(27 torr)	--	Insoluble	142	7.0
NONANAL $CH_3(CH_2)_7CHO$	142.24	0.8264 (22/4°C)	--	23 (93.5)	--/ 190-192	--	--	172	5.8
3,7-Dimethyl-2,6-OCTADIENAL $(CH_3)_2CCH(CH_2)_2C(CH_3)CHCHO$	152.24	0.8888 (20°C)	--	5(91-95) 20(118-119)	--/ 229 (decomposes)	--	Insoluble	161	6.2
OCTANAL $CH_3(CH_2)_6CHO$	128.22	0.8211 (20/4°C)	--	20(72)	--/ 171	--	Slightly soluble	191	5.2
7-Hydroxy-3,7-dimethyl-octanal $C_{10}H_{20}O_2$	172.27	0.925-0.930[c] (15°C)	--	3(103) 0.3(91)	--/	--	Slightly soluble	142	7.0

314

Name	Formula	MW	Density		B.P./M.P. etc.	Flash Point	Solubility		
2-(Phenylmethylene)-octanal	$C_6H_{13}C(CHO)=CHC_6H_5$	216.33	0.953-0.959^c (25°C)	--	--	--	--	113	8.8
OXIRANECARBOXALDEHYDE	$C_3H_4O_2$	72.06	--	--	--	--	--	340	3.0
PENTANAL	$CH_3(CH_2)_3CHO$	86.14	0.8095 (20/4°C)	3.0	50(ca. 25)	-91.5/ 103	54(cl. cup)	Slightly soluble	284	3.5
2-Methylpentanal	$CH_3CH_2CH_2CH(CH_3)CHO$	100.16	0.8092^c	--	--	--/ 116(737 torr)	68	Soluble	244	4.1
1,5-PENTANEDIAL	$OCH(CH_2)_3CHO$	100.13	0.72^c	3.4	17(20) 10(71-72)	-109.21/ 187-189 (decomposes)/ (F.P. -14)	None	Miscible	244	4.1
PROPANAL	CH_3CH_2CHO	58.08	0.8058 (20/4°C)	2.0	687(45) 300(ca. 25)	-81/ 48.8/ (F.P. -81)	15-19	Soluble	421	2.4
Benzenepropanal	$C_6H_5CH_2CH_2CHO$	134.18	--	--	13 (104-105)	47/ 223(745 torr)	--	Insoluble	182	5.5
4-(1,1-Dimethylethyl-α-methylbenzene propanal	$C_{14}H_{20}O$	204.32	--	--	--	--	--	--	120	8.4
2,2-Dimethylpropanal	$(CH_3)_3CCHO$	86.14	0.7923 (17°C)	--	--	6/ 77-78	--	--	284	3.5

TABLE A-1. Continued

Name and Formula	Mol. Wt.	Density, relative to H_2O or g/ml	Vapor Density, (air=1)	Vapor Pressure, mm Hg (°C)	Melting Pt./ Boiling Pt., °C	Flash Point, °F	Solub. in H_2O	Conversion Factors 1 mg/L (ppm)[a]	1 ppm (mg/m³)[b]
(PROPANAL – Contd.)									
α-Methyl-4-(methylethyl)benzenepropanal $C_{13}H_{18}O$	190.29	0.951 (15°C)	--	5(115)	--/133-137 (99 torr)	--	Insoluble	129	7.8
2-Methylpropanal $(CH_3)_2CHCHO$	72.12	0.794[c] (20/4°C)	2.48	170(ca. 20)	-65.9/ 65	-40 (cl.cup) -11 (open)	Insoluble	339	3.0
3-(Methylthio)propanal C_4H_8OS	104.18	--	--	--	--	--	--	235	4.3
2-Oxopropanal $CH_3(CO)CHO$	72.06	1.0455 (24°C)	--	--	--/72	--	--	340	3.0
PROPANEDIAL $C_3H_4O_2$	72.06	--	--	--	--	--	--	340	3.0
2-PROPENAL $CH_2=CHCHO$	56.07	0.8410 (20/4°C)	1.94	214 (20)	-86.95/ 52.5-53.5	<0	Very soluble	436	2.3
3(4-Hydroxy-3,5-dimethoxyphenyl)-2-propenal $C_{11}H_{12}O_4$	208.22	--	--	--	--	--	--	117	8.5
3-(4-Hydroxy-3-methoxyphenyl)-2-propenal $C_{10}H_{10}O_3$	178.19	--	--	2.5 (157)	84/ --	--	Slightly soluble	137	7.3

Name	Formula weight	Density		Melting point	Boiling point	Flash point	Solubility		
3-(2-Methoxyphenyl)-2-propenal $(CH_3O)C_6H_4(CH_3)_2CHO$	162.19	--	--	--	--	--	--	151	6.6
2-Methyl-3-phenyl-2-propenal $C_6H_5CH=C(CH_3)CHO$	146.19	1.0407 (17/4°C)	--	--	--/150(100 torr)	--	--	167	6.0
2-Methyl-2-propenal $CH_2=C(CH_3)CHO$	70.09	0.837 (20/4°C)	2.4	--	--/68.4	5(open cup)	Miscible	349	2.9
3-Phenyl-2-propenal $C_6H_5CH=CHCHO$	132.17	1.0497 (20/4°C)	4.6	16(127)	-7.5/253(slightly decomposes)	--	Very slightly soluble	185	5.4
3-(3,4,5-trimethoxyphenyl)-2-propenal $C_{12}H_{14}O_4$	222.24	--	--	--	--	--	--	110	9.1
2-PYRIDINECARBOXALDEHYDE C_6H_5NO	107.11	--	--	--	--	--	--	229	4.4
STREPTOMYCIN $C_{21}H_{39}N_7O_{12}$	581.58	--	--	--	--	--	--	42	23.8
Streptomycin sulfate $2(C_{21}H_{39}N_7O_{12})\cdot 3H_2SO_4$	1457.40	--	--	--	--	--	>20 mg/ml (ca. 28°C)	17	59.6
Tylosin $C_{45}H_{77}NO_{17}$	904.11	--	--	--	128-130/--	--	5 mg/ml	27	37.0

TABLE A-1. Continued

Name and Formula	Mol. Wt.	Density, relative to H_2O or g/ml	Vapor Density (air=1)	Vapor Pressure, mm Hg (°C)	Melting Pt./ Boiling Pt., °C	Flash Point, °F	Solub. in H_2O	Conversion Factors 1 mg/L (ppm)[a]	1 ppm (mg/m^3)[b]
n-UNDECANAL $CH_3(CH_2)_9CHO$	170.30	0.8251 (23/4°C)	—	18(117)	—/	235	Insoluble	144	7.0
2-Methylundecanal $CH_3(CH_2)_8CH(CH_3)CHO$	184.33	0.830 (15/4°C)	—	10(114)	—	—	Slightly soluble	133	7.5
9-UNDECENAL $CH_3CH=CH(CH_2)_7CHO$	168.28	—	—	—	—	—	—	145	6.9
10-Undecenal $CH_2=CH(CH_2)_8CHO$	168.28	0.840–0.850 (25/25°C)	—	—	—	—	—	145	6.9

---

[a] One milligram per liter in concentration of aldehyde is equivalent to the number of parts per million of aldehyde in air (25°C, 1 atm) shown in this column.

[b] One part per million of the aldehyde in air (25°C, 1 atm) is equivalent to a concentration in milligrams per cubic meter shown in this column.

[c] Specific gravity.

TABLE A-2. Uses and Synonyms of Selected Aldehydes

CAS Number	Name of Compound	Synonyms	Uses
75-07-0	ACETALDEHYDE	Acetic aldehyde Ethanal Ethyl aldehyde Methyl formaldehyde	Manufacture of acetic acid and acetic anhydride, ethyl aldehyde, n-butanol, 2-ethylhexanol, peracetic acide, pentaerythritol, pyridines, chloral, 1,3-butylene glycol, trimethylolpropane; also intermediate
122-78-1	Benzene acetaldehyde	Phenylacetaldehyde Benzeneacetaldehyde Hyacinthin Phenylacetic aldehyde α-Tolualdehyde α-Toluic aldehyde Phenylethanal	Perfumes; flavoring; laboratory reagent
99-72-9	α,4-Dimethylbenzene-acetaldehyde	p-Methylhydratropaldehyde p-Methylhydratropicaldehyde 2-(p-Methylphenyl)propionaldehyde	
7492-67-3	3,7-Dimethyl-b-octenyl-oxyacetaldehyde	Citronelloxyacetaldehyde 6,10-Dimethyl-3-oxa-9-undecanal	Perfumes; flavoring
59216-85-2	α-Formylbenzene acetic acid	α-Formylphenylacetic acid	
93-53-8	α-Methylbenzene-acetaldehyde	Hydratropaldehyde 2-Phenylpropanal 2-Phenylpropionaldehyde α-Phenylpropionaldehyde Hydratropa aldehyde Hydratropic aldehyde Cumene aldehyde Hyacinthal α-Methyl-α-toluic aldehyde	

TABLE A-2. Continued

CAS Number	Name of Compound	Synonyms	Uses
	(ACETALDEHYDE - Contd.)		
104-09-6	4-Methylbenzene-acetaldehyde	p-Tolylacetaldehyde p-Methylphenylacetaldehyde (4-Methylphenyl)acetaldehyde	
4395-92-0	4-(1-Methylethyl)benzene-acetaldehyde	(p-Isopropylphenyl)-acetaldehyde Cortexal p-Cymene-7-carboxaldehyde	
123-63-7	Paraldehyde (trimer of acetaldehyde)	2,4,6-Trimethyl-1,3,5-trioxane p-Acetaldehyde	Substitute for acetaldehyde; rubber accelerators; rubber antioxidants; synthetic organic chemicals; dyestuff intermediates; medicine; solvent for fats, oils, waxes, gums, resins, leather; solvent mixtures for cellulose derivatives; sedative
75-87-6	Trichloroacetaldehyde	Chloral Anhydrous chloral Trichloroethanal	Manufacture of chloral hydrate and DDT
84-83-3	1,3,3-Trimethyl-Δ-(2,α)-indoline-acetaldehyde	(1,3-Dihydro-1,3,3-trimethyl-2H-indol-2-ylidene)-acetaldehyde Fisher's aldehyde 1,3,3-Trimethyl-2-(formylmethylene)indoline 2-(Formylmethylene)-1,3,3-trimethyl-indoline	
100-52-7	BENZALDEHYDE	Benzenecarbonal Benzenecarboxaldehyde Benzoic aldehyde Phenylmethanal Artificial almond oil Benzaldehyde FFC Phenylformaldehyde	Chemical intermediate for dyes, flavoring materials, perfumes and aromatic alcohols; solvent for oils, resins, some cellulose ethers, cellulose acetate and nitrate; flavoring compounds; synthetic perfumes; manufacture of cinnamic acid, benzoic acid, pharmaceuticals; photographic chemicals.

89-98-5	2-Chlorobenzaldehyde	o-Chlorobenzaldehyde	
104-88-1	4-Chlorobenzaldehyde	p-Chlorobenzaldehyde	
92-10-4	4-((2-Chloroethyl)ethyl-amino)-2-methylbenzaldehyde	4-((2-Chloroethyl)ethylamino)-o-tolualdehyde	
94-31-5	4-((2-Chloroethyl)methyl-amino)benzaldehyde	p-((2-Chloroethyl)methylamino)-benzaldehyde	
94-21-3	4-((2-Cyanoethyl)methyl-amino)benzaldehyde	--	
83-38-5	2,6-Dichlorobenzaldehyde	--	
120-21-8	4-(Diethylamino)benz-aldehyde	p-(Diethylamino)benzaldehyde p-Formyl-N,N-diethylaniline	
93-02-7	2,5-Dimethoxybenzaldehyde	--	
120-14-9	3,4-Dimethoxybenzaldehyde	Veratraldehyde 3,4-Dimethoxybenzenecarbonal Protocatechualdehyde dimethyl ether Protocatechuic aldehyde dimethyl ether Vanillin methyl ether Veratric aldehyde Veratral Veratryl aldehyde Methylvanillin	Organic synthesis; dyestuff intermediate
939-97-9	4-(1,1-Dimethylethyl)-benzaldehyde	p-tert-Butylbenzaldehyde 4-tert-Butylbenzaldehyde	

321

TABLE A-2. Continued

CAS Number	Name of Compound	Synonyms	Uses
(BENZALDEHYDE - Contd.)			
10031-82-0	4-Ethoxybenzaldehyde	p-Ethoxybenzaldehyde Ethoxybenzaldehyde	
121-32-4	3-Ethoxy-4-hydroxy-benzaldehyde	Bourbonal Ethavan Ethovan Ethylvanillin Vanillal Vanirom Ethylprotal Quantrovanil Protocatechuic aldehyde ethyl ether 4-Hydroxy-3-ethoxybenzaldehyde	
120-25-2	4-Ethoxy-3-methoxy-benzaldehyde	4-Ethoxy-m-anisaldehyde Vanillin ethyl ether	Intermediate
88-39-1	4-Formyl-1,3-benzene-disulfonic acid	4-Formyl-m-benzenedisulfonic acid 2,4-Disulfobenzaldehyde Benzaldehyde-2,4-disulfonic acid	
91-25-8	2-Formylbenzenesulfonic acid	2-Sulfobenzaldehyde o-Formylbenzenesulfonic acid o-Sulfobenzaldehyde	
1008-72-6	2-Formylbenzenesulfonic acid sodium salt	o-Formylbenzenesulfonic acid sodium salt Benzaldehyde-o-sulfonic acid sodium salt 2-Sulfobenzaldehyde sodium salt Sodium benzaldehyde-2-sulfonate Sodium 2-formylbenzenesulfonate Sodium o-formylbenzenesulfonate Sodium o-benzaldehyde sulfonate	

90-02-8	2-Hydroxybenzaldehyde	Salicylaldehyde o-Hydroxybenzaldehyde Salicylal Salicylic aldehyde o-Formylphenol 2-Formylphenol	Analytical chemistry; perfumery (violet); synthesis of coumarin; gasoline additives; auxiliary fumigant
100-83-4	3-Hydroxybenzaldehyde	m-Hydroxybenzaldehyde m-Formylphenol 3-Formylphenol	
123-08-0	4-Hydroxybenzaldehyde	p-Hydroxybenzaldehyde p-Formylphenol p-Oxybenzaldehyde	Pharmaceuticals
134-96-3	4-Hydroxy-3,5-dimethoxybenzaldehyde	3,5-Dimethoxy-4-hydroxybenzene carbonal Gallaldehyde 3,5-dimethyl ether Syringaldehyde Syringic aldehyde Syringylaldehyde 3,5-Dimethoxy-4-hydroxybenzaldehyde	
148-53-8	2-Hydroxy-3-methoxybenzaldehyde	2-Hydroxy-m-anisaldehyde 3-Methoxysalicylaldehyde 3-Methoxy-2-hydroxybenzaldehyde 6-Formylguaiacol o-Vanillin 6-Formyl-2-methoxyphenol	
621-59-0	3-Hydroxy-4-methoxybenzaldehyde	3-Hydroxy-p-anisaldehyde 5-Formylguaiacol Isovanillin	

TABLE A-2. Continued

CAS Number	Name of Compound	Synonyms	Uses
	(BENZALDEHYDE - Contd.)		
121-33-5	4-Hydroxy-3-methoxybenz-aldehyde	Vanillin 3-Methoxy-4-hydroxybenzaldehyde Vanillic aldehyde Vanillaldehyde Lioxin p-Hydroxy-m-methoxybenzaldehyde 2-Methoxy-4-formylphenol 4-Hydroxy-m-anisaldehyde 4-Formyl-2-methoxyphenol	Perfumes; flavoring; pharmaceuticals; laboratory reagent; source of L-dopa
135-02-4	2-Methoxybenzaldehyde	o-Anisaldehyde o-Methoxybenzaldehyde o-Anisic aldehyde 2-Anisaldehyde 2-Methoxybenzenecarboxaldehyde	Intermediate
591-31-1	3-Methoxybenzaldehyde	m-Anisaldehyde m-Methoxybenzaldehyde	
123-11-5	4-Methoxybenzaldehyde	p-Anisaldehyde Crategine p-Methoxybenzaldehyde Aubepine p-Anisic aldehyde 4-Anisaldehyde p-Formylanisole Obepin	Perfumery; intermediate for antihistamines, electroplating, flavoring
529-21-4	2-Methylbenzaldehyde	o-Tolualdehyde o-Methylbenzaldehyde o-Toluic aldehyde o-Toluylaldehyde o-Tolylaldehyde 2-Tolualdehyde 2-Formyltoluene	

620-23-5	3-Methylbenzaldehyde	m-Tolualdehyde m-Methylbenzaldehyde 3-Formyltoluene	
104-87-0	4-Methylbenzaldehyde	p-Tolualdehyde p-Methylbenzaldehyde p-Tolylaldehyde p-Toluyl aldehyde p-Formyltoluene 4-Tolualdehyde	Perfumes; pharmaceutical and dyestuff intermediate, flavoring agent
122-03-2	4-(1-Methylethyl)benz-aldehyde	p-Isopropylbenzaldehyde 4-Isopropylbenzaldehyde p-Cuminaldehyde Cumaldehyde Cumic aldehyde Cuminal Cuminic aldehyde Cuminyl aldehyde	Perfumery; flavoring
61931-79-1	1-Methyl-4-isohexyl-hexahydrobenzaldehyde	1-Methyl-4-isohexylcyclohexane-1-carboxaldehyde 1-Methyl-4-(4-methylpentyl)cyclohexane-1-carboxaldehyde	
99-61-6	3-Nitrobenzaldehyde	m-Nitrobenzaldehyde	Synthesis of dyes, pharmaceuticals, surface active agents; vapor phase corrosion inhibitor; antioxidant for chlorophyll; mosquito repellent
626-19-7	1,3-BENZENEDICARBOX-ALDEHYDE	Isophthalaldehyde m-Phthalaldehyde	

TABLE A-2. Continued

CAS Number	Name of Compound	Synonyms	Uses
	(BENZALDEHYDE - Contd.)		
623-27-8	1,4-Benzenedicarbox-aldehyde	Terephthalaldehyde Terephthalic aldehyde 1,4-Diformylbenzene 4-Formylbenzaldehyde p-Benzenedicarboxaldehyde p-Phthalaldehyde p-Formylbenzaldehyde	
120-57-0	1,3-BENZODIOXOLE-5-CARBOXALDEHYDE	Piperonal Piperonaldehyde Piperonylaldehyde Geliotropin 3,4-(Methylenedioxy)benzaldehyde Heliotropin Heliotropine Protocatechuic aldehyde methylene ether 3,4-Dihydroxybenzaldehyde methylene ketal 3,4-Dimethylenedioxybenzaldehyde 5-Formyl-1,3-benzodioxole	Medicine; perfumery, suntan preparations; mosquito repellent; laboratory reagent; flavoring
123-72-8	BUTANAL	Butyraldehyde Butal Butaldehyde Butyl aldehyde n-Butyl aldehyde Butyral n-Butyraldehyde Butyric aldehyde Butyrylaldehyde n-Butanal Butanaldehyde	Manufacture of rubber accelerators, synthetic resins, solvents, plasticizers
27762-78-3	3-Ethoxy-2-oxo-butanal	Kethoxal β-Ethoxy-α-ketobutyraldehyde	Antiviral properties

CAS No.	Name	Synonyms	Uses
97-96-1	2-Ethylbutanal	2-Ethylbutyraldehyde; Diethylacetaldehyde; α-Ethylbutyraldehyde; 2-Ethylbutyric aldehyde; 3-Formylpentane	Organic synthesis; pharmaceuticals; rubber accelerators; synthetic resins
107-89-1	3-Hydroxybutanal	β-Hydroxybutanal; Aldol; Acetaldol	Synthesis of rubber accelerators and age resisters, perfumery; engraving; ore flotation; solvent; solvent mixtures for cellulose acetate; fungicides; organic systhesis; printer's rollers; cadmium plating; dyes; drugs; dyeing assistant; synthetic polymers
96-17-3	2-Methylbutanal (dl)	2-Methylbutyraldehyde; α-Methylbutanal; α-Methylbutyraldehyde; Methylethylacetaldehyde; 2-Methylbutyric aldehyde; α-Methylbutyric aldehyde; 2-Formylbutane	Flavoring
590-86-3	3-Methylbutanal	Isovaleraldehyde; Isovaleral; Isovaleric aldehyde; Isopentanal; β-Methylbutanal; 3-Methylbutyraldehyde; 2-Methylbutanal-4	Flavoring; perfumes; pharmaceuticals; synthetic resins; rubber accelerators
2439-44-3	3-Methyl-2-phenyl-butanal	α-Formylisobutylbenzene	

TABLE A-2. Continued

CAS Number	Name of Compound	Synonyms	Uses
	(BUTANAL - Contd.)		
4170-30-3	2-BUTENAL	Crotonaldehyde Crotonal Crotonic aldehyde Crotylaldehyde trans-2-Butenal β-Methyl acrolein	Intermediate for n-butyl alcohol and 2-ethylhexyl alcohol; solvent; preparation of rubber accelerators; purification of lubricating oils; insecticides; tear gas warning agent; organic synthesis, leather tanning; alcohol denaturant; manufacture of resins, rubber antioxidants
497-03-0	2-Methyl-2-butenal (E)	2-Methylcrotonaldehyde trans-2-Methyl-2-butenal Tiglaldehyde Tiglic aldehyde Tiglic acid aldehyde E-2-Methyl-2-butenal	
106-23-0	CITRONELLAL (d isomer)	d-Rhodinal 3,7-Dimethyl-6-octenal	Soap perfumery; manufacture of hydroxycitronellal; flavoring.
20691-52-5	4-(1,1-Dimethylethyl)-CYCLOHEXANECARBOXALDEHYDE	4-tert-Butylcyclohexanecarboxaldehyde 4-tert-Butylhexahydrobenzaldehyde	
100-50-5	3-CYCLOHEXENE-1-CARBOXALDEHYDE	1,2,3,6-Tetrahydrobenzaldehyde 3-Cyclohexen-1-aldehyde 4-Formylcyclohexene 1-Formyl-3-cyclohexene Cyclohexene-4-carboxaldehyde Δ^3-Tetrahydrobenzaldehyde	
31906-04-4	4-(4-Hydroxy-4-methylpentyl)3-cyclohexene-1-carboxaldehyde	Lyral 4-(4-Hydroxy-4-methylpentyl)-Δ^3-tetrahydrobenzaldehyde	

CAS	Name	Synonyms	Uses
112-31-2	DECANAL	Capraldehyde; Capric aldehyde; Caprinaldehyde; Caprinic aldehyde; Decaldehyde; Decyl aldehyde; n-Decanal; n-Decyl aldehyde; n-Decaldehyde; n-Decylic aldehyde; Decanaldehyde; Aldehyde C10	Perfumery, flavoring
112-54-9	Dodecanal	Lauraldehyde; n-Lauraldehyde; Lauric aldehyde; Lauryl aldehyde; Dodecanaldehyde; Dodecyl aldehyde; n-Dodecanal; 1-Dodecanal; n-Dodecyl aldehyde; n-Dodecylic aldehyde; Aldehyde C-12	Perfumery; flavoring agent
124-25-4	Tetradecanal	Myristaldehyde; Myristylaldehyde; Tetradecylaldehyde	
107-22-2	ETHANEDIAL	Glyoxal; Glyoxal aldehyde; Glyoxylaldehyde; Biformal; Biformyl; Diformyl; Ethanedione; 1,2-Ethanedione; Oxal; Oxalaldehyde	Permanent-press fabrics; dimensional stabilization of rayon and other fibers, insolubilizing agent for compounds containing polyhydroxyl groups (polyvinyl alcohol, starch, cellulosic materials); insolubilizing of proteins (casein, gelatin, animal glue); embalming fluids; leather tanning; paper coatings with hydroxyethylcellulose; reducing agent in dyeing textiles

TABLE A-2. Continued

CAS Number	Name of Compound	Synonyms	Uses
463-51-4	ETHENONE	Ketene Carbomethene	Acetylating agent, generally reacting with compounds having an active hydrogen atom; reacts with ammonia to give acetamide; starting point for making various commercially important products, especially acetic anhydride and acetate esters
50-00-0	FORMALDEHYDE	Formalin Formic aldehyde Formalith Formol Fyde BVF Methanal Methyl aldehyde Methylene oxide Morbicid Oxymethylene Oxomethane Lysoform Superlysoform Fannoform Formaldehyde solution Ivalon	Urea and melamine resins; polyacetal and phenolic resins; ethylene glycol; pentaerythritol; hexamethylenetetramine; fertilizer; dyes; medicine (disinfectant, germicide); embalming fluids; preservative; hardening agent; reducing agent, as in recovery of gold and silver; corrosion inhibitor in oil wells; durable-press treatment of textile fabrics; possible condensation to sugars and other carbohydrates for food use (experimental); industrial sterilant; treatment of grain smut
30525-89-4	Paraformaldehyde	Paraform Flo-Mor	Fungicides, bactericides, disinfectants; adhesives; hardener and waterproofing agent for gelatin
110-88-3	1,3,5-Trioxane	s-Trioxane sym-Trioxane Triformol Trioxan Troxane Formaldehyde trimer Triformol Trioxymethylene Trioxin Metaformaldehyde	Organic synthesis; disinfectant; nonluminous, odorless fuel

CAS	Name	Synonyms	Uses
98-01-1	2-FURANCARBOXALDEHYDE	2-Furaldehyde Furfural Furfuraldehyde Furfurole Furole Furale Fural Furancarbonal 2-Furancarbonal Pyromucic aldehyde α-Furole 2-Furanaldehyde 2-Formylfuran 2-Furfural 2-Furfuraldehyde 2-Furylaldehyde Furfurylaldehyde Furaldehyde Artificial ant oil	Solvent refining of lubricating oils, butadiene, rosin, other organic materials; solvent for nitrocellulose, cellulose acetate, shoe dyes; intermediate for tetrahydrofuran and furfuryl alcohol; phenolic and furan polymers; wetting agent in manufacture of abrasive wheels and brake linings; weed killer; fungicide; adipic acid and adiponitrile; road construction production of lysine; refining of rare earths and metals; flavoring; manufacture of varnish; accelerating vulcanization; insecticide, germicide; reagent in analytical chemistry; synthesis of furan derivatives
111-71-7	n-HEPTANAL	Enanthal Enanthaldehyde Enanthic aldehyde Enanthole Heptaldehyde n-Heptaldehyde n-Heptylaldehyde Heptanaldehyde Heptanal Oenanthal Oenanthol Oenanthole Oenanthic aldehyde Oenanthaldehyde Aldehyde C-7	Manufacture of 1-heptanol; organic synthesis; perfumery; pharmaceuticals; flavoring

TABLE A-2. Continued

CAS Number	Name of Compound	Synonyms	Uses
	(HEPTANAL - Cont.)		
122-40-7	2-(Phenylmethylene)-heptanal	α-Pentylcinnamaldehyde α-Amyl-β-phenylacrolein α-Amylcinnamaldehyde Amylcinnamaldehyde Flomine Jasminaldehyde	Perfumery; flavoring
106-72-9	2,6-Dimethyl-5-HEPTENAL	2,6-Dimethyl-5-hepten-1-al	Perfumery
142-83-6	2,4-HEXADIENAL	Sorbaldehyde Sorbic aldehyde 1,3-Pentadiene-1-carboxaldehyde 2,4-Hexadien-1-al	
66-25-1	HEXANAL	Caproaldehyde Caproic aldehyde Capronaldehyde n-Caproaldehyde n-Hexanal Hexaldehyde	Organic synthesis of plasticizers, rubber chemicals, dyes, synthetic resins, insecticides
1335-39-3	HEXENAL	Hexylenic aldehyde	
505-57-7	2-Hexenal	trans-2-Hexenal trans-Hex-2-enal trans-2-Hexen-1-al 2-trans-Hexenal (E)-2-Hexenal Leaf aldehyde	
3304-28-7	5-Methyl-2-(1-methylethylidene)4-hexenal	Isodihydro lavandulyl aldehyde 2-Isopropylidene-5-methyl-4-hexenal	

541b-80-8	2-Methyl-1H-INDOLE-3-CARBOXALDEHYDE	2-Methylindole-3-carboxaldehyde 3-Formyl-2-methylindole 2-Methyl-3-formylindole	
708-06-5	2-Hydroxy-1-NAPHTHA-LENECARBOXALDEHYDE	2-Hydroxy-1-naphthaldehyde 2-Hydroxynaphthaldehyde 2-Hydroxy-α-naphthaldehyde 2-Hydroxy-1-naphthylaldehyde 2-Hydroxynaphthaldehyde 1-Formyl-2-naphthol 2-Naphthol-1-carboxaldehyde β-Naphthol-1-aldehyde	
124-19-6	NONANAL	Nonaldehyde n-Nonaldehyde n-Nonylaldehyde Nonylic aldehyde Nonanoic aldehyde Pelargonaldehyde Pelargonic aldehyde	Perfumery; flavoring agent; n-Nonanal production
5392-40-5	3,7-Dimethyl-2,6-OCTA-DIENAL	Citral Geranial Geranialdehyde	Perfumes; flavoring agent, intermediate for other fragrances
124-13-0	OCTANAL	Caprylaldehyde Caprylic aldehyde n-Caprylaldehyde Octaldehyde Octanaldehyde Octanoic aldehyde Octylaldehyde n-Octanal n-Octylal n-Octaldehyde Aldehyde C-8 Antifoam-LF	Perfumery; flavors

TABLE A-2. Continued

CAS Number	Name of Compound	Synonyms	Uses
	(OCTANAL - Contd.)		
107-75-5	7-Hydroxy-3,7-dimethyl-octanal	Hydroxycitronellal 3,7-Dimethyl-7-hydroxyoctanal 7-Hydroxycitronellal Citronellal hydrate Cyclalia Cyclosia Fixol Laurine Lilyl Muguet synthetic Muguettine principle Phixia	Perfumery (fixative, muguet odor); flavoring; soap and cosmetic fragrances
101-86-0	2-(Phenylmethylene)octanal	α-n-Hexylcinnamaldehyde Hexyl cinnamic aldehyde α-n-Hexyl-β-phenylacrolein 1-Phenyl-1-octene-2-carboxaldehyde	Flavoring agent; fragrance
765-34-4	OXIRANECARBOXALDEHYDE	Glycidaldehyde Glycidal Epihydrinaldehyde 2,3-Epoxypropanal 2,3-Epoxypropionaldehyde	
110-62-3	PENTANAL	Valeraldehyde n-Valeraldehyde Valeral Valeric aldehyde Valerianic aldehyde Valerylaldehyde Valeric acid aldehyde n-Pentanal	Flavoring; rubber accelerators
123-15-9	2-Methylpentanal	2-Methylvaleraldehyde α-Methylpentanal 2-Formylpentane	Intermediate for dyes, resins, pharmaceuticals

111-30-8	1,5-PENTANEDIAL	Glutaraldehyde Glutaral Glutardialdehyde Aldesan Glutaric dialdehyde Hospex Sonacide 1,5-Pentanedione	Intermediate; fixative for tissues; crosslinking protein and polyhydroxy materials; tanning of soft leathers
123-38-6	PROPANAL	Propionaldehyde Propional Propaldehyde Propionic aldehyde Propylic aldehyde Propylaldehyde Methylacetaldehyde	Manufacture of propionic acid, polyvinyl and other plastics, synthesis of rubber chemicals; disinfectant; preservative
104-53-0	Benzenepropanal	Hydrocinnamaldehyde Hydrocinnamic aldehyde Dihydrocinnamaldehyde Benzylacetaldehyde 3-Phenylpropanal 3-Phenyl-1-propanal 3-Phenylpropionaldehyde 3-Phenylpropyl aldehyde β-Phenylpropionaldehyde	
80-54-6	4-(1,1-Dimethylethyl)-methylbenzene-propanal	p-tert-Butyl-α-methyl-hydrocinnamaldehyde Lilial Lilyal α-Methyl-p-(tert-butyl)-hydrocinnamaldehyde	Perfumery; flavors
630-19-3	2,2-Dimethylpropanal	Pivalaldehyde Pivalic aldehyde Trimethylacetaldehyde	

TABLE A-2. Continued

CAS Number	Name of Compound	Synonyms	Uses
	(PROPANAL - Contd.)		
	2,2-Dimethylpropanal (contd.)	2,2-Dimethylpropionaldehyde α,α-Dimethylpropionaldehyde α,α-Dimethylpropanal Neopentanal t-Butylcarboxaldehyde t-Butylformaldehyde	
103-95-7	-Methyl-4-(1-methylethyl)benzenepropanal	p-Isopropyl-α-methylhydrocinnamaldehyde Cyclamal Cyclamen aldehyde Aldehyde B p-Isopropylphenyl-α-methylpropyl aldehyde α-Methyl-p-isopropylhydrocinnamaldehyde 2-Methyl-3-(p-isopropylphenyl) propionaldehyde 3-(4-Isopropylphenyl)-2-methylpropanal	Perfumery; soap perfumes; flavoring
78-84-2	2-Methylpropanal	Isobutyraldehyde Isobutanal Iso-butyraldehyde 2-Methylpropionaldehyde Isopropyl aldehyde Isopropyl formaldehyde α-Methylpropionaldehyde Dimethyl acetaldehyde	Intermediate for rubber antioxidants and accelerators, for neopentyl glycol; organic synthesis; perfumes, flavors, plasticizers, resins, gasoline additives
3268-49-3	3-(Methylthio)propanal	Methional 3-(Methylthio)propionaldehyde β-(Methylthio)propionaldehyde 3-Methylthiopropionaldehyde β-(Methylmercapto)propionaldehyde 3-(Methylmercapto)propionaldehyde	

78-98-8	2-Oxopropanal	Pyruvaldehyde Pyruvic aldehyde Acetylformaldehyde Acetylformyl α-Ketopropionaldehyde 2-Ketoproprionaldehyde Methylglyoxal Pyroracemic aldehyde	Organic synthesis of complex chemical compounds such as pyrethrins; tanning leather; flavoring
542-78-9	PROPANEDIAL	Malonaldehyde Malondialdehyde Malonyldialdehyde Malonic dialdehyde 1,3-Propanedione 1,3-Propanedialdehyde	
107-02-8	2-PROPENAL	Acrolein Propenal Acrylaldehyde Acrylic aldehyde Allyl aldehyde Prop-2-en-1-al 2-Propen-1-one Aqualin NSC 8819	Intermediate for synthetic glycerol, polyurethane and polyester resins, methionine, pharmaceuticals; herbicide; tear gas; perfumes; warning agent in methyl chloride refrigerant; manufacture of colloidal forms of metals; organic syntheses
4206-58-0	3-(4-Hydroxy-3,5-dimethoxyphenyl)-2-propenal	4-Hydroxy-3,5-dimethoxy-cinnamaldehyde Sinapaldehyde Sinapic aldehyde Sinapyl aldehyde	
458-36-6	3-(4-Hydroxy-3-methoxyphenyl)-2-propenal	4-Hydroxy-3-methoxy-cinnamaldehyde Coniferaldehyde p-Coniferaldehyde Ferulaldehyde Coniferyl aldehyde	

TABLE A-2. Continued

CAS Number	Name of Compound	Synonyms	Uses
	(PROPENAL -Cont.)		
1504-74-1	3-(2-Methoxyphenyl)-2-propenal	o-Methoxycinnamaldehyde 2-Methoxycinnamaldehyde o-Methoxycinnamic aldehyde	
101-39-3	2-Methyl-3-phenyl-2-propenal	α-Methylcinnamaldehyde α-Methylcinnamic aldehyde 2-Methyl-3-phenylacrolein 2-Methyl-3-phenylacrylaldehyde	
78-85-3	2-Methyl-2-propenal	Methacrylaldehyde Methacrolein Isobutenal 2-Methylpropenal Methacrylic aldehyde α-Methylacrolein 2-Methylacrolein Methylacrylaldehyde	Copolymers; resins
104-55-2	3-Phenyl-2-propenal	Cinnamaldehyde 3-Phenylpropenal β-Phenylacrolein 3-Phenylacrolein Cinnamal Cinnamic aldehyde Cinnamyl aldehyde Cassia aldehyde Benzylideneacetaldehyde	Flavors; spice perfumes
34346-90-2	3-(3,4,5-trimethoxyphenyl)-2-propenal	3,4,5-Trimethoxycinnamaldehyde	
1121-60-4	2-PYRIDINECARBOXALDEHYDE	Picolinaldehyde 2-Picolinaldehyde Pyridine-2-aldehyde 2-Pyridaldehyde Picolinal 2-Picolinealdehyde Picolinic aldehyde	

	2-PYREDINECARBOXALDEHYDE (Contd.)	o-Nicotinaldehyde 2-Formylpyridine 2-Pyridylcarboxaldehyde	
57-92-1	STREPTOMYCIN	Streptomycin A o-2-Deoxy-2-(methylamino)-α- 1-glucopyranosyl-(1-4)N,N'-bis- (aminoiminomethyl)D-streptamine	Medicine (usually as sulfate)
298-39-5	Streptomycin sulfate	Sesquisulfate AgriStrep Streptobrettin Streptorex Vetstrep	Antibacterial (tuberculostatic)
1401-69-0	Tylosin	Tylosin A Tylosine Tylocine Tylan Tylon	Veterinary medicine
112-44-7	UNDECANAL	n-Undecanal Hendecanal Hendecanaldehyde Undecyl aldehyde n-Undecylic aldehyde	
110-41-8	2-Methylundecanal	Methyl n-nonyl acetaldehyde Methyl nonyl acetic aldehyde Methylnonylacetaldehyde 2-Methyl-1-undecanal Aldehyde M.N.A.	
143-14-6	9-UNDECENAL	Undecenoic aldehyde 9-Undecylene aldehyde	Perfumery; flavoring
1337-83-8	10-Undecenal	Undecylenic aldehyde	Perfumery; flavoring

339

REFERENCES

1. CHEMLINE Data Base. Bethesda, Md.: National Library of Medicine, April 1980.
2. CHEMNAME. DIALOG Database. Palo Alto, Cal.: Lockheed Missiles & Space Co., Inc., June 1980.
3. Fassett, D. W. Aldehydes and acetals, pp. 1959-1989. In Patty, F. A., Ed. Industrial Hygiene and Toxicology. 2nd rev. ed. D. W. Fassett and D. D. Irish, Eds. Vol. II. Toxicology. New York: Interscience Publishers, 1963.
4. Hawley, G. G., Ed. The Condensed Chemical Dictionary. 9th ed. New York: Van Nostrand Reinhardt, 1977.
5. Weast, R. C., and M. J. Astle, Eds. Handbook of Chemistry and Physics. 59th Edition. 1978-1979. Cleveland, Ohio: The Chemical Rubber Co., 1978.
6. Windholz, M., S. Budavari, L. Y. Stroumtsos, and M. N. Fertig, Eds. The Merck Index. An Encyclopedia of Chemicals and Drugs. 9th ed. Rahway, N.J.: Merck & Co., Inc., 1976.